E. T. BROWNE MONOGRAPH
OF THE
MARINE BIOLOGICAL ASSOCIATION OF
THE UNITED KINGDOM

THE
MEDUSAE OF THE BRITISH ISLES

VOLUME II

THE
MEDUSAE OF THE BRITISH ISLES

II. PELAGIC SCYPHOZOA
WITH A SUPPLEMENT TO THE FIRST VOLUME
ON HYDROMEDUSAE

BY

F. S. RUSSELL, F.R.S.

FORMERLY DIRECTOR OF THE PLYMOUTH LABORATORY OF
THE MARINE BIOLOGICAL ASSOCIATION OF
THE UNITED KINGDOM

CAMBRIDGE
AT THE UNIVERSITY PRESS
1970

Published by the Syndics of the Cambridge University Press
Bentley House, 200 Euston Road, London N.W.1
American Branch: 32 East 57th Street, New York, N.Y.10022

Library of Congress Catalogue Card Number: 69–15571
Standard Book Number 521 07293 X

Printed in Great Britain
at the University Printing House, Cambridge
(Brooke Crutchley, University Press)

TO MY WIFE

whose devotion has made this possible

CONTENTS

CONTENTS

CONTENTS

PREFACE

When *The Medusae of the British Isles* was published in 1953 I started research on the pelagic Scyphomedusae in the hope that I might be able to make the original monograph more complete by adding a volume on the larger jellyfish. This work continued slowly during the years of my Directorship of the Plymouth laboratory until my retirement in 1965, since when I have been able to devote most of my time to its completion. At the same time it has been possible for me to include a supplement to the first volume on Hydromedusae to bring it up to date.

The British species of Scyphomedusae were only treated cursorily by Forbes in his monograph of 1848, and indeed little was known about their biology at that time. Towards the end of the nineteenth century, however, as a result of the discovery of their remarkable life history and owing to the interest shown by comparative physiologists in their rhythmic pulsation, much attention was paid to this interesting group of animals.

The first modern systematic monograph on the Scyphomedusae was that of E. Haeckel (1880). This was his *System der Acraspeden* which formed the second half of his *System der Medusen*. Between then and the early years of the twentieth century our knowledge was much advanced by the work of H. B. Bigelow, H. Claus, O. Maas, and E. Vanhöffen. When in 1910 A. G. Mayer produced his third volume of *The Medusae of the World* devoted to the Scyphomedusae a new milestone was erected.

Since then there have been a number of summarizing works among which that of Th. Krumbach (1925) in the *Handbuch der Zoologie* is the most noteworthy, followed by his useful section in *Die Tierwelt der Nord- und Ost-see* in 1930. But the most active original research was being pursued by H. B. Bigelow, P. L. Kramp, G. Stiasny, M. E. Thiel and T. Uchida.

In 1936 the first part of the monumental work by M. E. Thiel on the Scyphomedusae in *Bronn's Tierreich* appeared. This was followed by five further contributions, the last of which was published in 1962. It remains only for him to complete the Rhizostomeae and the Bibliography. This work is of the greatest possible value to students, treating all aspects of the biology of the group in detailed historical sequence and including the reproduction of many representative published illustrations.

Up-to-date diagnoses of all known species are given by P. L. Kramp (1961) in his 'Synopsis of the Medusae of the World', which includes a bibliography of papers published since 1910. A monograph on the Scyphomedusae of waters of the U.S.S.R. was published by D. V. Naumov (1961) in Russian.

In the present monograph I have referred to all research done on the biology of our British species. Owing to their wide distribution in European and North Atlantic waters more has been published on the results of observations and experimental research on these species than on almost any other and the literature on them is very large. As in all works of this nature it is difficult to decide what degree of detail to include and a line has to be drawn somewhere. I have aimed, however, at making it possible for the student to find the necessary literature on any aspect of the research that has been done and have included brief statements indicating some of the more interesting results. As in the previous volume, unless stated otherwise, I have made the drawings myself.

PREFACE

I have received help from many people. The following have either sent me specimens, given information on local distribution, or helped me with their knowledge and experience: J. A. Adams, Miss S. M. van der Baan, J. H. Barrett, H. O. Bull, M. Cabioch, D. M. Chapman, J. S. Colman, E. D. S. Corner, J. H. Crother, C. Edwards, F. Evans, J. H. Fraser, E. Ghirardelli, R. Hamond, J. P. Hillis, G. A. Horridge, C. Burdon Jones, N. S. Jones, P. L. Kramp, J. R. Lewis, J. Mauchline, the late G. E. Newell, Miss E. J. Pope, H. T. Powell, J. E. G. Raymont, the late W. J. Rees, T. G. Skinner, M. E. Thiel, J. Verwey, B. Werner, V. C. Wynne-Edwards.

The beautiful photographs reproduced in Plates IX, X, and XII–XV were taken by Mrs Elizabeth Nybø: those in Plate VI by C. I. D. Moriarty, and in Plate I s by M. S. Laverack.

J. S. Alexandrowicz helped me in many ways, especially in the translation of foreign languages. Miss L. M. Serpell and A. Varley have given library assistance. The careful typing has been done by Miss M. L. Taylor and Mrs Wendy Kennedy.

I am especially grateful to J. S. Alexandrowicz, D. M. Chapman, and Elaine M. Robson for their very helpful suggestions and critical reading of the proofs of the whole work; and to C. Edwards for reading the manuscript and proofs of the supplement.

The Council of the Royal Society gave me a grant from the Browne Research Fund which enabled me to work at marine biological laboratories in Scotland and visit many parts of the Scottish coasts; and have also given a grant from the Scientific Publications Fund towards the cost of production of the monograph.

The Natural Environment Research Council have made it possible for me to continue my research since my retirement.

To all of these I extend my grateful thanks, and especially to the Presidents and Council of the Marine Biological Association and Dr J. E. Smith, F.R.S., Director of the Plymouth laboratory, for their encouragement and for so graciously according me working space in the Plymouth laboratory.

Finally, my thanks are also due to the Cambridge University Press for the great care they have taken in the printing and excellent production of the book.

In view of their size and abundance the Scyphomedusae must play a large part in the economy of the sea as predators and competitors of fish. I hope that this monograph will stimulate further investigations into their biology.

STRUCTURE AND CHARACTERS OF
PELAGIC SCYPHOMEDUSAE

The total number of known pelagic Scyphomedusae as listed by Kramp (1961) in his Synopsis of the Medusae of the World is of the order of 160. Of these only six species occur in coastal waters round the British Isles and seven deep-sea species may be found over deep water off the continental shelf along the western coasts of Ireland and Scotland, and off the western approaches to the English Channel.

There are four orders, Cubomedusae, Coronatae, Semaeostomeae and Rhizostomeae. The Cubomedusae are not represented in the British fauna and no reference will be made to them in this monograph. The Coronatae are all deep-water forms and only the Semaeostomeae and Rhizostomeae are represented in coastal waters.

The basic structure of a pelagic scyphomedusan consists of an umbrella in the centre of the subumbrellar side of which is a manubrium with a mouth opening. Marginal sense organs are present, and there may or may not be marginal tentacles.

The Scyphomedusae differ essentially from the Hydromedusae in that they have no velum, there are gastric filaments in the stomach, and the gonads are endodermal in origin. The gastrovascular system is also quite different, as is the nervous system.

As in Hydromedusae the four mouth lips are regarded as perradial, and there are four interradii and eight adradii (Text-fig. 1).

THE UMBRELLA

The umbrella is typically saucer-shaped or hemispherical, except in the Coronatae in which it may vary from the flat disk-like form of *Atolla* to the high conical bell of *Periphylla*. The mesogloea is usually thick and of solid consistency, being thickest in the centre of the umbrella and tapering off in thickness to the umbrella margin which is always interrupted by a number of clefts which result in the formation of separate lobes, the *marginal lappets*. According to whether the marginal lappets are situated next to a marginal sense organ or marginal tentacle they are known as *rhopalar* or *tentacular* lappets respectively. In *Rhizostoma* which has no marginal tentacles the marginal lappets lying between pairs of rhopalar lappets are known as *velar* lappets.

In coronate medusae the thickness of the mesogloea does not taper evenly towards the umbrella margin, but the marginal zone of the umbrella is separated from its central disk by a circumferential furrow known as the *coronal groove*. The marginal zone of the umbrella beyond the coronal groove is also grooved by a number of radial furrows which divide the margin into a number of thickened areas known as *pedalia*. The pedalia may be rhopalar or tentacular according to whether they are situated at the base of a marginal sense organ or of a marginal tentacle respectively.

The exumbrellar surface may be quite smooth, or it may be roughened by the presence of numerous raised *nematocyst warts*. These may be comparatively few and large, as in *Pelagia*; or they may be so numerous that they give the exumbrellar surface the appearance of frosted glass when out of water, as in *Rhizostoma*. The nematocyst warts may also occur on the manubrium and its basal thickenings, but usually not on the subumbrellar surface of the umbrella itself. The subumbrellar surface of the umbrella is characterized by the presence of the ectodermal musculature, composed of circular and radial muscle systems. The circular or *coronal muscle* may be

STRUCTURAL CHARACTERS

continuous and very massive, as in *Atolla*, or divided radially into a number of separate fields, as in *Rhizostoma*. The radial muscles, sometimes known as the *deltoid* muscles, may be very strongly developed, as in *Periphylla*, or absent, as in *Rhizostoma*.

MARGINAL TENTACLES

The marginal tentacles which may be solid or hollow are usually very extensile and capable of contraction by direct shortening or spiral coiling. Distributed throughout the ectoderm there are numerous aggregations of nematocysts to form warts or clasps. The ectodermal musculature consists of longitudinal muscle fibres. These are distributed round the circumference of the tentacle. In *Cyanea* there is, in addition, a sunken muscle strand on the adaxial side of the tentacle, while in *Pelagia* there are a number of sunken muscles on either side of the adaxial furrow.

In the coronate medusae, such as *Periphylla*, the muscle on the adaxial side is strongly developed and has two conical root muscles.

The marginal tentacles in *Aurelia* are not typical, nor are the marginal lappets. In other semaeostome medusae the marginal tentacles are situated either just on the margin of the umbrella or on the subumbrella. Their hollow cavities are continuations of the gastrovascular pouches on the subumbrellar side. In *Aurelia* the hollow tentacles are very small and situated on the exumbrellar surface of the margin of the umbrella between small vertical lappet-like structures (see Text-fig. 75, p. 144). Their cavities are continuations from the marginal ring canal. The umbrella margin is turned inwards and terminates in a thin velum-like fringe, the *velarium* (Text-fig. 76, p. 144).

GASTROVASCULAR SYSTEM

Basically the gastrovascular system consists of a central stomach in which the *gastric filaments* are situated, and a peripheral marginal zone, the *gastrovascular sinus*, or coronal sinus.

The umbrella can be regarded as having the form of two inverted saucers of which the lower or subumbrellar saucer is the flatter. These saucers are fused together round their margins and they enclose a space between them, equivalent to the gastrovascular cavity. In the centre of the lower saucer is an aperture leading to the exterior. The upper saucer is covered on its outer convex surface with ectoderm, and on its inner concave surface with endoderm. The solid substance between the ectoderm and endoderm layers is the mesogloea. Conversely the inner convex surface of the lower saucer is covered with endoderm and its outer concave surface with ectoderm.

At different places the two endoderm surfaces of the upper and lower saucers are fused together, as well as round their margins. It is the positions of these areas of fusion which determine the form of the gastrovascular system (Text-fig. 1).

In the Coronatae the central stomach cavity is separated from the peripheral gastrovascular sinus by four interradial crescent-shaped areas of fusion, alternating with four unfused areas allowing access between the stomach and gastrovascular sinus. These four passages between the stomach and gastrovascular sinus are known as the *gastric ostia*. The crescent-shaped areas of fusion have been known as cathammal plates or septal nodes. Neither of these two designations indicates the position of these areas of fusion and I prefer to call them *gastric septa* to distinguish them from the *radial septa* which are radial strips of fusion occurring at intervals in the gastrovascular sinus. These radial septa are regularly disposed round the peripheral region of the umbrella so as to divide the gastrovascular sinus into a number of partitioned areas or *pouches*. According to whether

2

these pouches lie in the radius of a marginal sense organ or of a marginal tentacle they are known respectively as *rhopalar* or *tentacular pouches*.

In semaeostome and rhizostome medusae the gastric septa are absent and the stomach can communicate with the gastrovascular sinus all round its periphery.

In some Semaeostomeae and in all Rhizostomeae the radial septa may be replaced by numerous island areas of fusion whose arrangement may give rise to a simple unbranched and branching *radial canal* system, as in *Aurelia*, or the complicated network of anastomosing canals typical of

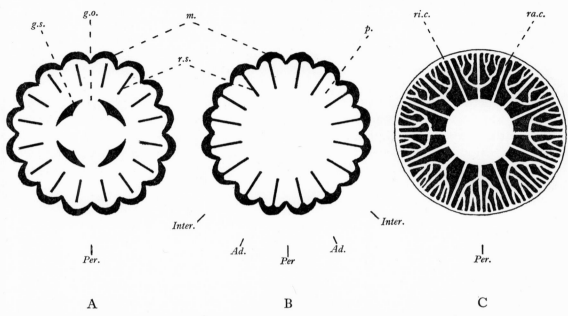

A   B   C

Text-fig. 1. Diagrams to show areas of fusion (black) of endodermal surfaces which determine the form of the gastrovascular system. A, Coronate; B & C, Semaeostome. *Per.* perradial; *Ad.* adradial; *Inter.* interradial; *g.o.* gastric ostium; *g.s.* gastric septum; *m.* umbrella margin; *p.* pouch; *r.s.* radial septum; *ra.c.* radial canal; *ri.c.* ring canal.

the Rhizostomeae. In such systems there are always a number of straight radial canals running from the stomach to the umbrella margin from which the branches arise. These are perradial, interradial, or adradial according to their position in relation to the perradial mouth lips. In some medusae the canals communicate with a continuous *ring canal* round the umbrella margin. In coronate medusae the rhopalar and tentacular gastrovascular pouches communicate with each other at the umbrella margin by a *festoon canal*. In those semaeostome medusae in which the gastrovascular sinus is divided by radial septa the resulting pouches do not communicate with their neighbouring pouches at the umbrella margin. They may be simple as in *Pelagia* and *Chrysaora*, or branching towards the umbrella margin, as in *Cyanea*.

The central stomach leads to the exterior through a mouth or oral opening in the centre of a *manubrium*. The manubrium may be no more than four perradial well-formed lips as in the coronate medusae, but in all other scyphomedusae it consists of a shorter or longer pillar-like fused basal portion of solid consistency, the *oral tube*, which divides at its end into four elongated structures, the *oral arms*. The margins of these oral arms are often thin and much folded and frilled, or even curtain-like, and form mouth lips bordering an axial groove running along the middle of the inner surface of the oral arm lined with endoderm. In *Rhizostoma* there is no, or only a rudimentary, central mouth opening and the manubrium has a complicated system of canals leading to innumerable mouth openings bordered by much-frilled lips.

Gastric filaments. The gastric filaments are arranged in four interradial groups, the *phacelli*. In the coronate medusae they are situated along the centripetal sides of the crescent shaped gastric septa, giving the impression that they form eight groups. In other medusae they are present as four continuous interradial groups, each of which may be several rows deep.

A gastric filament is a tapering structure with a solid mesogloeal core covered by an epithelium of endoderm cells which may include gland cells, nematocysts and muscle cells. Thus they may serve to paralyse prey further after entry into the stomach, and are the site of digestion. They may be very long and can then be seen extending into the gastrovascular sinus (see e.g. Text-fig. 20, p. 43).

The most detailed investigation on the part they play in digestion is that of Smith (1936) on *Cassiopea*, who found the following enzymes: protease, lipase, glycogenase, and an amylase originating from the zooxanthellae. Nutriment is carried to various parts of the body by wandering cells.

The part played by amoebocytes was studied by Wetochin (1930) in *Aurelia* (see p. 161), and in the same species and *Cyanea* Henschel (1935) examined the reactions of the oral arms to food (see pp. 121, 161).

A historical summary of research on digestion is given by M. E. Thiel (1959 *b*).

I have seen the gastric filaments in *Cyanea capillata* moving with a serpentine motion.

GONADS

In the coronate medusae the gonads are eight in number, each situated approximately on an adradius. They arise as endodermal proliferations of the gastrovascular endoderm of the coronal sinus, and are thus peripheral to the central stomach. They may vary in shape from simple round or oval plates (*Atolla*) to narrow U-shaped organs (*Periphylla*) or W-shaped (*Paraphyllina*). A histological study of gonad development in *Atolla* was made by Vanhöffen (1902) and in *Periphylla* by Maas (1897). Claus (1883) studied the development of the gonad in *Nausithoë punctata* from the Mediterranean.

In semaeostome and rhizostome medusae there are four gonads each situated interradially within the cavity of the central stomach. Detailed descriptions of their histology and development have been given by Claus (1877, *Chrysaora* and *Rhizostoma*), O. & R. Hertwig (1879, *Pelagia*), Claus (1883, *Pelagia*, *Discomedusa* and *Aurelia*), Hamann (1883, *Pelagia*), Paspalev (1938, *Rhizostoma*), Tsukaguchi (1914, *Aurelia*), and Widersten (1965, *Chrysaora*, *Cyanea*, *Aurelia*, *Rhizostoma*).

The gonads develop as ribbon-like thickenings in the subumbrellar endoderm of the stomach just peripheral to the gastric filaments. These thickenings are due to proliferation of the endoderm cells into the mesogloea. Cavities develop within these thickened cushions of cells which run together to form a *genital sinus* which is traversed by occasional trabeculae uniting the genital epithelium with the subumbrellar endodermal epithelium. The gonads become much folded. They may be supported by the wall of the subgenital cavity formed by thickened mesogloea growing outwards from the basal pillars of the manubrium as in *Aurelia* and *Rhizostoma*; or they may hang freely down beneath the subumbrella in a thin-walled pouch formed by the folded and convoluted subumbrellar wall of the stomach as in *Pelagia* and especially so in *Cyanea*.

4

Claus (1883, p. 41) concluded that the early process of formation of the genital sinus is the same in the coronate as in the semaeostome and rhizostome medusae, but that they develop to face in the opposite direction, the opening of the genital sinus being towards the centre of the stomach in the Coronatae instead of towards the periphery.

The sexes are distinct except in *Chrysaora* which is hermaphrodite. Here the male elements develop in small vesicles which may occur almost anywhere on the endodermal epithelium (see p. 93).

MARGINAL SENSE ORGANS

The marginal sense organ is a composite structure which may obviously serve more than one function. In its most developed form (Text-fig. 2), as found in some of the Semaeostomeae,

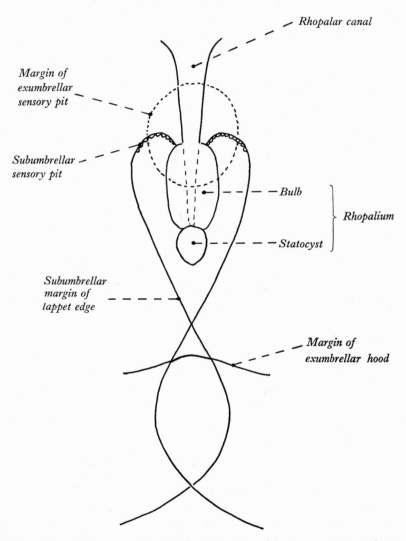

Text-fig. 2. Diagram of marginal sense organ of a semaeostome medusa,
viewed from exumbrellar side.

e.g. *Aurelia*, it is composed of a club-like body, the *rhopalium*, projecting from the umbrella margin and into which a small diverticulum from the gastrovascular system runs, the *rhopalar canal*. The rhopalium is situated in a protective *sensory niche* formed by a roof-like extension of the umbrella margin, or *hood*, and the lateral edges of the adjacent rhopalar lappets which form the sides and floor of the niche. On the exumbrellar surface above the basal stem of the rhopalium is a small depression, the *exumbrellar sensory pit*. Within the sensory niche itself, at the base of the rhopalium, there are two lateral depressions rich in nervous tissue known as the *subumbrellar sensory pits*.

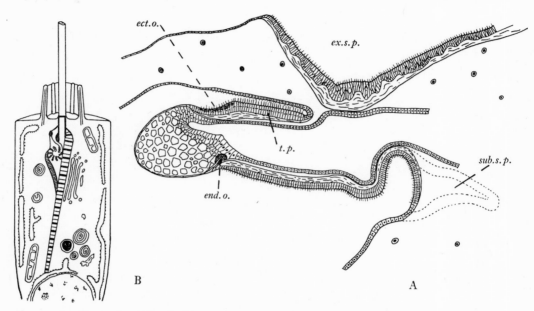

Text-fig. 3. *Aurelia aurita*. Marginal sense organ. A, radial section; B, sensory cell from touch-plate (electron microscope observation). *ect. o.* ectodermal ocellus; *end. o.* endodermal ocellus; *ex. s. p.* exumbrellar sensory pit; *sub. s. p.* subumbrellar sensory pit (in a different plane); *t. p.* touch-plate. (Unpublished drawing by D. M. Chapman.)

The rhopalium (Text-fig. 3) consists of a solid terminal body whose endoderm is filled with crystals, the *statocyst*, and a hollow basal stem receiving the rhopalar canal. The stem itself is clasped by a subumbrellar *sensory cushion* or *bulb* richly endowed with nervous tissue. On the exumbrellar side there is an ectodermal pigment spot or *ocellus*, and there is an endodermal ocellus on the subumbrellar side. (For a more detailed description see p. 145).

Other less-developed marginal sense organs may lack ocelli, but the general form is essentially the same throughout the Semaeostomeae and Rhizostomeae.

In the Coronatae on the other hand the marginal sense organ differs in that the rhopalium projects from a basal cushion on the umbrella margin and is not protected by a hood formed by an extension of the umbrella margin itself. The statocyst is, however, roofed over by a small spoon-shaped hood at the end of the rhopalium itself (Text-fig. 4).

Thus it must be realized that the term 'marginal sense organ' includes a complex of sense organs each of which may serve a different function, e.g. the statocyst sensitive to gravity, the ocellus sensitive to light, and the exumbrellar and subumbrellar sensory pits whose functions are not known, but which may perhaps be sensitive to chemical substances.

In the region of these organs the sensory epithelium is thickened, and contains spindle-shaped sensory cells interspersed between the ectoderm epithelium cells, each with a fibre running inwards and branching into a deeper layer of nerve fibres and ganglion cells.

Horridge (1968) made a study with the electron microscope of marginal sense organs of certain medusae, including *Nausithoë punctata*. He suggested that statocysts might have evolved from vibrator receptors each consisting of a single non-motile kinocilium around which stereocilia had developed, whose vibration reception was improved by the presence of calcareous concretions.

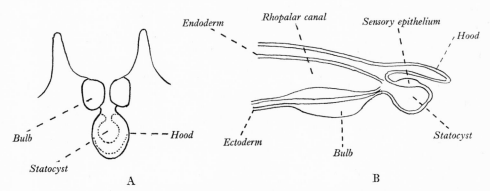

Text-fig. 4. Diagram of marginal sense organ of coronate medusa.
A, exumbrellar view; B, radial sectional view.

Extension of the bodies in which the concretions are enclosed to form pendants would give the receptors the additional sense of the direction of gravity.

The chemical composition of the concretions has been variously described as calcium carbonate, calcium oxalate, and calcium sulphate (see p. 145, and Spangenberg & Beck, 1968).

NERVOUS SYSTEM

Little is known about the nervous system of the Coronatae. Schäfer (1878) was the first to demonstrate the existence of a subumbrellar network of nerve fibres and their associated cells in *Aurelia*, thus complementing Romanes' physiological experiments. In the same year Eimer published the results of his investigations on locomotion and the nervous system, and the Hertwigs produced their studies on sense organs. But, although many physiological studies were made later on Scyphomedusae, and their sense organs were further described, it was not until 1927 that Bozler demonstrated that there were several kinds of neurones and suggested that the larger fibres were motor. More recently our knowledge of the general arrangement of the nervous system has been much advanced by the work of Horridge (1953, 1954*a*, *b*, 1956*b*).

Horridge demonstrated the existence and functions of at least two nerve nets, one of which he called the giant fibre system and the other the diffuse network. The giant fibre system is concerned with the propagation of the contraction wave leading to the pulsation of the umbrella. It may be spread evenly over the subumbrella as in *Aurelia*, or more restricted to the muscle fields as in *Cyanea* and *Rhizostoma*. The diffuse network spreads over the whole exumbrellar and subumbrellar surfaces, the manubrium and the marginal tentacles. It is concerned with local movements such as for feeding, and can evidently inhibit the action of the giant fibre network.

A detailed and up-to-date summary of the subject has been given by Bullock & Horridge (1965).

STRUCTURAL CHARACTERS

LOCOMOTION

Swimming is effected by pulsation of the umbrella in which the coronal and radial muscles are concerned. The radial muscles are the first to contract and this tends to arch the summit of the umbrella. This is followed after a slight interval by the contraction of the coronal muscle by which the subumbrellar space is reduced and water is forced out through the now diminished subumbrellar opening. In this way the umbrella margin acts in much the same capacity as the velum of the Hydromedusae.

This is probably typical of all Scyphomedusae, but the proportionate degree of development of coronal and radial muscles varies much from species to species. Bozler (1926a) studied the muscle contractions in *Cotylorhiza*, *Rhizostoma* and *Pelagia*. In *Rhizostoma* the radial muscles are absent, and Bozler thought that perhaps the strongly arched form of the umbrella rendered them superfluous.

The elucidation of the role played by the marginal sense organs in controlling the swimming movements of Scyphomedusae formed part of the classic experiments of Romanes and Eimer. Their discovery that removal of all the marginal sense organs resulted in complete cessation of the rhythmic pulsation led them to assume that the marginal sense organs were the stimulating centres. Both Romanes and Eimer, however, also noticed that in medusae with all sense organs removed rhythmic pulsation could begin again after a time, or could be effected by different external stimuli. This observation led to controversy as to whether the origin of pulsation was myogenic or neurogenic, and to investigations to decide which part of the marginal sense organ was concerned with regulation of pulsation.

Authors concerned in these studies were: Loeb (1899, *Aurelia*); v. Uexküll (1901, *Rhizostoma*); Bethe (1903, *Rhizostoma*); Mayer (1906, *Aurelia*, *Dactylometra*); Mayer (1909, *Cassiopea*); Jordan (1912, *Rhizostoma*); Morse (1921, *Cyanea*); Fränkel (1925, *Cotylorhiza*, *Rhizostoma*); Wetochin (1926b, *Aurelia*); Bozler (1926a, b, *Pelagia*, *Cotylorhiza*, *Rhizostoma*); Bauer (1927, *Cotylorhiza*, *Rhizostoma*); Horstmann (1934a, *Cyanea*, *Aurelia*).

It is now generally considered that the aggregation of nerve elements at the base of the rhopalium near the subumbrellar sensory pit and its immediate neighbourhood is concerned with the regulation of rhythmic pulsation.

It has been found by many investigators that, although pulsation is apparently simultaneous, the impulses set up by the sense organs are not synchronous but start from one only of the organs. This is not, however, a constant pace-maker, because the initiating role for successive pulsations can be taken over apparently at random by any one of the other organs.

During these investigations it was noticed that small young specimens could continue to pulsate after removal of all marginal sense organs, while larger specimens of the same species ceased all movement. The fact that in larger specimens pulsation can be later regained may be due to regeneration of the sense organs (C. W. Hargitt, 1909, *Rhizostoma*; Wetochin, 1926b, *Aurelia*).

The rate of pulsation is increased by raising the temperature, and is always greater in small specimens than in large ones of the same species.

Fränkel (1925) working with *Cotylorhiza*, *Rhizostoma* and *Pelagia* elucidated the mechanism of compensatory movement. If the medusa is in its normal upright position contraction of the coronal muscle is symmetrical. If the medusa is on its side, that part of the coronal muscle situated uppermost goes into a state of tonic contraction and does not fully relax. This enables the lower portion of the umbrella margin, in which the coronal muscle contracts and relaxes fully, to make a more

8

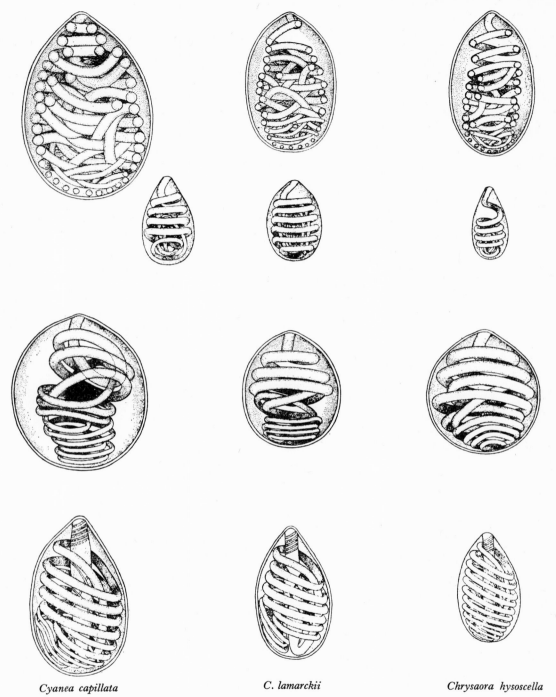

Cyanea capillata   *C. lamarckii*   *Chrysaora hysoscella*

Text-fig. 5. Undischarged nematocysts of some semaeostome medusae (from Papenfuss, 1936). Left-hand column, *Cyanea capillata*; middle column, *C. lamarckii*; right-hand column, *Chrysaora hysoscella*.

Top two rows, atrichous haplonemes, large and small; third row, holotrichous haplonemes; bottom row, heterotrichous microbasic euryteles (× 2,500).

9

effective beat than the upper portion. The medusa is thus brought again on to an even keel. According to Bozler (1926a) Fränkel observed similar compensatory movements in *Chrysaora* and *Cyanea*.

NEMATOCYSTS

Compared with the Hydromedusae, the Scyphomedusae only have a few types of nematocysts. These are atrichous haplonemes, holotrichous haplonemes, and microbasic heterotrichous euryteles (Text-figs. 5, 6).

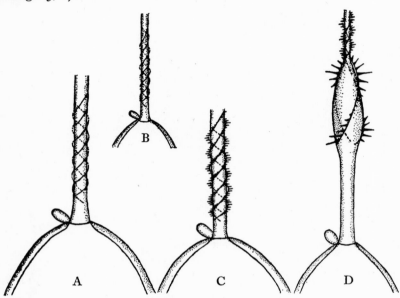

Text-fig. 6. Discharged nematocysts of *Cyanea capillata*. A and B, large and small atrichous haplonemes; C, holotrichous haploneme; D, heterotrichous microbasic eurytele (× 3,750) (from Papenfuss, 1936).

All three types occur in the semaeostome medusae, *Pelagia*, *Chrysaora*, and *Cyanea*. Only two types are found in *Aurelia* and *Rhizostoma*, namely atrichous haplonemes and microbasic heterotrichous euryteles. The coronate medusae also have only two types, holotrichous haplonemes and microbasic euryteles (Werner, 1967).

The fact that *Aurelia* and *Rhizostoma* differ from other semaeostome medusae in having only two types of nematocyst supports the evidence of other structural characters indicating that there may be a close phylogenetic relationship between the Ulmaridae and Rhizostomatidae (see von Lendenfeld, 1882; Uchida, 1926, 1934b).

DISTRIBUTION AND SEASONAL OCCURRENCE
ROUND THE BRITISH ISLES

DISTRIBUTION

There are six species of scyphomedusae which occur in British coastal waters, *Pelagia noctiluca*, *Chrysaora hysoscella*, *Cyanea capillata*, *Cyanea lamarckii*, *Aurelia aurita* and *Rhizostoma octopus*.

Of these *Pelagia* is an oceanic species with direct development. It only occurs in coastal waters regularly off the mouth of the English Channel, off the south-western and western shores of Ireland, and round the north-west and north of Scotland and the Faeroe–Shetland region, where deep oceanic water is near at hand. Its occurrence in other areas is spasmodic and due to unusual hydrographic conditions when it may occur in the English Channel, Irish Sea and off north-eastern coasts of England. *Chrysaora* and *Aurelia* occur commonly round all coasts, but *Rhizostoma* is absent, or extremely rare down the east coast of Scotland and of England to the Wash.

Of the two species of *Cyanea*, *C. capillata* is a northern species and *C. lamarckii* more southern. *C. lamarckii* probably occurs all round the British Isles, but *C. capillata* does not apparently occur so far south as the English Channel (see pp. 107 and 128).

To summarize: *Pelagia* is oceanic, *Aurelia* is coastal cosmopolitan; *Cyanea capillata* coastal northern boreal; *C. lamarckii*, *Chrysaora* and *Rhizostoma* are southern boreal.

In addition there are seven deep-sea species of the genera *Nausithoë*, *Atolla*, *Paraphyllina* and *Periphylla*, none of which appears in coastal waters, but only live in deep water off the continental shelf west of the British Isles.

SEASONAL OCCURRENCE

In general Scyphomedusae are noticed in waters round British coasts especially in summer and autumn. In fact most are present all through the year in various stages of growth.

For instance, ephyrae of *Aurelia* and *Cyanea* may often be taken in the plankton as early as January, though the peak period of abundance is nearer March and April. The ephyrae of *Chrysaora* and *Rhizostoma* probably appear first a bit later in the year.

The adults which are common during the summer months tend to disappear in September and October when many dying specimens are washed ashore. But some in fact, certainly *Aurelia*, *Cyanea* and *Rhizostoma*, live on over the winter in deep water. *Aurelia* and *Rhizostoma* are taken sometimes in great numbers in fishermen's trawls in winter (see pp. 167, 169, and 196) and large *Cyanea* and *Rhizostoma* may be washed ashore in November, December and January, while *Rhizostoma* has been recorded as late as February and March.

In Scotland the first appearances may be noticeably later on the east coast than on the west coast. For instance in early May 1968 I saw many *Aurelia* up to 130 mm. in diameter and *Cyanea capillata* of 75 mm. or more in the neighbourhood of Oban. At St Andrews, however, these species were not apparent; ephyrae and young stages of *Cyanea* up to 7 mm. in diameter only were collected. A similar state of affairs was noted in June 1967.

SWARMING AND STRANDING

There are many records of the occurrence of British and other species of Scyphomedusae in great abundance and swarms. These swarms extend often over large areas, and are sometimes in the form of windrows. They can be a nuisance to fishermen and to bathers. Jellyfish have been reported as blocking the condenser tubes and intakes of ships and power stations. They may also be abundant enough to impede the progress of small boats (see e.g. p. 167).

A number of observers have suggested that the swarms may be the result of active congregation for spawning purposes. There can be no doubt that windrows are produced passively by wind, wave and tidal action, but it should not be assumed that this is necessarily entirely fortuitous. By their vertical migrations to the surface the jellyfish may put themselves in a position to become aggregated by physical forces. Maaden (1942a) noted a tendency for shoals to consist of individuals from the same origin (see p. 195).

There are no observations on British species correlating swarming with the act of spawning, but in this respect Conklin (1908) recorded the habits of the coronate medusa *Linerges mercurius* (= *Linuche unguiculata* (Schwartz)) at Tortugas, Nassau and Bermuda. There the medusae suddenly swam up from deeper layers to the surface in countless numbers, remaining there only for a few days at the end of which time the sexual products were liberated. The medusae then sank downwards again to the bottom and disintegrated. There was a restricted period for the act of spawning which took place at 8 a.m. In *Linerges* spawning was completed in a day or two and the animal died thereafter. From our knowledge of the development of the egg in the ovary of our British species it seems likely that release of eggs or planulae may take place at intervals over fairly extended periods. There might thus be intermittent spawning at the sea surface.

Scyphomedusae are very buoyant and may as a result be more in evidence near the surface when the sea is calm.

It is noteworthy that mature specimens are cast ashore in great numbers, especially in the autumn. These are often in a moribund state, but their eggs or planulae will thus have been liberated in suitable sites in shallow water for the development of the scyphistoma. It seems possible that the change which takes place in the thickening and toughening of the mesogloea towards death (see p. 168) may affect the flotation of the medusae and bring them to the surface or perhaps sink them to the bottom where they may become aggregated by tidal action.

The stranding of medusae has been the subject of a special investigation by Maaden (1942a) who studied their frequency on the Dutch coast in relation to the wind. The effects of wind will of course depend upon offshore hydrographical conditions and whether the medusae are deep down in the water or near the surface. It is remarkable how the situation changes; one day a beach may be littered with hundreds of medusae or their fragments and the next day there will be none to be seen.

The extent of the autumn stranding may be illustrated in Text-fig. 67 on p. 129. This shows for instance that *Cyanea capillata* may be stranded almost simultaneously along the whole coastline. Specimens were seen stranded in various places on the north-west, north and east coasts of Scotland and north-east coast of England all within a period of ten days in September 1967. In 1968, they were relatively scarce on the Scottish coasts compared with 1967, when they were extremely abundant.

It is also noteworthy that while in 1967 Scyphomedusae were exceedingly abundant in the north hardly a single specimen was seen off Plymouth.

Photographs of stranded specimens of *Chrysaora* and *Cyanea* are reproduced in M. E. Thiel (1962a, figs. 498, 499).

LIFE HISTORY

THE DISCOVERY OF THE SCYPHISTOMA

A fully documented account of the early history of the discovery of the life cycle of *Aurelia* and *Cyanea* is given by L. Agassiz (1860, **3**, p. 29). In 1829 Michael Sars published an account of a new genus of polyps and a peculiar genus of jellyfish, both distinct from all known polyps and medusae at that time. He named the former genus *Scyphistoma* and the latter *Strobila*. He described as one species, *Scyphistoma filicorne*, having 24–32 tentacles, a retractile and protractile mouth, and the body annulate. The other he called *Strobila octoradiata*, and he thought that it might be a form transitional between a fixed zoophyte and a medusa. Ehrenberg (1836) mistook it for a *Lucernaria* in the process of transverse fission, while Thienemann, who supplied an abstract of Sars' paper for 'Isis', suggested that Sars should see whether it might not be a stage in the early development of a medusa.

In 1835 Sars proved that the *Scyphistoma* was an early stage of the *Strobila*, and that free disks from the *Strobila* were closely allied to medusae described by Eschscholtz (1829) under the genus *Ephyra*. In his paper published in 1837 Sars, without furnishing evidence, was satisfied that the *Ephyra*-like medusa produced by the *Strobila* was a young stage of *Aurelia aurita*, and he clinched the matter in 1841 when he stated that he had reared scyphistomas from the eggs of both *Aurelia aurita* and *Cyanea capillata*.

Meanwhile in Germany Siebold (1839) had reared the eight-tentacled polyp from the eggs of *Aurelia*, while in Scotland J. G. Dalyell (later Sir John) (1834, 1836) described under the name *Hydra tuba* his experiments on keeping the polyps alive. He was the first to describe the reproduction by budding and kept colonies and their descendants alive for six years; one specimen had 83 descendants in 13 months. He noted that they bred successive generations at all times of the year, while in February and March they became elongated and gradually developed into 20 or 30 successive strata which broadened and were liberated. Dalyell's observations were remarkable but somewhat unsystematic. Nevertheless, in his *Rare and Remarkable Animals of Scotland* (1847) it is quite probable that he has depicted the Scyphistoma of *Aurelia* in his pl. XIII, of *Chrysaora* in pl. XIV, and of *Cyanea* in pl. XIX.

THE CORONATE SCYPHISTOMA

Brief mention only is given of the scyphistoma stage of coronate medusae as they will not be found in British coastal waters. Some further information is given on p. 28 and detailed descriptions will be found in Komai (1935) and Werner (1967).

The following are some of the chief characters in which it differs from the semaeostome and rhizostome scyphistoma.

1. It has a complete chitinous tube.
2. The tube is blocked at its distal end by the oral end of the polyp before strobilation.
3. Ephyrae are produced from the lower end and upper end of the strobila.
4. The chain may be released from the tube before the ephyrae separate, or the ephyrae may swim out of the tube.
5. Very great numbers of ephyrae are produced.

The coronate scyphistoma *Stephanoscyphus* was first described as a hydrozoan genus by Allman (1874) from the Mediterranean (see p. 28). An up-to-date list of literature is given by Werner (1966).

THE SEMAEOSTOME AND RHIZOSTOME SCYPHISTOMA

The typical full-grown scyphistoma (Text-fig. 7) is a small sessile polyp whose external appearance is that of a short inverted cone or beaker, the *calyx*, which merges into a narrower short or long cylindrical *stalk* whose base or *pedal disk* is attached to the substratum. At its oral

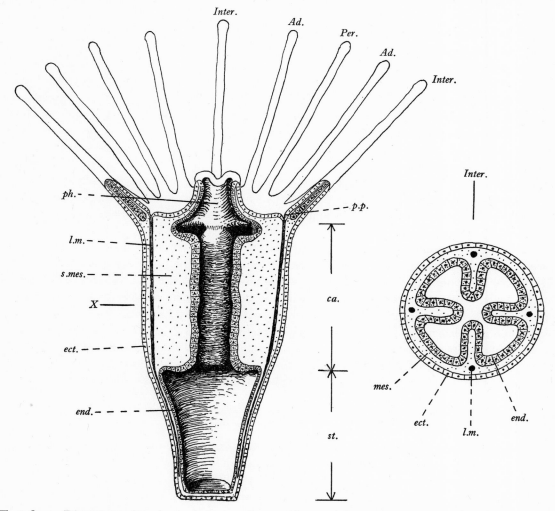

Text-fig. 7. Diagram to show basic structure of a scyphistoma polyp. *ca.* calyx; *ect.* ectoderm; *end.* endoderm; *l.m.* longitudinal muscle; *mes.* mesogloea; *ph.* pharynx; *p.p.* peristomial pit; *s.mes.* septal mesogloea; *st.* stalk. *X*, level of section shown on the right.

end is a flat circular oral disk with a protuberant mouth or *proboscis* in its centre. Around the margin of the oral disk is a whorl of sixteen to twenty-four or more tentacles. Four tentacles are perradial, four interradial, and eight adradial, this being usually the succession of their formation and growth. The lips of the mouth when closed are cross-shaped, the four arms of the cross being perradial.

The mouth leads into the gastric cavity which extends to the aboral base of the polyp. In the stalk region this cavity is cylindrical and entire. The portion of the cavity above the stalk is, however, divided into four peripheral compartments by *longitudinal septa* or taenioles extending along the wall from the oral disk to the oral end of the basal stalk, leaving a narrow undivided central cavity. These septa are interradial, a transverse section in this region showing the cross-shaped gastric cavity whose arms are perradial.

Running through the whole length of the scyphistoma from the oral disk almost to the pedal disk itself are four interradial *longitudinal muscles* buried in the mesogloea, one at the base of each longitudinal septum. Situated on the oral disk just above the insertion of each longitudinal muscle is a small depression known as the *peristomial pit*; these four pits are therefore interradial in position.

Externally the scyphistoma is covered with a single layer of ectoderm cells. Internally there is a single layer of endoderm cells, the interradial longitudinal septa being formed as folds in this layer. Between the two cell layers lies a thin mesogloea. The tentacles have a solid core of cylindrical cells. The four longitudinal muscle strands are ectodermal in origin, arising from the ectoderm of the oral disk and calyx (Widersten, 1967). It is thought by some that the lining of the oesophageal region of the proboscis is partly ectodermal in origin, but D. M. Chapman (1966, 1968) regarded the epithelial layer covering the inner surface of the pharynx and the free edges of the septa as a type of mesoderm distinct from the ectoderm and endoderm, and this area he called the scyphopharynx-filament complex. At the base of the stalk there may be a vestigial cuticle (D. M. Chapman, 1968).

DEVELOPMENT FROM EGG TO SCYPHISTOMA

While in all coastal species of semaeostome and rhizostome medusae it is probable that normally there is a sessile fixed scyphistoma stage in the life cycle, direct development from egg to medusa is known to occur in the oceanic semaeostome *Pelagia noctiluca* (Text-fig. 44, p. 81).

In all coastal species it is probably the general rule that the eggs are fertilized in situ in the medusa or in its gastric cavity; and that they develop either in the ovary or in special pockets in the stomach wall or manubrium, after which they are liberated to the exterior as free-swimming planulae.

All species so far described have separate sexes, except for *Chrysaora*, which is a protandrous hermaphrodite.

The eggs vary much in size, even within the same species, and would seem mostly to be between 0·1 and 0·2 mm. in diameter; but in *Pelagia noctiluca* they reach 0·3 mm., possibly related to its direct development (Berrill, 1949*b*).

The histological development from egg to scyphistoma was studied by a number of workers at the end of the nineteenth and beginning of the twentieth centuries. The following are the references for British species: *Pelagia*, Kowalevsky (1873), Goette (1893), Metschnikoff (1886); *Chrysaora*, Claus (1877, 1883, 1891), Hein (1900), Friedemann (1902), Hadži (1909), Heric (1909); *Cyanea*, Hamann (1890), McMurrich (1891), Smith (1891), Hyde (1894), C. W. & G. T. Hargitt (1910); *Aurelia*, Claus (1877, 1883, 1891), Goette (1887), Smith (1891), Hyde (1894), Hein (1900), Friedemann (1902), C. W. & G. T. Hargitt (1910).

Observations made by different workers, even on the same species, gave different results. It is probable that most of these differences were genuine, but a few were possibly due to misinterpretation of sections.

LIFE HISTORY

It would seem that segmentation of the egg, which is total, may be equal or unequal. This gives rise to a blastula whose cavity in those of the smallest size may be very small. The first differences of opinion arose over the process of gastrulation. Some observed normal invagination (*Pelagia*, *Chrysaora*), and others ingression of cells (*Cyanea*). One or more methods were observed in the same species (*Aurelia*). Cells which had ingressed into the blastula cavity were observed by some to degenerate and play no part in gastrulation.

Berrill (1949b), in a general analysis of development in Scyphozoa, concluded that the method of gastrulation appeared to be correlated with the size of the egg and resulting proportional differences between the size of the blastula and the thickness of its wall. Thus, in blastulae resulting from the smallest eggs unipolar ingression of cells occurs, and in those from the largest eggs invagination is the rule. From eggs of intermediate sizes combinations of ingression and invagination occur to varying degrees; and Berrill noted that it is significant that there is much variation in the size of the egg of *Aurelia* for which both methods of gastrulation have been recorded.

The gastrula, however formed, gives rise to a free-swimming planula which soon develops dissimilar oral and aboral poles. It soon attaches to some suitable substratum by its aboral end at a point slightly to one side of the apex. After attachment the oral end flattens to form the oral disk, the mouth develops and rudiments of the first four perradial tentacles appear, two opposite tentacles sometimes developing in advance of the others. At this stage the first signs of the interradial longitudinal septa appear as cushion-like swellings on the underside of the oral disk and extending to the wall of the gastric cavity. Very soon the ectodermal longitudinal muscle strands develop from ingrowing plugs of oral disk ectoderm cells, giving rise to the first signs of the peristomial pits, and from the ectoderm of the calyx (Widersten, 1967, *Cyanea*).

It is during this stage of development that the most significant differences of opinion have been held, namely whether the inner lining of the proboscis originates from ectoderm or endoderm (summarized by Gohar & Eisawy, 1960).

METHODS OF REPRODUCTION AND BIOLOGY OF THE SCYPHISTOMA

Scyphistomas can increase their numbers by a variety of methods of which the more normal are:

1. Buds are developed from the side of the polyp near the junction of the calyx with the stalk: these buds may (*a*) grow out and downwards towards the base, becoming attached to the substratum and constricting off at about the same time, the distal end forming the pedal disk; (*b*) grow out and upwards and form a stolon before liberation.

2. Buds develop on stolons growing out from the parent just above the junction of the calyx with the stalk. A bud may arise proximally, distally, or midway along the stolon, and there may be two on one stolon.

3. Hydra-like buds may grow up from just above the junction of the calyx with the stalk as wide outgrowths with mouth and tentacles at the distal end, which may or may not separate from the parent.

4. In some species it has been recorded that buds form as wide diverticulae from the polyp wall and swim away by ciliary activity before the formation of the oral and pedal disks.

5. The polyps strobilate and form free-swimming ephyrae which grow into medusae which give rise once more to planulae by sexual reproduction.

6. Resistant podocysts are formed beneath the pedal disk from which polyps are later developed.

Good detailed descriptions of budding are given by Perez (1922) and Halisch (1933). Gilchrist (1937), using scyphistomas presumed to be those of *Aurelia* from Pacific Grove and San Diego, described in detail the budding and stolonization. As regards budding he differentiated between the fig-type bud of Perez and Halisch, the stolonic bud, and the hydra-type bud. He also described non-budding stolons.

The pedal stolons typically grow outwards from the upper portion of the stalk and slightly upwards away from the base. The end then bends down and attaches to the substratum. By contraction these stolons enable the polyp to move.

Gilchrist did a number of experiments on regeneration, disclosing a polarity in the power to regenerate complete hydranths, the upper part of the polyp being the most strongly hydranth-forming, the tendency decreasing towards the aboral parts of the body.

This regenerative polarity is paralleled in the budding of the scyphistoma, buds being produced from the upper part of the body and non-budding stolons from the stalk.

Gilchrist found that the ectoderm alone could regenerate a whole polyp, but that endoderm alone rounds up and does not regenerate. Steinberg (1963) found that the endoderm in such polyps was formed by amoeboid cells which probably arose by de-differentiation of the ectoderm.

In Text-fig. 8 are reproduced four successive drawings I made at intervals showing the development and separation of buds from their parent polyps kept in the laboratory.

The formation of the podocyst has been investigated by D. M. Chapman (1968) in *Aurelia* (see p. 152).

Almost all the published work on the biology of scyphistomas has been the result of observations made on animals kept in the laboratory or aquaria for purposes of studying the histological details of development. Nevertheless all the above methods of reproduction have been recorded under such conditions and the various processes have been noted as occurring in a certain sequence. This has given rise to speculation about the factors of the environment which may influence the method of reproduction. Owing to the seasonal nature of the sequence it was natural that temperature should have been regarded as having some influence, but it was frequently noted that the amount of food eaten had a marked effect. For instance it was noticed that just before strobilation started the animal refused to eat. Tchéou-Tai-Chuin (1930*b*) in his histological study of various aspects of the biology of the scyphistoma of *Chrysaora* recorded the occurrence of reserve food materials and their transfer by migrating cells to regions of active cell proliferation during strobilation. Since feeding ceases at the onset of strobilation it could be assumed that reproduction by strobilation required the building up of food reserves to a threshold level first. The amount of food available would obviously therefore be an important environmental factor.

Lambert (1936*a*, *b*), a naturalist who had made a special study of living scyphistomas, concluded that the necessary conditions leading to strobilation were abundant food, a certain critical low temperature, and a rich oxygen supply at a critical period. But in the laboratory natural conditions can never be fully satisfactorily produced and such hypotheses need to be tested with observations in the field.

The first really detailed field observations were made by Hjalmar Thiel (1962) on natural populations of scyphistomas of *Aurelia aurita* in Kiel Bay. These observations were accompanied by instructive experiments made on test plates in the field. The situation chosen proved most suitable since the environmental conditions differed between the inner and outer harbours.

Adult medusae were liberating their planulae in the middle of August and within two to three weeks the primary polyps had sixteen tentacles, having lost their planula coloration and become

white. From September to mid-December daughter polyps were produced by lateral buds connected by a thin stalk to the parent (Type 1 above), but from the middle to the end of November
stolon buds began to appear (Type 2 above). Their numbers increased in proportion to those of
the lateral buds until at the beginning of December they formed about 95% of all buds. During

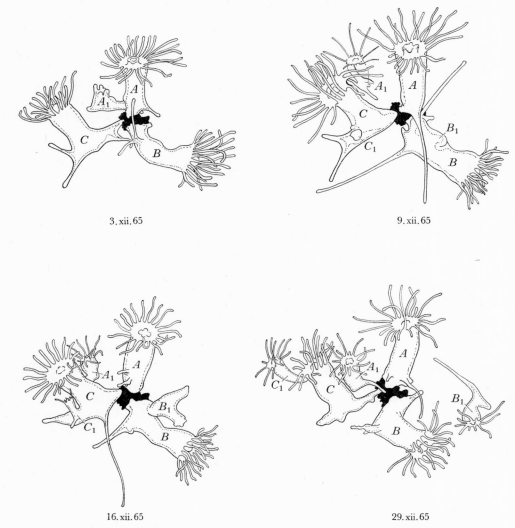

3. xii. 65

9. xii. 65

16. xii. 65

29. xii. 65

Text-fig. 8. Successive stages in development of buds on the dates given on polyps kept in the laboratory
at Plymouth. *A*, *B* and *C*, parent polyps; A_1, B_1 and C_1, buds.

December the rate of budding fell away and strobilation, which was already apparent in isolated
individuals in November, began to increase. In February and the beginning of March only club-
shaped reduced polyps remained. These later redeveloped and began to produce daughter polyps
and podocysts. Occasionally Hydra-like budding (Type 3 above) occurred in the spring.

While the above was the general sequence of events from autumn to spring, strobilation
continued among a small proportion of the population throughout the summer, there being a

pronounced smaller secondary peak in June. Thus in Kiel Bay strobilation can occur throughout the year, but in 1960/1 at any rate there were two peaks, the main one of which occurred in the winter.

Thiel found certain differences between the scyphistoma populations in the inner and outer harbours. He noted for instance the following differences in the sizes of the ephyrae and numbers produced per polyp, in the inner and outer harbours respectively.

Ephyra	Averages	
	Inner mm.	Outer mm.
Size*	1·6	3·6
Number per polyp	1·8	6·0

* (Measured between opposite sense organs).

He found also a slight increase in abnormality among those from the inner harbour in which the percentage was 1·5 (2,051 examined) compared with 0·7% (1,756 examined) in the outer harbour. The number of polyps strobilating compared with those not strobilating was also much higher in the outer than in the inner harbour.

On the assumption that these differences might be due to differences in the amount of food available he inserted a pair of test plates with attached scyphistomas into each harbour. Every three days the polyps on one plate of each pair were fed in the laboratory with *Artemia* larvae, each of the other two plates remaining unfed as controls. At the end of January and beginning of February the polyps on *both* plates in the outer harbour began to strobilate. But strobilation only appeared in polyps on that plate on which they had been fed in the inner basin. This situation continued throughout February and March.

This is the first really conclusive evidence that food supply is an effective controlling factor in nature.

By examination of individual polyps on his test plates Thiel found that it was possible for the same polyp to strobilate two, three, or even four times in a year. The ability to do this is also something that might be expected to be influenced by food supply.

Under natural conditions in the outer harbour the average duration of the strobilation of a polyp was 70 days, with a variation of 35 to 98 days, during the first winter strobilating period. At the time of the second peak of strobilation in the summer it was only 25 days, with a variation of 14 to 56 days. The shorter duration in summer must presumably be due to higher temperature. The average period between the first and second strobilation of a polyp was 58 days, with the very wide variation of 23 to 112 days.

Among possible factors which might determine the length of these periods Thiel suggests that the number of ephyrae produced per polyp must play a part.

There appears to be a period of arrested development or stagnation when the temperature falls to about 2° C. Thiel found that in a cold room the development of the strobilae ceased at 0–2° C, and that it also prevented strobilation from starting.

Other workers had made comparisons of the first appearance of ephyrae in the plankton with the temperature conditions. On this basis Verwey (1942), for instance, suggested a temperature region of 4–11° C as a necessary condition for strobilation. Thiel however points out that by this indirect method it is not possible to establish the time of onset of strobilation. But he noted that the late appearance of ephyrae in the eastern Baltic compared with that in the western Baltic is evidently temperature-dependent, since in the east from January to April the temperature remains below 1–2° C, which he had shown prevents the onset of strobilation.

Thiel concluded that since (1) strobilation can occur throughout the whole year; (2) there are two peak periods of strobilation; and (3), an individual polyp may strobilate more than once in the same year, temperature could not be a very important factor in nature. He drew attention to two other conditions of the environment that altered, like temperature, in the course of the year, namely plankton blooms and light intensity.

On the importance of food Thiel had two definite indications: (1) his experiments, already mentioned above; and (2) the occurrence of rich copepod and cladoceran swarms coinciding with the appearance of many new strobilae in the outer harbour at Kiel in 1961 and 1962. He cited also a record given by Korringa (1953) of the occurrence in the Oosterschelde in October and November of many strobilating scyphistomas on oysters when food conditions were known to be good for the oysters; the first strobila was seen as early as 30 August 1943 at a temperature of over 17° C.

That light intensity can also have an effect on strobilation was shown experimentally by Custance (1964) who found that too strong a light inhibited strobilation in *Aurelia*. Scyphistomas do of course tend to occur in shaded situations in nature, and it is in the winter months when light intensity is lowest. Kakinuma (1965) found 15° C in light to be most favourable.

To sum up, our present knowledge indicates that while the temperature and light intensity cycles play their part in conditioning the onset of winter strobilation, it would seem that throughout the year the food supply may play a predominating role, a threshold of certain food reserves being necessary before strobilation can begin. In any latitude the life cycle is probably geared to the conditions of the environment. Thus, in British and northern waters strobilation occurs mainly in the cold period of the year. That a cold temperature is however not necessary under favourable conditions is shown by the work of Spangenberg (1965 b) who kept *Aurelia aurita* scyphistomas in artificial sea water for three years at Little Rock, Arkansas. The scyphistomas were fed twice weekly on newly hatched *Artemia* larvae at a room temperature between 21 and 24° C. During the three years strobilation occurred continually, with only sporadic mass strobilation. *Aurelia* is, of course, a species with very wide distribution and must have races conditioned to a wide range of different temperatures (see e.g. p. 166 below; Mayer, 1914).

Scyphistomas can be induced to strobilate by changing the temperature. For instance, Sugiura (1965, 1966) found that it was necessary to raise the temperature to 20° C before the scyphistomas of *Mastigias papua* would strobilate and to 28° C for *Cephea*. If *Mastigias* had not strobilated at 20–22° C a further rise in temperature had no effect, but they could be made to strobilate by cooling to 20° C for about a month and then raising the temperature again.

Custance (1966) found that scyphistomas could continue to strobilate in the laboratory if the temperature was gradually raised to 12–14° C, but that a sudden rise in temperature above 10° C caused reversal of development and the production of tentacles in place of ephyra lappets. Later (Custance, 1967) he found also, that while the amount of food available influenced the number of ephyrae produced, starving scyphistomas would still strobilate. (See also p. 118 below; Loomis, 1961).

It is well known, however, that shock effects may induce spawning in marine animals and such conditions are not normally met with in the sea.

In laboratory experiments strobilation has also been induced by treating with iodine by Paspalev (1938) and Spangenberg (1967). (See below, pp. 152, 186.)

STROBILATION

Just before strobilation the upper portion of the scyphistoma starts to elongate, but the degree of elongation varies between species and according to circumstances. The two extremes are the monodisk form, producing only a single ephyra and usually having a narrow stalk, and the polydisk form with few or many annulae, which may produce up to as many as thirty ephyrae. The elongation is not due to proliferation of cells but to flattening of the existing cells (D. M. Chapman, 1966).

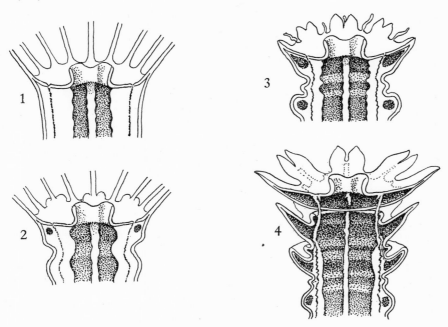

Text-fig. 9. Sectional diagrams to show stages in the process of strobilation.

In the polydisk scyphistoma shortly before the terminal ephyra is ready to be liberated the tentacles of the polyp are still present while the marginal lappets of the ephyra are well developed.

The processes whereby the changes in the structure of the scyphistoma take place have been the subject of detailed investigations by a number of authors, the most recent and complete being those of Percival (1923) on *Aurelia* and Tchéou-Tai-Chuin (1930) on *Chrysaora*.

The first external indication of strobilation (Text-fig. 9) is the appearance of a circular groove round the body of the scyphistoma a short distance below the tentacles. This is due to a proliferation of a circle of both ectoderm and endoderm cells immediately above and below the region of the groove which form protruding rings. Similar proliferations occur at intervals down the body of the polyp so that it appears to be encircled by a series of equidistant grooves. At the same time ostia appear in each gastric septum just above the position of each groove. In this way the neighbouring gastric compartments become connected to form a series of what have been called ring-sinuses. As the future ephyra increases in diameter the septal ostia increase in area peripherally so that the main peripheral region of the stomach cavity is continuous.

21

LIFE HISTORY

Simultaneously with these changes two other processes are taking place. First, additional grooves appear between the first formed grooves, again due to proliferation of circles of ectoderm and endoderm cells. These secondary zones do not however proliferate to the same extent as the primary zones, so that lesser protruding rings are formed between the primary rings. These rings are the future manubria of the ephyrae, the proboscis of the scyphistoma itself forming the manubrium of the terminal ephyra.

Secondly, the margins of the future ephyrae are becoming lobed. In the terminal ephyra, whose subumbrella is the original peristome of the scyphistoma, the tentacles become reduced by degeneration, cutting off at the base, or by fragmentation. In a typical 16-tentacled scyphistoma lobes grow outwards on either side of and at the bases of a pair of tentacles. These are the future lappets of the lobe of the ephyra. By the time the tentacles are completely reduced the eight lobes, each with their two lappets, are fully formed. Between the lappets the marginal sense organs are beginning to develop.

Meanwhile circular fissures begin to appear between the primary and secondary rings, being the lines of eventual separation of the margins of the manubrium from the exumbrella of the ephyra immediately above it. When these have parted all that now joins the ephyra to the parent stock are the remnants of the longitudinal muscles, at whose subumbrellar junction the gastric filaments develop. The active movements of the ephyra finally break this connection.

THE IDENTIFICATION OF EPHYRAE

The newly liberated ephyrae of the British semaeostome and rhizostome medusae are very similar. It is likely that there may be differences in proportional lengths of marginal lobes and lappets, as for instance drawn by Delap who reared specimens from the scyphistoma. But the shape and length of the marginal lappets is not a useful diagnostic character as it depends so much on their degree of contraction or that of the central disk in preserved specimens. Undoubtedly in their perfect state each marginal lappet is of an asymmetrical spatulate shape ending in a sharp point.

The size of the ephyra at liberation is also probably variable within the same species according to the size of the scyphistoma; and ephyrae grow very quickly when they start to feed.

There are, however, structural characters already present in the newly liberated ephyra which can be used as distinguishing features and others very soon develop (Text-fig. 10).

The first character to look at is the end of the radial canal. In *Aurelia* the perradial and inter-radial canals have somewhat square-cut ends. In *Cyanea* they have lateral horns which extend some way into the lappet on either side; similar horn-like diverticula are also present on the adradial canals projecting into the bases of the marginal lobes.

Pelagia, *Chrysaora* and *Rhizostoma* have rounded processes at the corners of the ends of the radial canals intermediate between those of *Aurelia* and *Cyanea*, but *Chrysaora* can be distinguished by pairs of nematocyst clusters at the bases of the marginal lobes.

The number of gastric filaments may also be diagnostic, e.g. in *Cyanea* there is never more than one in each quadrant, in *Aurelia* there are two or three.

In *Cyanea* one marginal tentacle develops first by the side of a perradial marginal lobe, to be followed quickly by another in the opposite position. In *Aurelia* all eight primary marginal tentacles develop more or less simultaneously. In *Pelagia* and *Chrysaora* eight marginal tentacles also develop but their ephyrae can be distinguished from those of *Aurelia* now by the development of the radial canals. In *Aurelia* outgrowths develop at the bases of the perradial and interradial canals and on the sides of the adradial canals which soon join up to form the marginal ring canal. In *Pelagia*, *Chrysaora* and *Cyanea* there are no such outgrowths. *Rhizostoma* has similar outgrowths to *Aurelia* but it has no marginal tentacles, and paired marginal velar lappets soon develop between the rhopalar lobes. Small tentacles also appear on the margins of the oral lips in *Aurelia* and *Rhizostoma*, but in *Rhizostoma* the lips start to divide to form the eight oral arms.

At a very early stage *Pelagia* ephyrae develop prominent nematocyst warts on the exumbrella arranged in a distinctive pattern. In *Pelagia* also the tentacular gastric pouches remain simple, but in *Chrysaora* protrusions develop from the corners of these pouches from which the additional marginal tentacles will arise to form the eight groups of three tentacles as opposed to the eight single tentacles in *Pelagia*.

There are other characters which distinguish living and freshly preserved specimens. For instance, the ephyrae of *Cyanea capillata* are a rich yellow ochre in colour, while those of *Pelagia* are golden yellow.

A diagram illustrating these different distinguishing characters during the course of development of *Chrysaora*, *Cyanea*, *Aurelia* and *Rhizostoma* is given in Text-fig. 10. *Pelagia* has not been included as it will only rarely be found in coastal waters.

Text-fig. 10. Diagrammatic key to the identification of ephyrae.

SUBJECT INDEX TO LITERATURE
OF SCYPHOMEDUSAE

This subject index is not complete and relates mainly to British species. It includes, however, sufficient references to enable the student to find what is known under the different headings.

ALIMENTARY SYSTEM

Digestion. Bodansky & Rose, 1922, enzymes in gastric filaments of *Stomolophus*; Metschnikoff, 1880, amoebocytes in *Pelagia*; Müller, Fr., 1859, gastric filaments of *Tamoya*; Smith, 1936, *Cassiopea*; M. E. Thiel 1959 *b*, historical summary; Wetochin, 1930, amoebocytes in *Aurelia*.

Food and feeding. Bozler, 1926 *b*, reactions of *Pelagia*; Delap, 1901, *Chrysaora*, 1905, *Cyanea* ephyra, 1907, *Pelagia* and *Aurelia*; Gemmill, 1921, ciliary currents in *Aurelia* ephyra; Gosse, 1853, *Chrysaora*; Hagmeier, 1933, *Cyanea* rearing; Hargitt, C. W. & G. T., 1910, *Cyanea* scyphistoma; Henschel, 1935, reactions of *Aurelia*; Lebour, 1922, 1923, *Chrysaora*, *Aurelia*; Orton, 1922, ciliary currents in *Aurelia*; Percival, 1923, ciliary currents in *Aurelia* scyphistoma; Southward, 1949, 1955, ciliary currents in *Aurelia*; Thiel, M. E., 1964, *Rhizostoma*, with review of literature; Wetochin, 1926, ciliary currents in *Aurelia*; Widmark, 1911, 1913, ciliary currents in *Cyanea* and *Aurelia*; Fraser, 1969, *Cyanea*, *Aurelia*.

Starvation, effects of. de Beer & Huxley, 1924, *Aurelia*; Hatai, 1917, *Cassiopea*; Mayer, 1914 *b*, *Cassiopea*; Steiner, 1935, *Aurelia* ephyra; Thill, 1937, *Aurelia*; Will, 1927, *Aurelia* scyphistoma and ephyra; Vernon, 1895, *Rhizostoma*.

NEUROMUSCULAR SYSTEM

Anatomical. Bullock & Horridge, 1965, general survey; Chapman, D. M., 1965, *Aurelia* scyphistoma; Eimer, 1878, *Aurelia*, *Rhizostoma*; Friedemann, 1902, *Aurelia* ephyra; Hesse, 1895, *Rhizostoma*; Horridge, 1953, 1954 *a*, *b*, 1956 *b*, *Cyanea*, *Aurelia* ephyra and adult; Horridge & Mackay, 1962, electron microscopy of *Cyanea* ganglion; Komai, 1942, *Pelagia* ephyra; Korn, 1966, *Cyanea*; Maaden, 1939, *Aurelia*; Schäfer, 1878, *Aurelia*; Schewiakoff, 1889, *Aurelia*; Woollard & Harpman, 1939, *Chrysaora*, *Cyanea*; Yamashita, 1951 *a*, *Aurelia*.

Experimental. Barnes, 1964, ion concentration, *Cyanea*; Barnes & Horridge, 1965, extracts of rhopalial region; Bauer, 1927, locomotion, *Pelagia*, *Rhizostoma*, *Cotylorhiza*; Bethe, 1903, 1908, neurophysiology, *Rhizostoma*; Bozler, 1926 *a*, *b*, *Pelagia*, *Rhizostoma*, *Cotylorhiza*; Bullock, 1943, facilitation, *Cyanea*, *Aurelia*, *Rhopilema*; Frankel, 1925, swimming movements, *Rhizostoma*, *Cotylorhiza*; Heymans & Moore, ionic content, *Pelagia*; Horridge, 1955, muscle physiology, *Rhizostoma*, 1959, rhythm in *Aurelia*, *Pelagia* habits; Horstmann, 1934 *a*, *b*, *Cyanea*, *Aurelia*; Jordan, 1912, *Rhizostoma*; Lehmann, 1923, *Chrysaora*, *Cyanea*; Loeb, 1899, *Aurelia*; Mayer, 1906, *Dactylometra*, *Aurelia*; Morse, 1910, *Cyanea*, *Aurelia*; Pantin & Dias, 1952, rhythm in *Aurelia*; Passano, 1958, nervous activity, *Cyanea*, 1965, rhythm in *Cyanea*, *Cassiopea*; Passano & McCullough, 1960, 1961, nervous activity; Romanes, 1876 *b*, 1877 *b*, *Aurelia*; Schaefer, 1921, *Chrysaora*, *Cyanea*, *Rhizostoma*; von Skramlik, 1945, *Pelagia*; Thill, 1937, *Aurelia*; v. Uexküll, 1901, *Rhizostoma*; Vernon, 1895, *Rhizostoma*; Wetochin, 1926 *b*, *Aurelia*; Yamashita, 1957 *b*, *Aurelia*.

CHEMICAL OBSERVATIONS

Chemical composition. Aarem et al. 1964, lipid in *Rhizostoma*; Danmas & Ceccaldi, 1965, amino acids in *Rhizostoma*; Denton & Shaw, 1962, ionic composition and buoyancy, *Pelagia*; Hattai, 1917, *Cassiopea*; Haurowitz, 1920, fats in *Rhizostoma*; Koizumi & Hosoi, 1936, *Cyanea*; Nicol, 1957, coronal muscle, of *Atolla*; Robertson, 1949, *Aurelia*; Stipos & Ackman, 1968, lipids in *Cyanea*; Teissier, 1932, planula of *Chrysaora*.

Fluorescence. Ries & Ries, 1924, *Aurelia*.

Mesogloea. Alexander, 1964, visco-elastic properties in *Chrysaora* and *Cyanea*; Astbury, 1940, collagen; Chapman, 1953, structure and composition in *Chrysaora*, *Cyanea* and *Aurelia*; Christomanos, 1954, composition; Macallum, 1903, composition, *Cyanea* and *Aurelia*.

Pigment. Christomanos, 1954, *Rhizostoma*; Fox & Pantin, 1944, *Rhizostoma*; Krukenberg, 1882, *Rhizostoma*; Millott & Fox, 1954, *Pelagia colorata* (as *P. panopyra*); von Zeynek, 1912, *Rhizostoma*.

Water content. Bateman, 1932, *Chrysaora*, *Cyanea*, *Aurelia*; Hyman, 1938, 1943, *Aurelia*; Krukenberg, 1880, *Chrysaora*, *Aurelia*; Lowndes, 1942, *Aurelia*; Möbius, 1880, 1882, *Aurelia*; Teissier, 1926, *Chrysaora*; Thill, 1937, *Aurelia*.

SUBJECT INDEX

HABITS AND BIOLOGY

Luminescence. *Atolla*, Nicol, 1958.

    Periphylla, Nicol, 1958.

    Pelagia, Dahlgren, 1916; Dubois, 1914; Heymans & Moore, 1923, 1924; Hunt, 1952; Hykes, 1928 (? also *Rhizostoma*); M'Andrew & Forbes, 1847; Moore, 1926; Newton-Harvey, 1952; Nicol, 1958; Panceri, 1872; Vanhöffen, 1902.

Pressure sensitivity. Digby, 1967, *Cyanea*; Ebbecke, 1935, *Cyanea*; Rice, 1964, *Aurelia* ephyra.

Rate of growth. Krüger, 1968, oxygen requirement and growth, *Rhizostoma*, *Cyanea* and *Chrysaora*; M'Intosh, D. C., 1910, *Aurelia*; Russell, 1931, *Cyanea*; Thiel, M. E., 1959 a, *Aurelia*, 1966 a, *Rhizostoma*.

Regeneration. Goldfarb, 1914, effects of salinity in *Cassiopea*; Hargitt, C. W., 1904, *Rhizostoma*; Kaufmann, 1957, scyphistoma of *Cyanea*; Stockard, 1908, *Cassiopea*; Wetochin, 1926 b, sense organ of *Aurelia*.

Stinging. Barnes, 1967, *Cyanea* poison; Bigelow, 1926, *Cyanea*; Cleland & Southcott, 1965, *Pelagia*; Ehrenberg, 1837, *Aurelia*, no sensation on tongue; Eysenhardt, 1821, *Rhizostoma*; Evans, Muir, 1943, *Rhizostoma*; Evans & Ashworth, 1909, *Cyanea*; Gosse, 1863, *Chrysaora*, *Cyanea*; Halstead, 1965, review; Högberg et al. 1956, 1957, histamine-releasing principle in *Cyanea*; Krumbach, 1925, *Rhizostoma*; Mitchell, 1902, stinging of cornea by *Cyanea*; Russell, F. E., 1965, review; Stadel, 1965, medical treatment, good list of references; Thiel, M. E., 1935, review; Wagner, 1841, *Pelagia*; Welsh, 1955, *Cyanea* extract on crab autotomy and mollusc heart activity; Wood, 1882, *Cyanea*; Zeynek, 1912, *Rhizostoma* causing sneezing.

Swarming. Agassiz, L., 1862, *Cyanea*; Bigelow, 1926, *Aurelia*; Bourne, 1890, *Pelagia*; Conklin, 1908, spawning of *Linuche* (as *Linerges*); Johnstone, 1908, *Aurelia*; M'Intosh, 1885, *Cyanea* habits; Möbius, 1880, *Aurelia*; Vanhöffen, 1896, 1902, *Pelagia*.

Vertical distribution. Russell, 1927, *Cyanea*; Vallentin, 1907, *Rhizostoma*; Verwey, 1966, vertical movements and state of tide, *Chrysaora*, *Rhizostoma*.

Weather effects. Bauer, 1927, *Pelagia*, *Rhizostoma*; Claus, 1877, *Aurelia*; Eysenhardt, 1821, *Rhizostoma*; Horstmann, 1934 a, *Cyanea*.

COMMENSALS AND PARASITES

Bowman et al. 1963, *Aurelia* and *Hyperia*; Dahl, E., 1959 a, b, nematocysts of *Cyanea* in stomach of *Hyperia*, 1961, *Cyanea* and young fish; Damas, 1909 a, b, *Cyanea* and young gadoids; Gosse, 1856, *Cyanea* and young whiting; Hjort & Dahl, 1900, *Aurelia* and young whiting; Hollowday, 1947, *Rhizostoma* and *Hyperia*; Künne, 1948, *Cyanea* and larva of *Peachia*; Lambert, 1936 a, *Aurelia*, *Cyanea* and *Hyperia*; Lawless, 1877, *Aurelia* and young fish; M'Andrew & Forbes, 1847, *Pelagia* and crustacean; Mansueti, 1963, review of association between jellyfish and young fish, with bibliography; Monticelli, 1897, cysts of *Pemmatodiscus* in *Rhizostoma*; Nagabhushanam, 1959, *Rhizostoma* and *Hyperia* and gadoids; Rees, 1966, *Cyanea* and young whiting; Romanes, 1876, 1877 a, *Aurelia* and *Hyperia*, 1877 c, *Aurelia* and young fish; Russell, 1928, *Cyanea* and young whiting, 1930, *Cyanea* and young *Caranx*; Scheuring, 1915, *Cyanea* and young fish, review of literature; Shojima, 1963, *Aurelia* and phyllosoma larvae; Stiasny, 1937, *Pelagia* and amphipods; Tattersall, 1907, *Pelagia*, *Chrysaora* and *Aurelia* and *Hyperia*; Thomas, 1963, *Pelagia* and phyllosoma larvae.

VARIATIONS AND ABNORMALITIES

Agassiz, A. & Woodworth, 1896, review to date; Ballowitz, 1899, *Aurelia*; Bateson, 1892, *Aurelia*; Brandt, 1870, *Rhizostoma*; Browne, 1889, 1895 b, 1901, *Aurelia* adult and ephyra; Ehrenberg, 1837, *Aurelia*; Hargitt, C. W., 1905, *Aurelia* adult and ephyra, *Rhizostoma*; Lambert, 1936 a, *Aurelia*; Low, 1921, *Aurelia* ephyra; M'Intosh, D. C., 1910, 1911, *Aurelia*; Romanes, 1876 a, 1877 a, *Aurelia*; Sorby, 1894, *Aurelia*; Stiasny, 1928, 1929, *Rhizostoma*, 1930, *Aurelia*; Thiel, M. E., 1959 b, review to date, 1960, *Cyanea*; Thiel, Hj., 1963 a, b, *Aurelia* ephyra; Unthank, 1894, *Aurelia*.

ECONOMICS

Fishing industry. Fraser, 1954, inconvenience by *Pelagia*; Kramp, 1937, bursting of nets and stinging by *Cyanea*; Hela, 1951, clogging of nets by *Aurelia*, inverse relationship of *Aurelia* to sprat and herring; Mauchline, this volume, p. 167, effect of *Aurelia* shoals on fishing in Faeroes; Netschaeff and Neu, 1940, clogging of nets by *Rhizostoma*.

Miscellaneous. Borlase, 1758, jellyfish eaten by man; Davidson & Huntsman, 1926, dead *Aurelia* and productivity; Pope, 1951, eaten by man; Weill, 1934, p. 243, *Aurelia* used for medical purposes.

ORDER

CORONATAE

Scyphomedusae with umbrella with coronal groove surrounding a central disk and with peripheral thickenings of jelly or pedalia corresponding to numbers of solid marginal tentacles and marginal sense organs, and with marginal lappets; with stomach attached to subumbrella over four inter-radial triangular gastric septal areas of exumbrellar and subumbrellar endodermal fusion; with radial septa in gastrovascular sinus; mouth with simple lips.

An attached scyphistoma stage is known for two non-British species, *Nausithoë punctata* and *Atorella vanhoeffeni*, but the life histories of all other species are still unknown.

In their structural features the coronate scyphomedusae are probably the most primitive of the group. Werner (1966) has, indeed, shown the relationship of the polyp with the fossil group *Conulata* which flourished from Cambrian to Triassic times.

The chief characteristics which together distinguish the medusae from those of other orders are:

1. The coronal furrow on the umbrella.
2. The pedalia round the umbrella periphery.
3. The triangular gastric septa fusing the subumbrellar wall of the base of the stomach with the exumbrellar wall.
4. The hood at the end of the rhopalium.

The *coronal groove* is a deep furrow encircling the umbrella, usually about midway between the umbrella margin and its apex, and dividing the umbrella into a central disk of lens-shaped jelly and a peripheral zone. It is in fact a region where the mesogloea is very thin in proportion to that of the bulk of the umbrella.

The *pedalia* are radial thickenings of jelly on the peripheral zone of the exumbrella between the coronal groove and the bases of the marginal lappets. Deep radial grooves between adjacent pedalia correspond to the positions of radial septa which fuse together the exumbrellar and subumbrellar walls of the gastrovascular coronal sinus. These septa extend outwards from near the proximal margin of the coronal muscle into the centres of the marginal lappets. Each pedalion occupies the space between adjacent radial grooves, and extends approximately from the coronal groove to the base of either a marginal tentacle or a rhopalium. They are therefore designated either as tentacular or rhopalar pedalia.

The digestive system is basically the whole of the volume contained between the exumbrellar and subumbrellar layers of endodermal epithelium. It is, however, divided into two regions by a ring of the four interradial triangular gastric septa where the exumbrellar and subumbrellar endoderm layers are fused. The central region proximal to these septa is the stomach, and the distal peripheral region is the gastrovascular coronal sinus. The two regions are connected through four perradial ostia each of which is situated between the bases of adjacent gastric septa. Along each of the two sides of each gastric septum there is a phacellus, or row of gastric filaments. The gastrovascular or coronal sinus is divided distally by the radial septa corresponding to the number of marginal lappets mentioned above. The gastrovascular sinus extends as a pouch into each lappet, and the radial septum divides each pouch into two. But as the septum only runs to about

27

the centre of the lappet the two sides of the pouch are connected beyond the distal end of the septum. In this way a simple peripheral canal system is formed appearing as a festoon round the umbrella margin.

Apart from the meristic variation in distribution and numbers of marginal tentacles and rhopalia which characterize genera and species, the coronate medusae show two trends in their shape. The average form is somewhat hemispherical and from this norm there has been a tendency either to increase in height, as in *Periphylla*, or to become almost completely flattened, as in *Atolla*. There have been certain characteristic developments in keeping with these two extremes.

In *Periphylla* the increase in height, coupled also probably with its large size, has resulted in the development of four additional areas of fusion of subumbrellar and exumbrellar endoderm in the gastrovascular sinus which form attachments for the four very strongly developed interradial deltoid muscles. The perradial deltoid muscles have their points of attachment in the thickened mesogloea of the margins of the gastric ostia.

In all genera except *Atolla* the marginal sense organs are situated approximately on the same circumference as that of the bases of the marginal tentacles. The rhopalar and tentacular pedalia are therefore ranged more or less in a single peripheral series. In *Atolla*, partly owing to the flatness of the umbrella and probably mainly due to the great number of alternating marginal tentacles and marginal sense organs, the rhopalar pedalia have been squeezed outwards so that they form a complete ring distal to an inner ring of tentacular pedalia. The base of each rhopalar pedalion is dovetailed between the outer margins of the two adjacent tentacular pedalia.

Until recently the only coronate medusa of which the complete life history was known was *Nausithoë punctata*, but Werner (1967) has now succeeded in rearing a deep-sea species, *Atorella vanhoeffeni*. Although these are not British species it will be useful to give a brief account here because it may be that other species may have a similar course of development. A good account of the life history of *Nausithoë punctata* in Japanese waters is given by Komai (1935). *Nausithoë* has a fixed polyp stage known as *Stephanoscyphus*. This has a well-developed periderm which forms branched tubes carrying thecae characterized by numerous lamellar rings. The polyps, which have 100–200 tentacles each, are typically scyphozoan in structure, but they have many taenioles instead of the usual four. The whole forms a branching colony which lives in sponges. Early descriptions were given by Allman (1874) and Schulze (1877).

Ephyrae are produced by strobilation, and when first liberated they have eight marginal lobes and eight marginal sense organs, but no marginal tentacles. There is one gastric filament in each interradius, attached at both ends. The ephyra is 1·5 to 2·0 mm. in diameter.

The eight marginal tentacles soon develop and when the ephyra is 3 mm. in diameter they are about as long as the marginal lappets. There are then two gastric filaments in each interradius and the rudiments of the gonads are apparent. Apart from increase in size, and assumption of the typical coronate form and proportions, there is very little further change as development proceeds. It is evident that the adult medusa is very ephyra-like in character. Komai (1935) figures a number of instances of abnormal development, this being especially prevalent in the disposition of the gonads.

Werner (1967) brought back living *Stephanoscyphus* from the Indian Ocean and reared the ephyrae to the adult stage which proved to be *Atorella vanhoeffeni*. The ephyrae were 0·6–0·8 mm. in diameter and had six marginal lobes, six sense organs and two interradial marginal tentacles. The numbers of ephyrae produced by a single polyp were very great, the maximum from one individual being 4,256. A species of *Nausithoë* with a solitary polyp, also reared by Werner from the Indian Ocean, produced a maximum of 1,069 ephyrae, 1·2–1·8 mm. in diameter.

The Coronatae are classified according to the number and disposition of the marginal tentacles and marginal sense organs.

The British species, all of which are deep-water, fall into four families characterized as follows:

Nausithoïdae: with four perradial and four interradial marginal sense organs and eight adradial marginal tentacles; without sac-like pouches on subumbrella.
Atollidae: with more than eight marginal sense organs and more than eight marginal tentacles, the sense organs and tentacles being equal in number.
Paraphyllinidae: with four perradial marginal sense organs and twelve marginal tentacles.
Periphyllidae: with four interradial marginal sense organs and 4–28 marginal tentacles.

Coronate medusae have two kinds of nematocyst, holotrichous haplonemes and microbasic euryteles (Werner, 1967).

FAMILY NAUSITHOÏDAE

Coronate Scyphomedusae with four perradial and four interradial marginal sense organs; with eight adradial marginal tentacles; without sac-like pouches on subumbrella; with four or eight gonads.

There are two genera in the family Nausithoïdae, *Nausithoë* Kölliker and *Palephyra* Haeckel. The distinguishing characteristic is that *Nausithoë* has eight gonads while *Palephyra* has four. One cannot help wondering whether specimens which have been placed in the genus *Palephyra* may not be abnormal specimens with a reduced number of gonads or specimens in which two contiguous gonads have been regarded as one.

The only British genus is *Nausithoë*.

Genus **Nausithoë** Kölliker, 1853

Nausithoïdae with umbrella with dome-shaped hemispherical or somewhat flattened central disk; with eight adradial gonads.

There are only two species of *Nausithoë* that are known to occur in deep water west and north of the British Isles, namely *N. atlantica* Broch and *N. globifera* Broch.

These two species can be immediately distinguished when freshly preserved by their coloration. *N. atlantica* has the whole umbrella uniformly coloured deep chocolate red, while *N. globifera* has a completely transparent umbrella through which can be seen the deep purple-red stomach. In specimens kept for a long time in preservative the colours fade, but the remarkable high dome-like central disk and the occurrence of the gonads almost entirely above the coronal groove in *N. globifera* are features which at once distinguish it from the more flattened *N. atlantica*.

The identity of some of the other described species of *Nausithoë* is uncertain. *N. punctata* Kölliker, the type species, is a well-known surface and shallow-water species of worldwide distribution in tropical and warm waters. It has never been recorded in the Atlantic as far north as the British Isles. A deep-sea species *N. rubra* Vanhöffen (1902) has possibly been confused with *N. atlantica*, but I have seen no specimens from collections in deep water off British coasts agreeing with Vanhöffen's description. Hartlaub (1909) described a species from near the east coast of Greenland, *N. limpida*, in which the rhopalium is shown in his drawing as apparently having an ocellus, though no mention of this is made in the text. *N. albatrossi* described by Maas

(1897) has the gastric filaments arranged in clusters. *N. challengeri* Vanhöffen, *N. clausi* Vanhöffen, and *N. picta* Agassiz & Mayer have been considered by some authors as possibly synonymous with *N. punctata*, but their identity remains in doubt.

It is known that *N. punctata* has a fixed scyphistoma stage. This is a branching polyp *Stephanoscyphus mirabilis* which lives in sponges. Other somewhat similar polyps have been described and attributed to the genus *Stephanoscyphus*. For details of these polyps see Komai (1935), Thiel (1936), Leloup (1937) and Kramp (1951b, 1959b).

It is possible that the deep-sea species *S. simplex* Kirkpatrick may be the polyp of one of the deep-sea species of *Nausithoë* (see Kramp, 1951b, p. 127, and 1959b).

The young stages of the two British species *N. atlantica* and *N. globifera* are not known. But the life history of *N. punctata* has been fully described and it is probable that the development will be similar in the other species.

Nausithoë atlantica Broch

Plate I, fig. 1; Plate VII, fig. 1; Text-figs. 11–13

Nausithoë atlantica Broch, 1913, *Rep. 'Sars' N. Atl. Deep Sea Exped.* **3**, pt. 1, p. 9, pl. I, figs. 1–4; text-fig. 5.
Ranson, 1945, *Résult. Camp. scient. Prince Albert I*, fasc. cvi, p. 25 and addendum.
Russell, 1956b, *J. mar. biol. Ass. U.K.* **35**, 363, pl. I, left; text-figs. 1, 2, 5.

SPECIFIC CHARACTERS

Umbrella smooth and somewhat flattened; eight marginal sense organs each with rhopalium with large, flat, wedge-shaped, basal cushion with straight exumbrellar carina, and with underside of hood and sensory bulb deeply coloured; about 160 simple gastric filaments in single rows; gonads extending nearly to distal margin of coronal muscle; whole medusa chocolate red.

DESCRIPTION OF ADULT

Umbrella smooth with flattened hemispherical central disk with coronal groove situated about midway between umbrella margin and apex; with fairly thick jelly; with sixteen pedalial thickenings. Well-developed coronal muscle on subumbrellar surface. Sixteen slightly elongate marginal lappets with evenly rounded margins. Eight solid adradial marginal tentacles. Eight marginal sense organs, four perradial and four interradial, each consisting of rhopalium with hood, statocyst and sensory bulb, but with no ocellus, with large flat wedge-shaped basal cushion with straight exumbrellar carina. Eight adradial gonads, at first somewhat shield shaped, eventually elongated distally nearly to distal margin of coronal muscle. Stomach attached interradially to subumbrellar summit over four flattened triangular gastric septa to form four pouches between manubrium and subumbrellar surface; eight phacelli, one along each side of each triangular septal plate, each with 20 or more simple gastric filaments making 160 or more in all; four perradial ostia leading from stomach cavity into gastrovascular sinus; gastrovascular sinus divided distally by sixteen radial septa running from near proximal margin of coronal muscle into middle of each marginal lappet to form simple peripheral system of marginal lappet canals. Manubrium slightly folded and not extending beyond umbrella margin. Colour of freshly preserved specimens uniform chocolate red over whole subumbrella, marginal tentacles, and stomach; gonads colourless; sensory bulb and underside of hood of marginal rhopalium deeply coloured. Size up to 35 mm. in diameter, 19 mm. at coronal groove.

DISTRIBUTION

Nausithoë atlantica has been recorded from the North Atlantic at three positions by Broch (1913), namely 48° 29′ N., 13° 55′ W.; 46° 58′ N., 19° 16′ W.; and 36° 53′ N., 29° 47′ W. Ranson (1945) recorded one specimen from 27° 43′ N., 18° 28′ W. near the Canary Isles. I have myself had specimens from deep water off the mouth of the English Channel (Russell, 1956*b*).

If it should be the same species as that identified by Bigelow (1928) as *N. rubra* Vanhöffen then it occurs also in the S. Atlantic, Pacific and Indian Oceans.

STRUCTURAL DETAILS OF ADULT

Umbrella

The somewhat flattened shape of the umbrella (Text-fig. 11) at once distinguishes *Nausithoë atlantica* from *N. globifera*. In some of the specimens I have seen, the jelly of the central disk was pitted; this was, however, caused by the impression of radiolarians present with the other plankton organisms in the vessels in which the catches were stored.

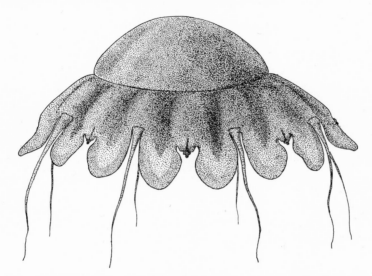

Text-fig. 11. *Nausithoë atlantica*. Adult medusa, 35 mm. in diameter.

Marginal sense organs

The rhopalia are rather distinctive (Text-fig. 13). Each has at its base an unusually large wedge-shaped cushion with a straight exumbrellar carina which extends along the rhopalium itself. This basal cushion is so wide that it gives rise to a distinct space between the bases of adjacent marginal lappets.

The deep coloration of the underside of the hood is also characteristic.

Gonads

The shape of the gonad changes as it develops. At first it is somewhat U-shaped with the inter-radial arm of the U thicker than the perradial arm. As the gonad develops its distal end grows outwards until it almost reaches the base of the marginal tentacle (Text-fig. 12).

NAUSITHOÏDAE

Sections of the gonad show that the mature ova are each completely enveloped by a layer of densely packed oval bodies lying immediately beneath the egg membrane. These bodies, the largest of which are 12–14 μ in length, appear in the smallest oocytes as minute granules. As the ova increase in size the granules grow and tend to move towards the periphery. Each body has a central core but no other structure is visible (Russell, 1956b). The mature eggs are white.

Ranson (1945) records an abnormal specimen with only seven gonads.

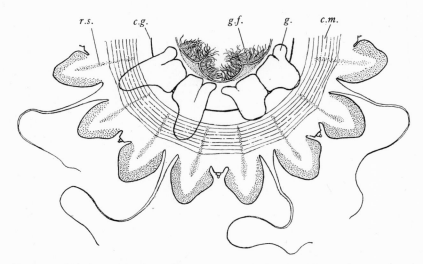

Text-fig. 12. *Nausithoë atlantica*. Semidiagrammatic view of subumbrella with buccal walls of stomach cut away. *c.g.* coronal groove; *c.m.* coronal muscle; *g.* gonad; *g.f.* gastric filaments; *r.s.* radial septum (from Russell, 1956b).

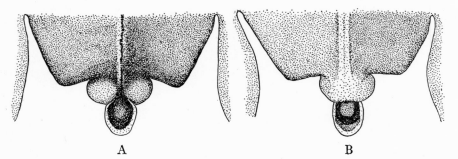

Text-fig. 13. *Nausithoë atlantica*. Marginal sense organ. A, exumbrellar view; B, subumbrellar view (from Russell, 1956b).

COLORATION

The colour of the whole medusa when freshly preserved in formalin is chocolate-red. After storage in formalin and sea water for several months the whole medusa becomes much more transparent and amber in colour.

Sections show that the coloration is limited to the mesogloea and forms a dark membranous layer immediately beneath the endoderm. The brownish yellow colour also occurs faintly throughout the whole mesogloea so that the jelly does not appear transparent. The stomach walls are especially densely coloured, both beneath the endoderm and the ectoderm; and the mesogloea of the marginal lappets is coloured throughout. The whole rhopalium is coloured and the underside of the hood is especially dark.

HISTORICAL

Nausithoë atlantica was named and described by Broch (1913) on the basis of eight specimens caught on the 'Michael Sars' Expedition to the North Atlantic. Subsequently a further description was given by Russell (1956*b*).

Nausithoë globifera Broch

Plate I, fig. 2; Plate VII, fig. 2; Text-figs. 14–17

Nausithoë globifera Broch, 1913, *Rep. 'Sars' N. Atl. Deep Sea Exped.* **3**, pt. 1, p. 10, pl. I, figs. 5–8; text-fig. 5.
Ranson, 1945, *Résult. Camp. scient. Prince Albert I*, fasc. cvi, p. 25.
Kramp, 1947, *Dan. Ingolf-Exped.* **5**, pt. 14, p. 16.
Russell, 1956*b*, *J. mar. biol. Ass. U.K.* **35**, 363, pl. I, right; text-figs. 3, 4, 6.

SPECIFIC CHARACTERS

Umbrella smooth, with high dome-shaped central disk; eight marginal sense organs each with small high wedge-shaped basal cushion with exumbrellar carina bent down at right angles, ventral bulb of rhopalium only slightly coloured; about eighty simple or compound gastric filaments in single rows; gonads situated almost entirely above coronal groove; umbrella colourless, stomach deep purple-red.

DESCRIPTION OF ADULT

Umbrella smooth, with high dome-shaped central disk, with coronal groove situated about midway between umbrella margin and apex and peripheral margin of umbrella beyond coronal groove forming nearly horizontal shelf; with thick jelly over central disk; with sixteen pedalial thickenings. Well-developed coronal muscle on subumbrellar surface. Sixteen broad marginal lappets with evenly rounded margins. Eight solid adradial marginal tentacles. Eight marginal sense organs, four perradial and four interradial, each consisting of rhopalium with hood, statocyst and ventral bulb but with no ocellus, with small high wedge-shaped basal cushion with exumbrellar carina bent downwards at right angles. Eight adradial oblong gonads situated almost entirely above coronal groove. Stomach attached interradially to subumbrellar summit over four flattened triangular gastric septa to form four pouches between manubrium and subumbrellar surface; eight phacelli, one along each side of each triangular gastric septum, each with ten or more simple or compound gastric filaments making eighty or more in all; four perradial ostia leading from stomach cavity into gastrovascular sinus; gastrovascular sinus divided distally by sixteen radial septa stretching from near the coronal groove into middle of each marginal lappet to form simple peripheral system of marginal lappet canals. Manubrium slightly folded, extending a short distance beyond umbrella margin. Colour of freshly preserved specimens limited to stomach, marginal tentacles, rhopalia and gonads; stomach deep purple-red, marginal tentacles orange-red, gonads white to reddish brown, ventral bulb of rhopalium faintly coloured; remainder of umbrella transparent. Size up to 22 mm. in diameter, 10 mm. at coronal groove.

NAUSITHOÏDAE

DISTRIBUTION

Nausithoë globifera is a deep-sea oceanic species likely to be caught anywhere in deep water west and north of the British Isles.

It has so far only been recorded in the North Atlantic from south of Iceland (Kramp, 1947) to the coast of Portugal (Ranson, 1945) and west to 45° 26′ N., 25° 45′ W. (Broch, 1913). I have recorded several specimens from deep water off the mouth of the English Channel (Russell, 1956*b*) and have seen other specimens from that area since.

STRUCTURAL DETAILS OF ADULT

Umbrella

The dome-like shape of the central disk (Text-fig. 14) is quite characteristic, and is implied in the specific name *globifera*; the manner in which the peripheral part of the umbrella, distal to the coronal groove, spreads out nearly at right angles to the upright walls of the central disk is also a distinctive feature.

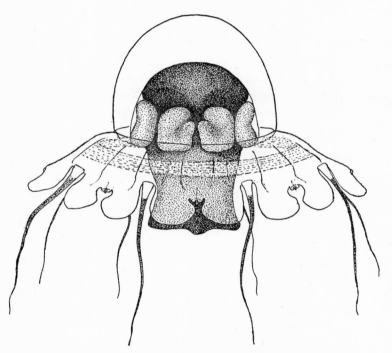

Text-fig. 14. *Nausithoë globifera*. Adult medusa 22 mm. in diameter.

Marginal sense organs

Each rhopalium has at its base a small thickened wedge-shaped cushion with an exumbrellar carina which bends down at right angles to the neck of the rhopalium (Text-fig. 16). Because of this bending the rhopalium is hidden from view when the medusa is looked at from above.

Gonads

The mature gonads (Text-fig. 17 C, D) may be so large that they are contiguous on both sides. They are situated almost entirely above the coronal groove and reach up to the bases of the gastric septa, forming an upright collar surrounding the stomach (Text-figs. 14, 15).

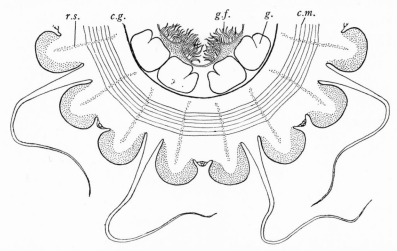

Text-fig. 15. *Nausithoë globifera*. Semidiagrammatic view of subumbrella with buccal walls of stomach cut away. *c.g.* coronal groove; *c.m.* coronal muscle; *g.* gonad; *g.f.* gastric filaments; *r.s.* radial septum (from Russell, 1956 b).

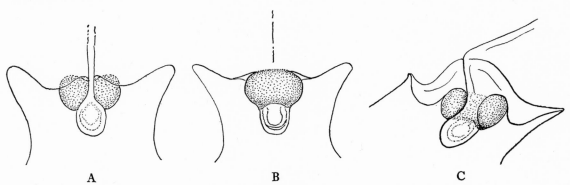

Text-fig. 16. *Nausithoë globifera*. Marginal sense organ. A, B, and C, exumbrellar, subumbrellar and lateral oblique views respectively.

Each gonad is oblong in outline with rounded corners, but with the perradial side shorter than the interradial side, giving the appearance of a flat boxing glove. This shape can be appreciated when the development of the gonad is considered. Each gonad starts as a crescent-shaped endodermal fold in the subumbrellar wall of the gastrovascular sinus with the concavity of the crescent on the perradial side (Text-fig. 17, A, B). The upper and middle portions of the crescent develop more quickly than the lower or distal portion. The middle portion increases in width so that the cavity of the crescent becomes filled. At the same time the two ends of the crescent grow towards one another. On the subumbrellar side the edges of the gonads are folded inwards.

3-2

NAUSITHOÏDAE

In the mature female gonad the eggs are contiguous. They are about 0·25 mm. in diameter and white in colour.

I have seen two specimens in which the gonads had developed very abnormally. In one specimen two gonads were abnormally large. It seems just possible that the medusa described by Vanhöffen (1902) as *Palephyra indica* may have been an abnormal specimen of *Nausithoë globifera*, but Stiasny (1940) has provisionally identified nine specimens under this name.

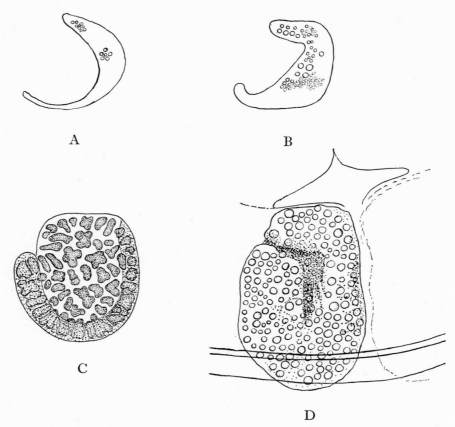

Text-fig. 17. *Nausithoë globifera*. Gonad. A and B, in specimens 17 mm. in diameter;
C, in adult male; D, in adult female.

COLORATION

The transparency of the umbrella is a striking feature, and through this the dark purple-red stomach is fully visible (Pl. I, fig. 2; Pl. VII, fig. 2). A short portion of the base of each marginal tentacle is also without colour, the remainder of the tentacle being orange-red. The ventral bulb of the rhopalium is only faintly coloured. The young gonads are white except for fine flecks of pigment and they appear in striking contrast against the dark stomach. When fully developed they may have a reddish brown coloration. After several years in formalin and sea water the dark pigment gradually fades through dark amber to an almost colourless or very pale flesh-pink tint.

Sections show that the coloration is limited to a membranous layer lying in the mesogloea immediately beneath the endoderm of the stomach. In the buccal walls of the stomach and the

36

gastric filaments the whole mesogloea is coloured. There are, in addition, pigment granules present in the endoderm cells lining the whole stomach and in the walls of the gonads. The mesogloea of the marginal tentacles is also coloured.

HISTORICAL

Nausithoë globifera was named and described by Broch (1913) on the basis of seven specimens collected on the 'Michael Sars' expedition to the North Atlantic. The species was recognized as a good species by Ranson (1945) and Kramp (1947), and a further description was given by Russell (1956*b*).

FAMILY ATOLLIDAE

Coronate Scyphomedusae with more than eight marginal sense organs alternating with an equal number of marginal tentacles; with eight adradial gonads.

The chief characteristics of the family are the very flattened umbrella and the disposition of tentacular pedalia and rhopalar pedalia in two concentric circles, the rhopalar pedalia being distal to the tentacular pedalia.

There is only one genus in the family, *Atolla* Haeckel.

Genus **Atolla** Haeckel, 1880

Coronate scyphomedusae with flattened disk-like umbrella; with marginal lappets with or without warts; with very strongly developed coronal muscle; with 18, 20, 22, 24, or more, marginal tentacles alternating with marginal sense organs.

Five species of *Atolla* are known, of which three are likely to occur in British deep waters, *A. wyvillei* Haeckel, *A. vanhoeffeni* Russell, and *A. parva* Russell. Of the other species *A. chuni* Vanhöffen has only been recorded from southern latitudes, and *A. russelli* Repelin off the coast of South-West Africa.

The three species are reasonably easy to identify, except when very small, and can be separated as follows.

1. Mature specimens large, over 40 mm. in diameter . . . . **wyvillei**
2. Mature specimens small, 10–30 mm. in diameter . . . . {**vanhoeffeni** **parva**

A. vanhoeffeni can be distinguished from *A. parva* by its pigmentation and the shape of the basal attachment of the stomach.

Base of stomach cross-shaped; eight pigment spots on either side of each gastric ostium . **vanhoeffeni**

Base of stomach in form of four-leaved clover; uniformly pigmented . . . . . **parva**

In small specimens without gonads *A. parva* can be distinguished from *A. wyvillei* as follows.

Radial septa diverging at centripetal ends and extending beyond inner margin of coronal muscle
wyvillei

Radial septa not diverging at centripetal ends and not extending beyond inner margin of coronal muscle
parva

In all three species one of the marginal tentacles is larger than the others. Repelin (1966) found that while in *A. parva* and *A. vanhoeffeni* this hypertrophied tentacle is situated at an interradius, in *A. wyvillei* it is just either to the right or left of an interradius (Text-fig. 26).

37

A. russelli is very like *A. parva*, but differs in that its radial septa diverge proximally as in *A. wyvillei* and that the gonads are bilobed. The southern species *A. chuni* appears to be very similar to *A. wyvillei*, the chief difference being that its marginal lappets are covered with warts (see Stiasny, 1934, p. 382).

It seems possible that *A. tenella* Hartlaub may be a distinct species if the pairs of pigment spots round the margin of the umbrella are a constant feature.

Atolla wyvillei Haeckel

Plate I, fig. 3; Plate VIII, fig. 3; Text-figs. 18, 19 A, B, 20–24, 25 B, 26 A, 28 B

Atolla wyvillei Haeckel, 1880, *System der Medusen*, p. 485.
    Haeckel, 1881 *a*, *Rep. Challenger Exped. Zool.* **4**, 113–23, pl. XXIX.
    Vanhöffen, 1902, *Wiss. Ergebn. dt. Tiefsee-Exped. 'Valdivia'*, **3**, 13, pl. V, fig. 22.
    Bigelow, 1909, *Mem. Mus. comp. Zool. Harv.* **37**, 39–41, pl. 8, fig. 1; pl. 9, fig. 3; pl. 10, figs. 8, 9.
    Mayer, 1910, *Medusae of the World*, **3**, 566, fig. 361.
    Stiasny, 1934, '*Discovery' Rep.* **8**, 366–79 (mostly), pl. XV, fig. 4; text-figs. 3–6.
    Bigelow, 1938, *Zoologica*, **23**, 160.
    Ranson, 1945, *Résult. Camp. scient. Prince Albert I*, fasc. CVI, pp. 32–43 (mostly), pl. II, fig. 13.
    Russell, 1959, *J. mar. biol. Ass. U.K.* **38**, 33, figs. 1, 2 *b*.
    Naumov, 1961, *Fauna of U.S.S.R.*, no. 75, p. 52, text-fig. 33, pl. I, fig. 2.
    Repelin, 1964, *Cah. O.R.S.T.O.M. océanogr.* **2**, 13, text-fig. 3, pl. I, II *a*.
    Repelin, 1966, *Cah. O.R.S.T.O.M. océanogr.* **4**, 22, fig. 3.
Collaspis achillis Haeckel, 1880, *System der Medusen*, p. 489, pl. XXVIII.
Atolla bairdii Fewkes, 1886, *U.S. Comm. Fish Fisheries*, pt. XII, report for 1884, pp. 936–9, pl. I–III.
    Maas, 1904, *Résult. Camp. scient. Prince Albert I*, fasc. XXVIII, pp. 49–53, pl. IV, figs. 29–34; pl. V, figs.
        39–43 (not 38).
    Mayer, 1910, *Medusae of the World*, **3**, 563, fig. 357.
    Broch, 1913, *Rep. Sars N. Atl. Deep Sea Exped.* **3**, 12.
    Kramp, 1924, *Rep. Dan. Oceanogr. Exped. 1908–10*, **2**, H1, pp. 44–6, fig. 36.
Atolla verrillii Fewkes, 1886, *U.S. Comm. Fish Fisheries*, pt. **12**, report for 1884, pp. 939–44, pl. IV–V.
Atolla gigantea Maas, 1897, *Mem. Mus. comp. Zool. Harv.* **23**, 80, pl. XII, figs. 2–4; pl. XIV, fig. 6.
    Mayer, 1910, *Medusae of the World*, **3**, 565, fig. 359.
Atolla alexandri Maas, 1897, *Mem. Mus. comp. Zool. Harv.* **23**, 81, pl. XI, fig. 2; pl. XIV, figs. 4, 5.
    Mayer, 1906 *a*, *U.S. Fish. Comm. Bull. for 1903*, pt. III, p. 1138; pl. II, fig. 7; pl. III, figs. 10, 11.
Atolla valdiviae Vanhöffen, 1902, *Wiss. Ergebn. dt. Tiefsee-Exped. 'Valdivia'*, **3**, 13–14, pl. I, fig. 3; pl. V,
    figs. 22, 23.
    Maas, 1903, *Siboga-Exped. Mon.* **11**, 17–18, pl. I, fig. 3 (not 4), pl. III, fig. 23 (not pl. XII, fig. 108).
?*Atolla tenella* Hartlaub, 1909, *Crois. océanogr. 'Belgica'*, pp. 17–18. pl. LXXVII, figs. 1, 2.
Atolla bairdii forma *valdiviae*, Mayer, 1910, *Medusae of the World*, **3**, 565, fig. 358.
Atolla wyvillei forma *verrillii*, Mayer, 1010, *Medusae of the World*, **3**, 567.

SPECIFIC CHARACTERS

Usually twenty-two marginal tentacles; marginal lappets without warts; base of stomach in form of four-leaved clover, fully pigmented. Radial septa diverging centripetally and extending beyond margin of thin part of coronal muscle.

DESCRIPTION OF ADULT

Umbrella flattened and disk-like; with deep coronal groove situated slightly nearer umbrella margin than centre of disk; with thick jelly; rim of central lens-shaped disk overhanging wide coronal groove, with or without slight marginal notches one less in number than tentacular pedalia. Usually forty-four pedalial thickenings arranged in two series of equal number, the innermost being tentacular and the outermost rhopalial. Tentacular pedalia may have longitudinal furrows. Very thick continuous coronal muscle on subumbrellar side with thinner proximal portion; thick and thin portions of equal width. Usually forty-four smooth marginal lappets, elongated and with evenly rounded margins. Usually twenty-two solid marginal tentacles. Usually twenty-two marginal sense organs, alternating with marginal tentacles, each consisting of rhopalium with hood and statocyst, but with no ocellus. Eight adradial oblong rounded gonads. Base of stomach in form of four-leaved clover; stomach wall attached interradially to umbrella disk over four flattened triangular gastric septa to form four deep triangular pockets externally. Eight phacelli, one along each side of triangular gastric septa, each with up to twenty-four or more simple gastric filaments, making 192 or more in all; four perradial horizontal ostia from stomach cavity into gastrovascular sinus, which is divided by usually forty-four radial septa, corresponding in number to that of marginal lappets; radial septa situated in pairs close together, leaving narrow passage or canal leading from gastrovascular sinus opposite each rhopalium and very broad passage or pouch opposite each marginal tentacle; septa tending to diverge at centripetal ends and extending beyond margin of thin part of coronal muscle; rhopalar canals and tentacular pockets united by peripheral canals. Manubrium extending some way below subumbrellar surface; with four perradial thickenings of jelly running from base to tip of each somewhat pointed lip. Colour when present reddish brown; stomach always deeply pigmented; pigmentation of umbrella variable from slight coloration through intermediate stages to uniformly dark; coronal muscle white to greenish yellow when exposed. Size up to 150 mm. in diameter.

DISTRIBUTION

Atolla wyvillei is a deep-sea medusa widely distributed in all oceans. It is likely to be found in any deep-water collection made off the west and north coasts of the British Isles.

STRUCTURAL DETAILS OF ADULT

Umbrella

Atolla wyvillei is generally described as having a smooth annular zone (Text-fig. 18) between the coronal groove and the bases of the tentacular pedalia. It is this zone that the rim of the central disk may overhang and the apparent presence or absence and the extent of a smooth zone was shown by Bigelow (1928) to depend on the state of contraction and preservation of the specimen. Vanhöffen (1902) described the arrangement of mesogloeal fibres; at the thin connection between the peripheral zone and the central disk there is a supporting band of fibres.

The notches round the rim of the central disk are also a variable character. These notches or grooves may or may not be evident and they vary in width. This was at first regarded as a specific character and, accordingly, authors designate three types as follows.

wyvillei type: with broad radial notches and central lens smooth.
verrillii type: central lens scored with narrow radial notches.
bairdii type: central lens smooth and without notches.

Complete intergradation of these three types has now been shown (Broch, 1913; Browne, 1916; Stiasny, 1934). Broch (1913) found that if the summit of the central lens was pressed upon, the furrows became apparent over the whole of the lens in otherwise smooth specimens. I have found the same effect in specimens which have not been very long preserved. Stiasny (1934) was unable to produce this effect in the material from the 'Discovery' collections and supposed that it was because the specimens had been for several years in preservative.

ATOLLIDAE

When these notches are obvious it can be seen that they are one less in number than the number of marginal tentacles. Stiasny (1934, p. 370, pl. XV, fig. 4) drew attention to this indication of bilateral symmetry which was first noticed by Vanhöffen (1902). Stiasny pointed out that the absence of a notch corresponded in position to that of a very broad tentacular pedalion which itself did not carry the longitudinal furrow which is often evident. In small specimens the marginal tentacle opposite this point of bilateral symmetry is often also very large (Russell, 1959, p. 39), and this feature persists in adult specimens (Repelin, 1966). Stiasny also states that these radial grooves, when present, are broad in large specimens and narrow in small specimens. He

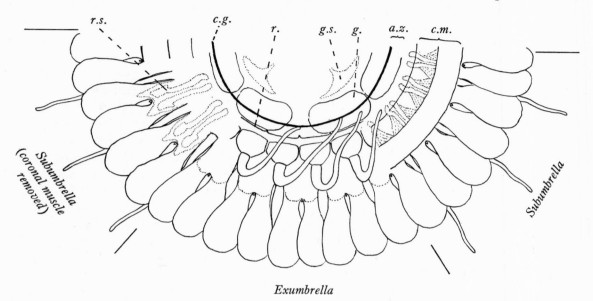

Exumbrella

Text-fig. 18. *Atolla wyvillei*. Diagram to show general morphology. Exumbrellar and subumbrellar views are given as indicated. *a.z.* annular zone; *c.m.* coronal muscle; *c.g.* coronal groove; *g.* gonad; *g.s.* gastric septum; *r.* rim; *r.s.* radial septum.

mentions specimens from the Antarctic in which the grooves are very conspicuous, being large, long, deep and darkly pigmented. There may or may not be grooves in the tentacular pedalia and these are not necessarily correlated with the notches on the central disk.

The mesogloea has many scattered cells and fibres mostly running vertically and branching to the upper and lower surfaces (Vanhöffen, 1902). Ranson (1945) draws attention to granular pigmented cells in the mesogloea which may be more or less abundant and whose number seems to bear some relation to the degree of pigmentation. These cells are $10-12\,\mu$ in diameter and are spherical, oval, and with or without prolongations (Ranson, 1945, pl. II, fig. 13).

Marginal tentacles

The number of marginal tentacles in fully grown specimens seems to be typically twenty-two. In fact, I have myself examined over 200 specimens and never found more than twenty-two (Russell, 1959). Ranson (1945, p. 38) also says that he found the number to be most usually twenty-two, but he records two specimens 25 mm. and 35 mm. in diameter which had thirty-six and thirty-two tentacles respectively. Maas (1903) and Vanhöffen (1902) each record a specimen with twenty-nine tentacles. Smaller specimens may have seventeen to twenty-one tentacles but the full number may be present in very young medusae; for instance, Ranson (1945) records a specimen only 4 mm. in diameter which already had twenty-two tentacles.

The occurrence of one unusually large tentacle in some young specimens as an indication of bilateral symmetry has been mentioned above. Whether this has any significance in the very early development of the medusa is not known as the first stages have not been described. Repelin (1966) points out that this hypertrophied tentacle is always present, and that its base has a rounded protrusion into the mesogloea as opposed to the wide concave intrusion of the other tentacles, and that it is situated just to the right or left of an interradius (Text-fig. 26 A).

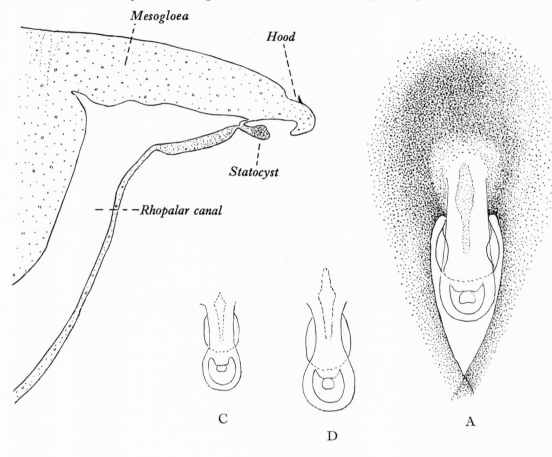

Text-fig. 19. Marginal sense organs. A and B *Atolla wyvillei*, 60 mm. in diameter; A, exumbrellar view; B, radial sectional diagram. C, *A. parva*, 20 mm. in diameter; D, *A. vanhoeffeni*, 25 mm. in diameter. All to same scale.

Marginal sense organs

The marginal sense organ (Text-fig. 19 A, B), whose number varies in agreement with the number of marginal tentacles, consists of a rhopalium with sensory bulb and statocyst which by comparison with that of other coronate medusae is rather weakly developed. The statocyst itself is small in relation to the hood, and the sensory bulb, which forms a half-collar round the under-side of the stalk of the rhopalium, is not very prominent. The whole organ is colourless, apart from the pigmentation of the endoderm in the rhopalar canal.

The rhopalium arises direct from the umbrella margin between adjacent marginal lappets without any basal cushion. It is situated on the exumbrellar side of the umbrella and the rhopalar canal runs up nearly vertically through the thickened mesogloea from the marginal canal system which is on the subumbrellar side (Text-fig. 19 B).

Microscopic sections of the rhopalium have been described and figured by Vanhöffen (1902) and Maas (1904). A large sensory cushion surrounds the underside of the rhopalar canal, and there is a sensory cushion on the exumbrellar side of the statocyst.

In specimens in which the marginal lappets are frayed the fragile rhopalia are generally broken off, or partially disintegrated with the statocysts and sensory bulbs missing.

Gastrovascular system

The basal attachment of the stomach (Text-fig. 25, p. 51) is a distinguishing feature by which *A. wyvillei* may be at once separated from *A. vanhoeffeni*. When viewed from above it is seen that the base has the appearance of a cross, but the sides of each arm of the cross con-verge distally along rounded curves, which gives the whole base the appearance of a four-leaved clover.

The gastric septa have small cavities and canals in them (Vanhöffen, 1902), and Repelin (1964) drew attention to periform protuberances in the centre of each gastric septum. The gastro-vascular system has been well described by Stiasny (1934). It can, perhaps, most easily be under-stood by comparison with the simplest system such as is to be found in the genus *Nausithoë*, in which the gastrovascular sinus is divided by equidistant radial septa. In *Nausithoë* the tentacular and rhopalar pedalia are the same distance from the centre of the central disk. But in *Atolla* the rhopalial pedalia are more distal than the tentacular pedalia; they form a ring nearer the umbrella margin than that of the tentacular pedalia which are thus enabled to close together and form a continuous and compact ring. As a result the radial septa which separate the tentacular pedalia from the rhopalar pedalia lie close to one another in pairs along the rhopalar radii. Very wide tentacular pouches are thus formed in the gastrovascular sinus, and the rhopalar pouches are reduced to straight, narrow canals (Text-figs. 18, 20).

The radial septa are straight and narrow until they reach the region of the rhopalar pedalia, where each septum broadens out into a spatula-like lobe leaving only narrow peripheral canals connecting with the adjacent tentacular gastrovascular pouches. The septa in each pair diverge at their centripetal ends (Text-fig. 28 B; Pl. VIII, fig. 3), and extend beyond the margin of the thin part of the coronal muscle (Russell, 1959).

The radial septa dividing the rhopalar canals from the tentacular pouches are the only septa in the gastrovascular sinus. Septa have been described by some authors which subdivide the tentacular pouches proximally; these are areas with little pigment which have the appearance of septa, but there is, in fact, no fusion of the subumbrellar and exumbrellar mesogloea in these areas. Stiasny (1934) has called them 'false septa' and says that they may vary in size and shape, being leaf-, egg-, or spoon-shaped.

The tentacular pouches and rhopalar canals are connected by wide peripheral sinuses which run along the whole of the vertical sides of the thick rhopalar pedalion (Text-fig. 23 A). As the pedalion jelly thins towards the marginal lappet these sinuses flatten out to form broad canals

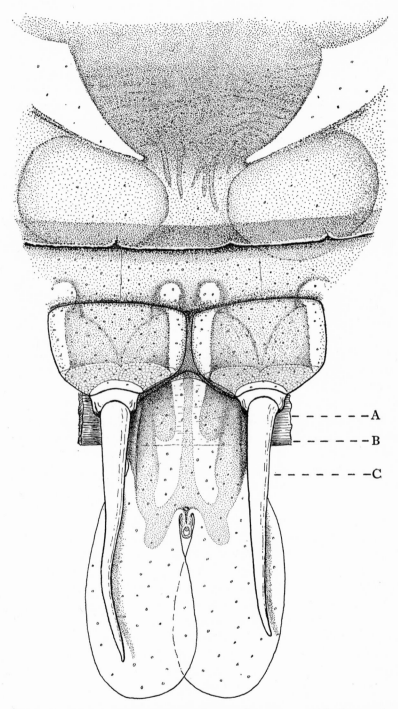

– – – A

– – – B

– – – C

Text-fig. 20. *Atolla wyvillei*. Marginal gastrovascular canal system, marginal tentacles, coronal muscle, coronal groove, gonads, gastric and radial septa and stomach. A, B, and C are levels at which sectional drawings in Text-fig. 23 were made.

running round the periphery of the pedalion and meeting in the centre where they join the rhopalar canal. At this point of junction a short canal runs upwards to the rhopalium (Text-fig. 19 B). Vanhöffen (1902) remarks on the absence of gland cells in the gastric filaments.

Manubrium

The manubrium (Text-fig. 21) has four leaf-shaped perradial lips with thick jelly of tough consistency separated by interradial areas of softer texture but in which nevertheless the jelly is just as thick. The external pockets resulting from the shape of the gastric septa are in this softer interradial region. In perfect specimens the whole outer surface of the manubrium is covered

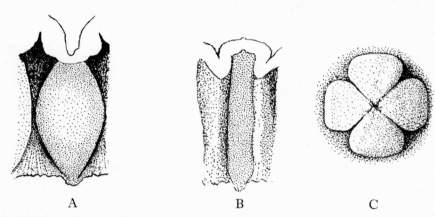

A   B   C

Text-fig. 21. *Atolla wyvillei.* Manubrium. A, external view of perradially thickened lip; B, internal view of same; C, subumbrellar view of closed mouth.

with pigmented epithelium. This ectodermal layer is however usually missing from the perradial lips in preserved specimens. The underlying endodermal pigment shows through the thick layer of transparent jelly and gives rise to a slaty blue velvety appearance. The four lips thus show up clearly, separated by the still pigmented interradial region. In section it can be seen that the interradial areas project into the cavity of the manubrium as pillar-like cushions.

Muscular system

The coronal muscle is remarkable for its great development and strength. When its coating of pigmented epithelium is rubbed off it forms a strikingly obvious whitish or greenish ring on the subumbrellar side 5 to 10 mm. in width. It is divided into two portions, an outer very thick ring and an inner much thinner ring of approximately the same width. The outer circumference of the coronal muscle lies peripheral to the bases of the marginal tentacles and runs across the rhopalar pedalia at about the point at which the latter thin down to the marginal lappets (Text-fig. 22 A). The inner circumference of the coronal muscle does not reach to the proximal ends of the radial septa. The coronal muscle is completely unattached in the region of the bases of the marginal tentacles (Text-fig. 22 B). Deep pockets are thus formed which penetrate up beneath the basal muscles of the tentacles. These pockets are lined with cushions of nematocyst-bearing tissue (see p. 45).

There are longitudinal muscles running along the abaxial and adaxial sides of the marginal tentacles. The adaxial muscle runs the whole length of the tentacle and ends in two very conspicuous root muscles attached to the subumbrellar surface of the gastrovascular tentacular pouch (Text-fig. 22). According to Haeckel (1881 *a*) the abaxial muscle is weaker and shorter and extends only along the basal third of the tentacle on to the exumbrellar surface of the tentacular pedalium.

The roots of the adaxial muscles, which lie under the thin part of the coronal muscle, form a saw-edge pattern along the inner edge of the thick portion of the coronal muscle.

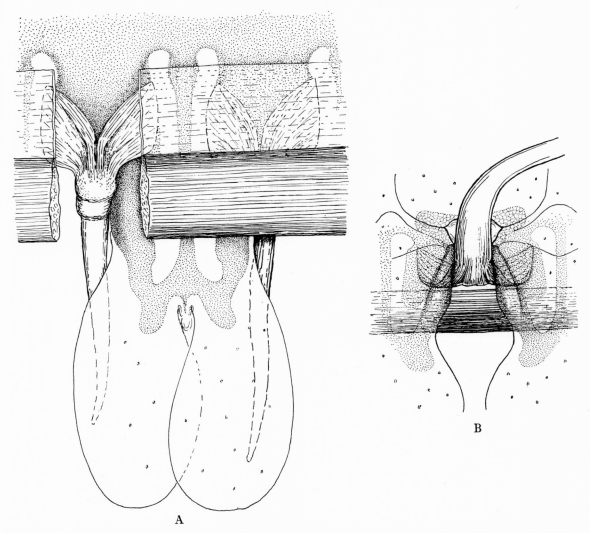

A

B

Text-fig. 22. *Atolla wyvillei.* A, subumbrellar view of marginal lappet and coronal muscle, part of which has been cut away. B, exumbrellar view; marginal tentacle turned upwards to show space between its base and the coronal muscle.

Nematocyst cushions

Special areas packed with nematocysts occur along the outer walls of the gastrovascular sinuses which run along the vertical walls of the rhopalar pedalia (Text-fig. 23 A, B). The upper halves of these nematocyst pads are protected by a hood of tissue which hangs over them like an eyelid.

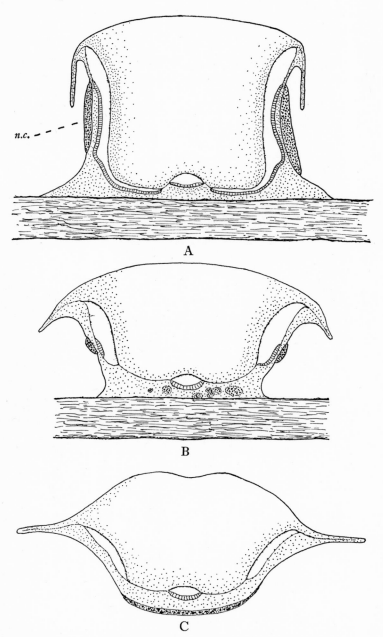

Text-fig. 23. *Atolla wyvillei.* Sections through marginal gastrovascular canal region at levels A, B, and C in Text-fig. 20. *n.c.* nematocyst cushion.

The nematocyst-bearing tissue is continued inwards along the walls of the pockets under the bases of the tentacles and into pockets beneath the triangular basal tentacular muscles.

Diverticula, presumably in which nematocysts are formed, also occur in the mesogloea immediately beneath the outer edge of the thick part of the coronal muscle.

Gonads

The gonads (Text-fig. 24 A–E) first appear when the medusa is *ca.* 15 mm. in diameter, as crescent-shaped endodermal thickenings on the subumbrellar wall of the gastrovascular sinus opposite the ends of the gastric septa at about the level of the coronal groove. They develop to form sack-like outgrowths with a mesogloeal pad lying between the gastric endoderm and the germinal epithelium. The eggs develop on the subumbrellar or genital sinus side of the mesogloeal

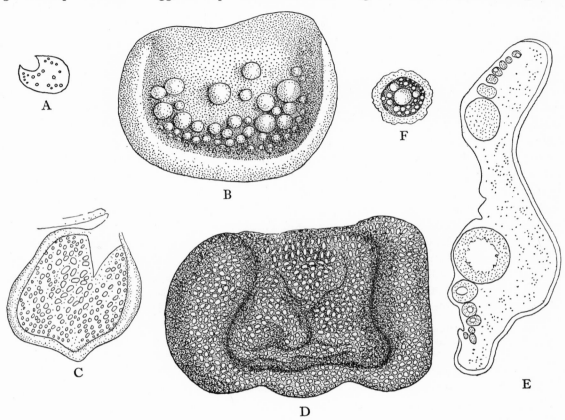

Text-fig. 24. *Atolla wyvillei*. Development of gonad. A, in specimen 33 mm. in diameter; B, female, 95 mm. in diameter; C, developing male 48 mm. in diameter; D, mature male, 72 mm. in diameter; E, section through female gonad; F, female gonad of *A. parva*. A, D, and F to same scale.

pad, the youngest ova being round the periphery and the oldest in the centre. The eggs may reach a size of about 1 mm. in diameter. According to Vanhöffen (1902) the young eggs are connected by stalks to the endodermal layer.

In the male gonads the mesogloeal pad is not so thick as in the female. The sperm follicles are not produced only round the periphery, but lie in a single layer next to one another. The follicles are somewhat oval in shape. Throughout the mesogloeal pad there are many irregularly distributed small eosinophil cells with lobated nuclei, resembling leucocytes.

In ripe females the gonads are regular in shape, being somewhat oval or bean-shaped, and the individuals in each pair are separated from one another by a small space, so that each gonad is quite distinct. In mature males, however, the gonads become irregularly folded and lengthened so that they touch each other and appear to form a more or less continuous circle of gonad tissue.

The gonads do not become full until the medusa is at least 40 mm. in diameter and in a large sample from the Bay of Biscay the majority of the mature specimens were between 50 and 80 mm. in diameter (Russell, 1959).

Fine structure

The finer structure and histological details of *Atolla* have been most fully described by Vanhöffen (1902).

COLORATION

In nature *Atolla wyvillei* is completely covered with a thin pigmented epithelial layer, dark reddish brown in colour. In perfect specimens the internal structure is thus almost completely hidden from view. The stomach walls are always intensely dark. Pigmentation is thus both ectodermal and endodermal.

As pointed out by Vanhöffen (1902) and Stiasny (1934), this epithelium is very delicate and easily rubbed off, small pieces being nearly always present at the bottoms of the jars in which specimens have been stored. In fact, it is rare that a perfect specimen is seen. In many specimens the coronal muscle appears as an intensely white or greenish opalescent ring, but the occurrence of tattered remains of pigmented epithelium shows that the muscle is normally completely covered on its subumbrellar side by pigment. In general the smaller specimens are less heavily pigmented and somewhat more transparent. When a specimen has been badly rubbed the pigmented epithelium is usually still intact in the coronal groove and in furrows and depressions between pedalia. The medusa then appears to have a definite pigmentation pattern.

Broch (1913) divided the degree of pigmentation into four groups:

I. Only the stomach, and occasionally the gonads, pigmented.
II. The ring muscle also pigmented.
III. Pigment covering other parts of the subumbrella, but gonads always visible from the exumbrellar side.
IV. Pigmentation so dense that the gonads are quite invisible from the exumbrellar side.

While it seems to me likely that groups III to IV are more an expression of the degree of loss of pigmented epithelium, there can be no doubt that occasional specimens are found which prove group I to be a definite type of pigmentation. I have seen a few specimens with this type of pigmentation. I have not seen any specimens agreeing with Broch's pigmentation group II, neither did Stiasny (1934, p. 374) see any in the 'Discovery' collections.

Nicol (1958) made some tests on the coronal muscle, showing that its white appearance is not due to deposits of purine materials such as guanine, pterine, or uric acid. He concluded that 'It is probably a characteristic of the physical structure of the muscle or its contained connective tissue.'

HABITS

A. wyvillei is an oceanic deep-sea medusa. Reports on its vertical distribution have been given by Bigelow (1913), Browne (1910), Broch (1913), Stiasny (1934), and Kramp (1968). In open waters in the daytime the level of maximum abundance appears to be somewhere in the region of 500 to 1,500 metres. In areas of upwelling, and at night, specimens may be caught nearer the surface.

Broch (1913) considered that there was some differentiation with depth according to pigmentation, the more intensely pigmented individuals tending to live somewhat deeper. Stiasny (1934) was inclined to agree with Broch on this.

Observations made in the Bay of Biscay indicate that the medusae may mature at any time of year (Russell, 1959).

The early development of *A. wyvillei* is not known, nor is it known whether there is a fixed or sessile stage in the life history, but it is very probable that it has a *Stephanoscyphus* stage.

Nicol (1958, p. 717) was the first to observe luminescence in *A. wyvillei*; this is a blue flash or glow which may last for two seconds. The luminescing area is a thin streak inside the peripheral edge of the coronal muscle, and the glow shines upwards through the rhopalar pedalion, the white coronal muscle possibly acting as a reflector. The source of luminescence may lie in a patch of tall columnar epithelial cells in the floor of the rhopalar canal and peripheral canals.

Nicol (1958, p. 744) estimated that luminescing *Atolla* might be seen up to a distance of about 30 metres under water.

HISTORICAL

Atolla wyvillei was first named and briefly described by Haeckel (1880, p. 488) preparatory to his full and detailed account in the Reports of the Challenger Expedition (1881 *a*). He had five specimens, three from 53° 55′ S., 108° 35′ E. between the Kerguelen Islands and Melbourne, and two from 42° 32′ S., 56° 27′ W. not far from the coast of Patagonia. As has happened so often with deep-sea medusae, for a number of years this description was followed by descriptions of occasional and sometimes poorly preserved specimens each of which was given a new name. In this way no less than nine species of *Atolla* were erected.

Browne (1916) considered that most of these species were the same and reduced their number to two, *A. wyvillei* and *A. chuni*. This conclusion was supported by most modern authors, e.g. Stiasny (1934) and Bigelow (1928). More recently, however, Russell (1958; 1959) has shown, on the basis of other characteristics than those usually examined, that there exist at least two other species, *A. vanhoeffeni* and *A. parva*; and Repelin (1962 *a*, *b*) has added a third, *A. russelli*.

Atolla vanhoeffeni Russell

Plate VIII, fig. 1; Text-figs. 19 D, 25 A, 26 C

Atolla, Vanhöffen, 1902, *Wiss. Ergebn. dt. Tiefsee-Exped. 'Valdivia'*, **3**, 21, pl. V, figs. 27–9.
Atolla valdiviae, Maas, 1903, *Siboga-Exped. Mon.* **11**, 10; pl. I, fig. 4; pl. XII, fig. 108.
Atolla bairdii, Maas, 1904, *Résult. Camp. scient. Prince Albert I*, fasc. XXVIII, p. 52, pl. V, fig. 38.
Atolla wyvillei, Stiasny, 1934, '*Discovery' Rep.* **8**, 373 (records, in part).
   Stiasny, 1940, '*Dana' Rep.* **4**, no. 18, p. 14 (records, in part).
   Ranson, 1945, *Résult. Camp. scient. Prince Albert I*, fasc. CVI, p. 32 (records, in part).
Atolla vanhöffeni Russell, 1957, *J. mar. biol. Ass. U.K.* **36**, 275, pl. I (above), Text-fig. 1 *a*.
   Repelin, 1964, *Cah. O.R.S.T.O.M. océanogr.* **2**, 22, pl. IV *b*, V.
   Repelin, 1966, *Cah. O.R.S.T.O.M. océanogr.* **4**, 26, figs. 2, 3.

SPECIFIC CHARACTERS

Usually twenty marginal tentacles; marginal lappets without warts; base of stomach cross-shaped, pigmented centrally and with eight pigment spots, one on each side of gastric ostia; radial septa not or hardly diverging centripetally, not or hardly extending beyond margin of thin part of coronal muscle.

ATOLLIDAE

DESCRIPTION OF ADULT

Umbrella flattened and disk-like; with deep coronal groove situated slightly nearer umbrella margin than centre of disk; with thick jelly; rim of central lens-shaped disk overhanging wide coronal groove, with marginal notches one less in number than tentacular pedalia. Usually forty pedalial thickenings arranged in two series of equal number, the inner being tentacular and the outermost rhopalial. Very thick continuous coronal muscle on subumbrellar side with thinner proximal portion; thick portion usually equal to or wider than thin portion. Usually forty smooth marginal lappets, elongated and with evenly rounded margins. Usually twenty solid marginal tentacles. Usually twenty marginal sense organs, alternating with marginal tentacles, each consisting of rhopalium with hood, statocyst and sensory bulb, but with no ocellus. Eight adradial circular gonads. Base of stomach in form of cross; stomach wall fused interradially to umbrella disk over four flattened triangular gastric septa to form deep triangular pockets externally. Eight phacelli, one along each side of triangular gastric septa, each with numerous simple gastric filaments; four perradial horizontal ostia from stomach cavity into gastrovascular sinus, which is divided by usually forty radial septa corresponding in number to that of marginal lappets; radial septa situated in pairs close together leaving narrow passage or canal leading from gastrovascular sinus opposite each rhopalium and very broad passage or pouch opposite each marginal tentacle; radial septa straight, not or hardly diverging at centripetal ends, and not or hardly extending beyond margin of thin part of coronal muscle; rhopalar canals and tentacular pouches united by peripheral canals. Manubrium extending some way below subumbrellar surface, with four perradial thickenings of jelly running from base to tip of each somewhat pointed lip. Colour, when present, reddish brown; stomach always deeply pigmented, with eight pigment spots, one on each side of gastric ostia; umbrella usually colourless, but muscles at bases of marginal tentacles may be pigmented; gonads orange brown. Up to *ca.* 50 mm. in diameter, usually less than 35 mm.

DISTRIBUTION

Atolla vanhoeffeni has been recorded for certain from the central North Atlantic, the Bay of Biscay, the Straits of Gibraltar, off the Portuguese coast, and off the west coast of Africa (Repelin, 1962b, 1966). It might be expected to occur in collections from deep water west of the British coasts. Its distribution is probably worldwide, since a specimen is figured by Maas (1903, pl. XII, fig. 108) from material collected in the Dutch East Indies area on the 'Siboga' Expedition. It occurred in collections of the 'Valdivia' Expedition (Vanhöffen, 1902, p. 21); in 'Discovery' collections off Cape Agulhas and the Cape of Good Hope (Stiasny, 1934, pp. 367–8, St. 405, 407 and 440); in the Prince of Monaco's collections off Cape Finisterre (Ranson, 1945, p. 34, St. 2870), and in Pacific and Indo-Pacific regions (Stiasny, 1940; Kramp, 1968).

STRUCTURAL DETAILS OF ADULT

In its general structural characters *A. vanhoeffeni* is essentially the same as *A. wyvillei*. Attention should be drawn to the following differences.

A. vanhoeffeni is a small species, not yet having been recorded larger than 50 mm. in diameter, or with more than twenty marginal tentacles. The basal attachment of the stomach (Text-fig. 25A) forms a cross whose arms slope downwards towards the umbrella margin. The sides of the arms are approximately parallel so that the arms are of fairly uniform width until they narrow suddenly to form the ostia into the gastrovascular sinus. The four interradial gastric septa are thus more triangular in shape than in *A. wyvillei* or *A. parva*, in both of which these areas are distinctly crescentic. As in *A. parva*, the radial septa separating the rhopalar and tentacular gastrovascular canals and pouches do not or hardly diverge at their centripetal ends and do not usually extend beyond the margin of the thin portion of the coronal muscle. Repelin (1964) measured the widths of the thick and thin parts of the coronal muscle and found that the proportion thick:thin = one or more, but in *A. parva* it was always less than one.

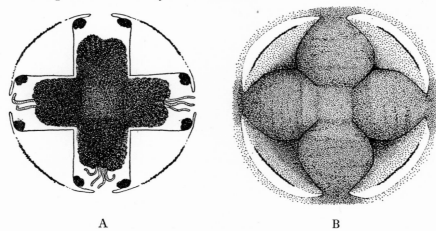

A   B

Text-fig. 25. Exumbrellar view of the base of the stomach of, A, *Atolla vanhoeffeni*, and B, *A. wyvillei*. For clarity the buccal portion of the stomach has been omitted (from Russell, 1957).

The remaining characteristic difference is in the pigmentation of the stomach, which is a more intense blackish purple than in *A. wyvillei* or *A. parva*, and is confined to the central part of the basal attachment of the stomach, stopping short before the end of each arm of the cross and leaving transparent areas through which some of the pigmented gastric filaments can be seen. The eight dark pigment spots in the distal corners of the arms of the cross show up strongly in this transparent area. There are narrow lines of intense pigment along the outer edges of the triangular interradial gastric septal areas.

Some specimens have no other pigmentation except on the gonads, which are yellowish brown on the exumbrellar side. In others there may be pigmentation in the region of the muscles at the bases of the marginal tentacles.

As in the other two species of *Atolla* there is one hypertrophied marginal tentacle; this is situated in an interradius (Text-fig. 26C) and its base projects into the mesogloea as a laterally compressed protrusion while in the other tentacles the protrusion has a wide base with two lateral swellings.

The eggs are large when ripe, being about 1·0 mm. in diameter, and there is room in each gonad for only one full-sized egg at a time.

The black pigment spots were first described by Vanhöffen (1902, p. 21) who thought that they were excretory pores. Sections do not, however, show any passage through the mesogloea so there are, in fact, no pores. Both the ectoderm and the endoderm cells are pigmented, but the ectoderm cells are the more darkly pigmented. The ectoderm and endoderm cells of these spots are higher than those in neighbouring areas.

The rhopalia are identical in form to those of *A. wyvillei*.

HABITS

Repelin (1964) found that in the Gulf of Guinea it lived nearer the surface than the other species of *Atolla*, being caught with 600 m. of wire out while the other species were not caught with less than 1000 m. of wire. Specimens I have seen in 'Discovery' collections were mostly taken between 500 and 1000 m. depth. As mature specimens have been taken in February, March, May, June, July and September, it probably breeds all the year round.

Nothing is known of its early development.

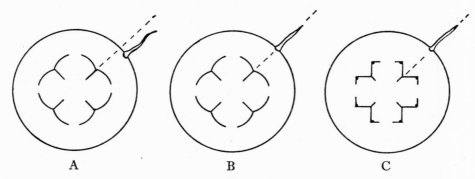

Text-fig. 26. Diagram showing positions of hypertrophied marginal tentacle in A, *Atolla wyvillei*; B, *A. parva*; and C, *A. vanhoeffeni* (from Repelin, 1966, fig. 3).

HISTORICAL

On the evidence of the presence of the eight pigment spots on the stomach *A. vanhoeffeni* was first recorded in a general description of the anatomy of *Atolla* given by Vanhöffen (1902). It was subsequently figured by Maas (1903) under the name of *A. valdiviae* (a synonym of *A. wyvillei*), and the spots were again mentioned by him (1904). Stiasny (1934) also mentioned the occurrence of these spots in some of his specimens of *A. wyvillei*, as did Ranson (1945).

The medusa was recognized as a distinct species by Russell (1957) and named by him after its first discoverer Ernst Vanhöffen. The type specimens are in the British Museum.

Atolla parva Russell

Plate VIII, fig. 2; Text-figs. 19C, 24F, 26B, 27, 28A

Atolla parva Russell, 1958, *Nature, Lond.* **181**, 1811.
    Russell, 1959, *J. mar. biol. Ass. U.K.* **38**, 33, figs. 1, 2a, 3.
    Repelin, 1962a, *Bull. I.F.A.N.* **24**, 664.
    Repelin, 1962b, *Bull. Inst. Res. scient. Congo*, p. 94.
    Repelin, 1964, *Cah. O.R.S.T.O.M. océanogr.* **2**, 19, text-fig. 4, pl. II b, III, IV a.
    Repelin, 1966, *Cah. O.R.S.T.O.M. océanogr.* **4**, 24, fig. 3.

SPECIFIC CHARACTERS

Usually twenty or twenty-four, sometimes twenty-six, marginal tentacles; marginal lappets without warts; base of stomach in form of four-leaved clover; radial septa not or hardly diverging centripetally, not or hardly extending beyond margin of thin part of coronal muscle.

DESCRIPTION OF ADULT

Umbrella flattened and disk-like; with deep coronal groove situated slightly nearer umbrella margin than centre of disk; with thick jelly; rim of central lens-shaped disk overhanging wide coronal groove, with marginal notches one less in number than tentacular pedalia. Usually forty or forty-eight, sometimes fifty-two, pedalial thickenings arranged in two series of equal number, the innermost being tentacular and the outermost rhopalial. Very thick coronal muscle on subumbrellar side, with thinner proximal portion; thick portion narrower than thin portion. Usually forty or forty-eight, sometimes fifty-two, smooth marginal lappets, elongated and with evenly rounded margins. Usually twenty or twenty-four, sometimes twenty-six, solid marginal tentacles. Usually twenty or twenty-four, sometimes twenty-six, marginal sense organs, alternating with marginal tentacles, each consisting of rhopalium with hood, statocyst and sensory bulb, but with no ocellus. Eight adradial circular gonads. Base of stomach in form of four-leaved clover; stomach wall fused interradially to umbrella disk over four triangular gastric septa to form four deep triangular pockets externally. Eight phacelli, one along each side of triangular gastric septa, each with numerous simple gastric filaments; four perradial horizontal ostia from stomach cavity into gastrovascular sinus, which is divided by usually forty or forty-eight, sometimes fifty-two, radial septa, corresponding in number to that of marginal lappets; radial septa situated in pairs close together leaving narrow passage or canal running from gastrovascular sinus opposite each rhopalium and very broad passage or pouch opposite each marginal tentacle; radial septa straight, not or hardly diverging at centripetal ends, and not or hardly extending beyond margin of thin part of coronal muscle; rhopalar canals and tentacular pouches united by peripheral canals. Manubrium extending some way below subumbrellar surface, with four perradial thickenings of jelly running from base to tip of each pointed lip. Colour, where present, deep reddish brown; stomach always deeply pigmented; pigmentation of umbrella variable, from slight coloration to uniformly dark. Size up to 63 mm. in diameter.

DISTRIBUTION

Atolla parva has been recorded over a wide area from the North Atlantic as far south as 1° 30′ N. and as far north as 67° 56′ N. It has been recorded in the South Atlantic off the African coasts between 0° 30′ N. and 14° 25′ S. (Repelin, 1962a, b). It is probably as widespread as *A. wyvillei* and is certain to occur in deep water off the western coasts of the British Isles. Repelin (1962b) noted that south of 14° 25′ S. it was replaced by *A. russelli*.

STRUCTURAL DETAILS OF ADULT

In its general structural characters *A. parva* is essentially the same as *A. wyvillei* and it resembles that species more closely than *A. vanhoeffeni*.

Attention should be drawn to the following differences. *A. parva* is a small species reaching 33 mm. in diameter in the North Atlantic, the largest yet recorded being 63 mm. in diameter in equatorial waters. The marginal tentacles are usually twenty or twenty-four in number, some specimens from the far north having twenty-six. The basal attachment of the stomach is the same as that in *A. wyvillei*, a character in which it differs from *A. vanhoeffeni*, though according to Repelin (1964) the stomach pouches are more rounded, and the gastric septa have less rigidity. The radial septa separating the rhopalar and tentacular gastrovascular canals and pouches are straight (Pl. VIII, fig. 2; Text-fig. 28A), and do not normally diverge to any marked degree at their centripetal ends, nor do they usually extend beyond the margin of the thin portion of the coronal muscle. In this respect they resemble *A. vanhoeffeni*. Repelin (1964) found that the proportion of the width of the thick to the thin part of the coronal muscle was always less than one (see p. 51). The general pigmentation seems to be much the same as that of *A. wyvillei*; some specimens are practically unpigmented except for the stomach. As in *A. wyvillei*, *A. parva* has one unusually large marginal tentacle; this is situated at an interradius (Text-figs. 26 B, 27), and its base has a convex intrusion into the mesogloea (Repelin, 1966).

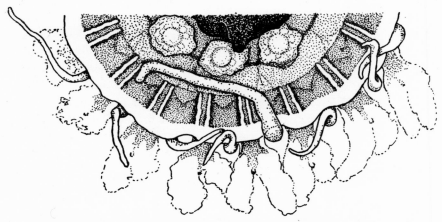

Text-fig. 27. *Atolla parva*. Portion of umbrella margin showing hypertrophied marginal tentacle.

A   B

Text-fig. 28. Diagrammatic drawings to show the shapes of the radial septa and their positions in relation to the thin part of the coronal muscle; A, *Atolla parva*; B, *A. wyvillei* (from Russell, 1959).

Repelin (1964) notes that the eggs, which have a yellow tint, are of unequal sizes and all together enveloped by very fine fibres; in one female 37 mm. in diameter there were eggs up to 1·3 mm. in diameter (Repelin, 1966). The rhopalia are identical in form to those of *A. wyvillei*.

The species is very like *A. russelli*, which differs in that radial septa separating the rhopalar and tentacular gastrovascular canals and pouches diverge at their centripetal ends, that the gonads are bilobed, and that the number of marginal tentacles is usually eighteen.

HABITS

There are not sufficient records on the occurrence of this species to give details on its habits, but Repelin (1966) in collections from the Gulf of Guinea found that the depth of most frequent occurrence was around 350 m. Specimens in 'Discovery' collections were mostly taken between 1000 and 2000 m. depth. Nothing is known of its early development. Specimens may mature at any time of year (Russell, 1959). This species has been shown by Nicol (1958) to be luminescent.

HISTORICAL

A. parva was first recognized as a distinct species and described by Russell (1958, 1959). There is no doubt that it must have been included in some past reports under the name of *A. wyvillei*. The type specimens are in the British Museum.

The species has since been recorded and described by Repelin (1962*a, b*; 1964; 1966). M. E. Thiel (1966*b*) thought that it was a precocious *A. wyvillei*, but there can be no doubt that it is a separate species.

Family PARAPHYLLINIDAE

Coronate Scyphomedusae with four perradial marginal sense organs; with twelve marginal tentacles in groups of three between adjacent marginal sense organs.

The distinctive feature of the family is the perradial position of the four marginal sense organs, between which the marginal tentacles are arranged in groups. The family is thus differentiated from the Periphyllidae in which the four marginal sense organs are interradial.

There is only one genus in the family, *Paraphyllina* Maas, and this is probably represented in British waters.

Genus **Paraphyllina** Maas, 1903

Paraphyllinidae with dome-shaped umbrella; with marginal sense organ with rhopalium with or without ocellus; with bean-shaped or W-shaped gonads.

There is only one species of *Paraphyllina* so far known which might be found in deep water west of the British Isles. It is *Paraphyllina ransoni* Russell and cannot really be confused with any other species. Care should, however, be taken not to confuse it with small specimens of *Periphylla periphylla* in which the marginal sense organs are interradial.

The type species of the genus is *Paraphyllina intermedia* Maas (1903) from the Malay Archipelago, the specimens of which I have seen (Russell, 1956*a*). Another medusa from the Mediterranean identified by Maas (1904, p. 48, footnote) as a *Periphylla* and figured by Lo Bianco (1903) was redescribed by Mayer (1910) as *Paraphyllina intermedia*, but its identity remains uncertain.

Paraphyllina ransoni Russell

Plate I, fig. 4; Pl. VII, fig. 3; Text-figs. 29–31

Paraphyllina intermedia, Ranson, 1936, *Bull. Mus. Hist. nat. Paris* (2nd sér.), **8**, no. 3, p. 269, fig.
Paraphyllina ransoni Russell, 1956*a*, *J. mar. biol. Ass. U.K.* **35**, 105, pl. I–II, text-figs. 1–3.
 Repelin, 1965, *Cah. O.R.S.T.O.M. océanogr.* **3**, 81, figs. 1–6.

SPECIFIC CHARACTERS

Marginal sense organ with rhopalium without ocellus or lens; gonad W-shaped, extending beyond proximal margin of coronal muscle; whole medusa coloured deep chocolate-red.

PARAPHYLLINIDAE

DESCRIPTION OF ADULT

Umbrella with hemispherical dome-shaped central disk, with deep coronal groove situated somewhat nearer to the umbrella margin than to the apex; with fairly thick jelly; with sixteen pedalial thickenings, twelve of which are tentacular and four rhopalar, the rhopalar pedalia being slightly narrower than the tentacular pedalia. Well-developed coronal muscle on subumbrellar surface. Sixteen marginal lappets with evenly rounded margins. Twelve solid marginal tentacles, in four groups of three, four being interradial and eight adradial. Four perradial marginal sense organs, each consisting of rhopalium with hood and statocyst, but with no ocellus. Eight adradial flattened elongated gonads, each forming an asymmetrical W whose interradially situated arms are coiled inwards and whose adradial arms bend outwards towards the perradii. Stomach attached interradially to subumbrellar summit over four flattened triangular gastric septa to form four deep pockets between manubrium and subumbrellar surface; eight phacelli, one along each side of triangular gastric septa, each with twenty or more simple gastric filaments making 160 or more in all; four perradial horizontal ostia leading from stomach cavity into gastrovascular sinus; gastrovascular sinus divided distally by sixteen radial septa running from just above proximal margin of coronal muscle into middle of each marginal lappet, the rhopalar septa being slightly the longer, to form simple peripheral system of tentacular and rhopalar lappet canals. Manubrium slightly folded and reaching about to bases of marginal tentacles, with four perradial thickenings and very slightly crenulated margin. Colour of freshly preserved specimens deep chocolate-red, uniform over whole subumbrella and marginal tentacles, darkest on stomach; gonads colourless. Size up to 75 mm. in diameter; 49 mm. at coronal groove.

DISTRIBUTION

Paraphyllina ransoni is evidently a deep-sea medusa, and must be very rare. Specimens have only been caught on two occasions and then they were evidently in swarms. Forty-five specimens were taken in a 2 m. ring-trawl with 900 m. wire out in 48° 26′ N., 9° 42′ W., over the continental slope west of the English Channel on 28 April 1955 (Russell, 1956a). Subsequently a catch of 610 specimens was made in an Isaacs Kidd Trawl with 700 m. wire out on 24 April 1963 off the Ivory Coast, 4° 58′ N., 4° 2′ W. (Repelin, 1965). One other specimen has been seen; this was washed ashore at Villefranche (Ranson, 1936) so that the species evidently occurs in the Mediterranean. Kramp (1959) also provisionally identified one young specimen off the west coast of Africa.

STRUCTURAL DETAILS OF ADULT

Umbrella (Text-fig. 29)

In most of the specimens so far seen the marginal lappets were somewhat frayed. Above the bases of the marginal tentacles and rhopalia there are six to eight or more radially directed dark stripes. Sections show that these are thickenings of the mesogloea lying under the coronal muscle, which probably play a part by their elasticity in the contraction and expansion of the umbrella margin. They are clearly shown in Pl. VII.

The approximate total diameters of the medusae taken off the mouth of the English Channel lying free in formalin and sea water were as follows:

Diameter (mm.)	11–14	15–19	20–4	25–9	30	35
No. of specimens	3	13	14	8	2	2

The specimens recorded by Repelin were much larger, ranging from 30 to 75 mm. in diameter with a mode between 45 and 60 mm. The largest, a female, was 75 mm. in diameter and 67 mm. high; the coronal muscle was 8 mm. wide and the marginal tentacles 55 mm. long.

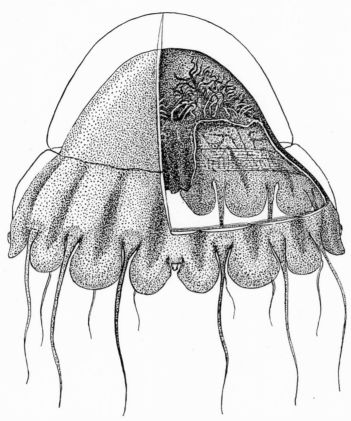

Text-fig. 29. *Paraphyllina ransoni*. Adult medusa with a section of the umbrella and a portion of the manubrium cut away to show the subumbrella and internal anatomy (from Russell, 1956 a).

Marginal sense organs

The rhopalium (Text-fig. 31) has an upper short transparent spoon-shaped hood covering the stalked statocyst, which is enveloped on its subumbrellar side by an opaque pigmented sensory bulb; this sensory bulb has ectodermal pigment. The hood and bulb fuse to form the main stalk of the rhopalium, the lower half of which is wide and rounded and the upper half of which narrows into a thin short carinate crest. At its base the rhopalium joins a wedge-shaped cushion-like thickening of the umbrella margin.

There is no ocellus with lens such as Maas (1903) described in *P. intermedia*.

Gonads

The W-shaped form of the gonad is very characteristic (Text-fig. 30). In some specimens with the most fully developed gonads the interradial arms of the W may recoil upon themselves and

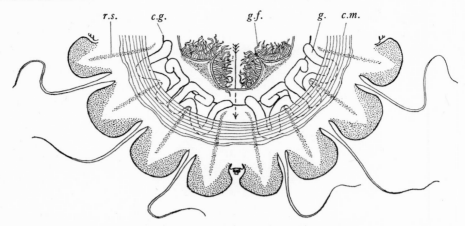

Text-fig. 30. *Paraphyllina ransoni*. Semi-diagrammatic view of subumbrella with buccal walls of stomach cut away. The arrow shows a perradial entrance from stomach to the gastrovascular sinus. *c.g.* coronal groove; *c.m.* coronal muscle; *g.* gonad; *g.f.* gastric filaments; *r.s.* radial septum (from Russell, 1956 *a*).

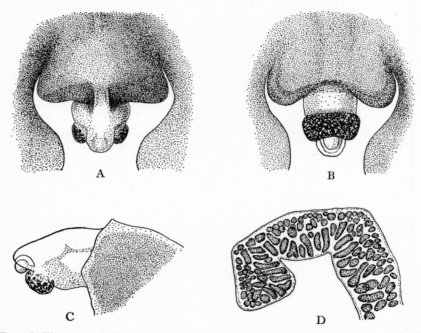

Text-fig. 31. *Paraphyllina ransoni*. Marginal sense organ; A, B, and C, exumbrellar, subumbrellar and lateral views respectively (A and B, reflected light; C, transmitted light); D, portion of male gonad (from Russell, 1956 *a*).

overlie, or even fuse with, their neighbours, so that their true outlines are not very clear, or they may be abnormal in shape.

In female gonads the ova are in various stages of development. In the males there are elongated oval or slightly branched seminiferous follicles in the gonad walls (Text-fig. 31 D). In the large collection recorded by Repelin (1965) 252 were males and 312 females.

58

COLORATION

Over the main body of the medusa the coloration is limited to the mesogloea lying immediately beneath the endoderm; here it forms a dark membranous layer which fades off gradually, but usually for only a short distance, towards the exumbrellar surface. As a result the colourless transparent thickened dome of jelly above the coronal groove and the transparent pedalia become striking features.

In the buccal walls of the stomach the mesogloea is coloured brownish yellow throughout, and is especially dark near the base; in these stomach walls the coloration may also be dense just beneath the ectodermal epithelium. The mesogloea of the marginal lappets is coloured throughout, and the thin dark zone is to be found here both on the exumbrellar and subumbrellar sides. In the marginal tentacles the coloration is also in the mesogloea.

The gonads are colourless, and appear white in contrast to the dark surrounding coloration. In fresh specimens the gonads cannot be seen through the exumbrella owing to the dark coloration; but after a few months in formalin and sea water they become clearly visible and in due course the coloration of the whole medusa is reduced through a dark clear amber tint to almost colourless.

YOUNG STAGE

Kramp (1959a) describes what might be a young stage of this medusa 3 mm. in diameter. It was colourless and only one marginal tentacle was developed, the remaining eleven being minute bulbs; there were twenty gastric filaments in all.

HISTORICAL

Paraphyllina ransoni was first described by Ranson (1936) from a specimen found washed up on the shore at Villefranche, but he regarded it as *P. intermedia* described by Maas (1903) from the Malay Archipelago. Subsequently Russell (1956a) redescribed the species from a collection of a larger number of specimens caught in one haul in deep water west of the English Channel, and named it after Ranson, who had given the first description. The type specimens are in the British Museum. Further specimens were recorded by Repelin (1965) off the Ivory coast of Africa.

FAMILY PERIPHYLLIDAE

Coronate Scyphomedusae with four interradial marginal sense organs and four, twelve, twenty or (?) twenty-eight marginal tentacles in groups of equal numbers between adjacent rhopalia.

The distinctive feature of the family is the interradial position of the four marginal sense organs, between which the marginal tentacles are arranged in groups. The family is thus differentiated from the Paraphyllinidae in which the four marginal sense organs are perradial. There is only one British genus in the family, *Periphylla* Haeckel.

Genus **Periphylla** Haeckel

Periphyllidae with hemispherical or pointed dome-shaped umbrella; with twelve marginal tentacles in four groups of three, four perradial and eight adradial.

There is only one species of *Periphylla* and it is likely to occur anywhere in deep ocean water off the west and north coasts of the British Isles. It is *Periphylla periphylla* (Péron & Lesueur), and it cannot really be confused with any other species. Although they are quite distinctive, care should be taken not to confuse it with *Paraphyllina* in which the marginal sense organs are perradial.

The validity of the generic name *Periphylla* is discussed in great detail by Kramp (1947) and in agreement with him this familiar name is here retained.

Periphylla periphylla (Péron & Lesueur)

Plate I, fig. 5; Plate IX; Text-figs. 32–7

Charybdea periphylla Péron & Lesueur 1809, p. 332 (pl. XI, figs. 19–21, pl. V, figs. 1–3, not printed).
  Kramp, 1947, *Dan. Ingolf-Exped.* **5**, pt. 14, p. 42, fig. 16.
  ?Medusa (Melitea) hyacinthina Faber 1829, *Naturgesch. Fische Islands*, p. 197.
Charybdea bicolor Quoy & Gaimard 1833, *Voy. Astrolabe, Zool.* **4**, 293, pl. 25, figs. 1–3 (umbrella only).
Periphylla hyacinthina Steenstrup 1837, *Acta Mus. Hafniensis* (never printed, fide Kramp, 1947).
  Haeckel, 1880, *System der Medusen*, p. 419, pl. XXIV.
  Vanhöffen, 1892, *Plankton Exped.* **2**, K.d. p. 6, pl. II, figs. 1, 2.
  Vanhöffen, 1902, *Wiss. Ergebn. dt. Tiefsee-Exped.* '*Valdivia*', **3**, 23, pl. II, fig. 9; pl. V, figs. 30–4.
  Maas, 1904, *Résult. Camp. scient. Prince Albert I*, fasc. XXVIII, p. 47, pl. V, fig. 35, pl. VI, figs. 45–6.
  Bigelow, 1909, *Mem. Mus. comp. Zool. Harv.* **36**, 26, pl. 1, fig. 3; pl. 9, fig. 2.
  Mayer, 1910, *Medusae of the World*, **3**, 544.
  Broch, 1913, *Rep. 'Sars' N. Atl. Deep Sea Exped.* **3**, pt. 1, p. 4, figs. 2–4.
  Bigelow, 1928, *Zoologica, N.Y.* **8**, no. 10, **3**, 495.
  Stiasny, 1934, *Discovery Rep.* **8**, 345, text-fig. 2; pl. XIV, figs. 1, 2; pl. XV, figs. 1–3.
  Bigelow, 1938, *Zoologica, N.Y.* **23**, 155.
  Stiasny, 1940, '*Dana' Rep.* **4**, no. 18, p. 6, chart I.
  Ranson, 1945, *Résult. Camp. scient. Prince Albert I*, fasc. CVI, p. 7, pl. II, fig. 12.
  Naumov, 1961, *Fauna of U.S.S.R.* no. 75, p. 54, text-fig. 35, pl. I, fig. 1.
Chrysaora (Dodecabostrycha?) dubia Brandt, 1838, *Mém. Acad. Petersb.* **4**, 387, pl. 29, 30.
Cassiopea dubia Lesson, 1843, *Acalèphes*, p. 408.
Quoyia bicolor, L. Agassiz, 1862, *Contrib. nat. Hist. U.S.A.* **4**, 173.
Periphylla bicolor, Haeckel, 1880, *System der Medusen*, p. 419.
Periphylla peronii Haeckel, 1880, *System der Medusen*, p. 420.
Periphylla regina Haeckel, 1880, *System der Medusen*, p. 421.
  Vanhöffen, 1902, *Wiss. Ergebn. dt. Tiefsee-Exped.* '*Valdivia*', **3**, 23, pl. II, figs. 6, 8; pl. V, fig. 35.
  Broch, 1913, *Rep. 'Sars' N. Atl. Deep Sea Exped.* **3**, pt. 1, p. 7, fig. 4.
Periphema regina Haeckel, 1881 *a*, *Rep. 'Challenger' Exped. Zool.* **4**, 85, pl. XXIV, XXV.
Periphylla mirabilis Haeckel, 1880, *System der Medusen*, p. 422.
  Haeckel, 1881 *a*, *Rep. 'Challenger' Exped. Zool.* **4**, 64, pl. XVIII–XXIII.
Periphylla humilis Fewkes, 1886, *Rep. U.S. Comm. Fish.* (1884), pl. XII, XXXVI, p. 931.
Periphylla dodecabostrycha, Haeckel, 1880, *System der Medusen*, p. 421.
  Vanhöffen, 1892, *Plankton Exped.* **2**, K.d. p. 10, pl. II, figs. 1, 2.
  Maas, 1897, *Mem. Mus. comp. Zool. Harv.* **23**, no. 1, XXI, pl. IX, figs. 1–6.
  Vanhöffen, 1902, *Wiss. Ergebn. dt. Tiefsee-Exped.* '*Valdivia*', **3**, 23.
  Maas, 1903, *Siboga-Exped. Mon.* **11**, 6, pl. II, fig. 15.
  Maas, 1904, *Résult. Camp. scient. Prince Albert I*, fasc. XXVIII, p. 47, pl. V, figs. 36, 37.
  Mayer, 1906, *U.S. Fish. Comm. Bull. for 1903*, pt. III, p. 1136, pl. II, figs. 5, 8.

Periphylla, Maas, 1897, *Mem. Mus. comp. Zool. Harv.* **23**, no. 1, XXI, p. 27, pl. IV–VIII.

Periphylla hyacinthina forma *dodecabostrycha*, Mayer, 1910, *Medusae of the World*, **3**, 546.

   Stiasny, 1934, 'Discovery' Rep. **8**, 345.

Periphylla hyacinthina forma *regina*, Mayer, 1910, *Medusae of the World*, **3**, 546.

   Stiasny, 1934, 'Discovery' Rep. **8**, 348.

SPECIFIC CHARACTERS

   As genus.

DESCRIPTION OF ADULT

   Umbrella hemispherical or somewhat conical, with deep coronal groove situated approximately midway between apex and margin of umbrella; with thick jelly. Sixteen pedalial thickenings, twelve of which are tentacular and four rhopalar, the rhopalar pedalia being slightly narrower than the tentacular pedalia. Well-developed coronal muscle on subumbrellar surface, the distal margin of which has sixteen crenulations the points of which are in line with the radial septa. Eight well-developed deltoid muscles. Sixteen marginal lappets with elongated rounded margins. Twelve solid marginal tentacles, in four groups of three, four being perradial and eight adradial. Four interradial marginal sense organs, each consisting of rhopalium with hood and statocyst but with no ocellus. Eight long U-shaped gonads situated adradially on either side of the perradial deltoid muscles, with the convexities distal; each gonad with horizontal folds directed outwards. Stomach attached interradially to subumbrellar summit over four elongated triangular gastric septa to form four deep pockets between manubrium and subumbrellar surface; apex of stomach often elongated to form narrow canal; eight phacelli, one along each side of triangular gastric septa, each with 140 or more simple gastric filaments making 1,120 or more in all, extending along margins of perradial ostia; four perradial vertical ostia leading from stomach cavity into gastrovascular sinus; gastrovascular sinus continuous around area between proximal edges of ostia down to proximal edge of coronal muscle, except for four short interradial septa situated a little above coronal groove; gastrovascular sinus divided distally by sixteen radial septa, running from proximal margin of coronal muscle into middle of each marginal lappet, to form peripheral festoon canal system of tentacular and rhopalar lappet canals. Manubrium with four broad perradial thickenings of jelly, not extending beyond umbrella margin. Colour usually deep reddish brown over whole stomach and gastrovascular sinus and canal system, sometimes less dense peripherally so that gonads are visible; or on stomach and manubrium only, leaving gastrovascular sinus and periphery of umbrella transparent except for pigment on rhopalia. Size up to 200 mm. in height.

DISTRIBUTION

   Periphylla periphylla is a deep-sea medusa widely distributed in all oceans, except the Arctic; it is likely to be found in any collections made from deep water west and north of the British Isles.

STRUCTURAL DETAILS OF ADULT

   Periphylla was described in detail by Haeckel (1881*a*), and by Maas (1897) who included its microscopic anatomy.

Umbrella

   The shape of the umbrella changes as the medusa grows (Text-fig. 32). Presumably in its very earliest stages it is flattened in form but it soon increases in height until it has a highly domed or conical shape. After the medusa has reached a diameter of about 40 mm. it begins to grow more quickly in circumference than in height. In this way the summit of the umbrella becomes more flattened and hemispherical and the height of the medusa approximates to its diameter.*

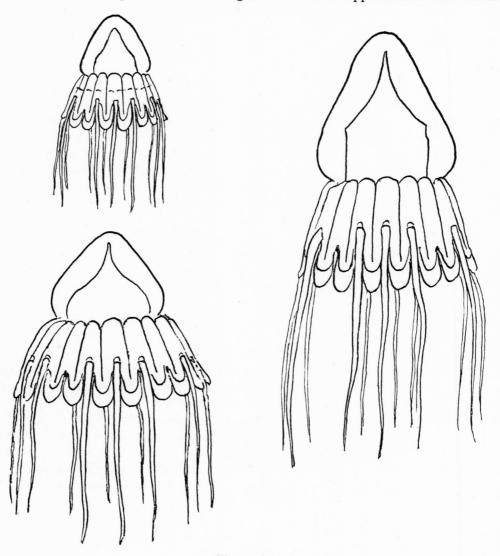

Fig. 32 (i)

This alteration in shape with growth, as well as individual variation in form and pigmentation, at first led to the erection of a number of species which have now been merged into one. It was already realized by Mayer (1906) that the shape of the medusa changed with age and Stiasny (1934, p. 343) and Bigelow (1938) divided the growth stages approximately according to their

* In preserved collections specimens are often laterally flattened so that the umbrella appears in outline to have a shallower summit than it really has. Many of the drawings and photographs in the literature are obviously made from such specimens.

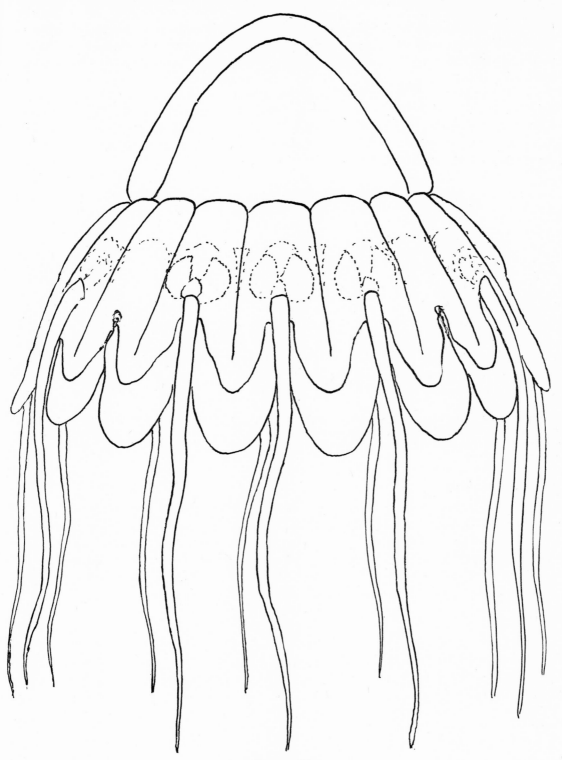

Fig. 32 (ii)

Text-fig. 32. *Periphylla periphylla*: changes in form of umbrella with growth (nat. size).

specific names, i.e. *dodecabostrycha* up to 35 mm.; *hyacinthina* up to 80 mm.; and *regina* up to 200 mm. diameter. This arrangement clarified the situation, but now that most authors agree that there is a single species and since growth is a continuous process it is unnecessary to fit specimens into such a scheme.

The mesogloea is very tough. It has scattered cells in it and numerous well-developed fibres clearly visible under a binocular microscope. These fibres run vertically through the mesogloea and branch towards the exumbrellar and subumbrellar surfaces. If the mesogloea is compressed these straight fibres fold like a concertina. Nowikoff (1912) confirmed that it was not cartilaginous in nature in the sense used by Haeckel.

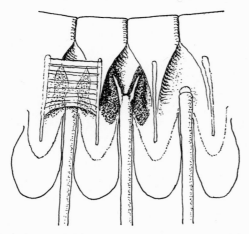

Text-fig. 33. *Periphylla periphylla*. Subumbrellar view of umbrella margin showing successive dissection away of coronal muscle, and tentacle base muscles.

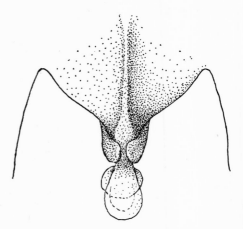

Text-fig. 34. *Periphylla periphylla*. Marginal sense organ, exumbrellar view.

Marginal sense organs

The marginal sense organ (Text-fig. 34), consists of a rhopalium with hood, statocyst, and sensory bulb. The statocyst itself is very well developed and appears large by comparison with the hood. The sensory bulb is fairly well developed. The base of the rhopalium, which has an exumbrellar carina, joins a broad wedge-shaped cushion on the margin of the umbrella.

A description and drawings of microscopic sections of the rhopalium are given by Vanhöffen (1900, 1902), who states (1892) that the rhopalium may be carrot-red.

Musculature

The coronal muscle is well developed and forms a ribbon-like band whose proximal circumference lies about midway between the coronal groove and the bases of the marginal tentacles just distal to the centripetal ends of the radial septa. The distal margin of the coronal muscle appears crenulated, the points of the crenulations being the attachments to the radial septa and the embayments where the muscle lies free and unattached so that pockets are formed between it and the bases of the marginal tentacles, and these run up under the root muscles of the tentacles (Text-fig. 33). The whole muscle is thrown into seven or eight circular folds (Haeckel, 1881*a* says ten to twelve) the proximal three or four of which are more strongly developed than the distal folds.

The most striking feature of the musculature is the strong development of the four interradial deltoid muscles, which extend as narrow strands from the bases of the gastric septa (Haeckel's intergenital muscles) and begin to fan out distally just above the coronal furrow where their main septal attachments lie, and end on a level of the proximal margin of the coronal muscle. The four perradial deltoid muscles are weaker and have their basal attachments at the level of the coronal groove on the lower edges of the gastric ostia (Text-fig. 35).

The musculature of the marginal tentacles, which are solid, has been described by Haeckel (1881 a, p. 68). The abaxial longitudinal muscle runs only along the proximal quarter or third of the tentacle, but the adaxial muscle runs the entire length of the tentacle and splits at its base into the two root muscles which are inserted on the radial septa near the proximal margin of the coronal muscle, lying in leaf-shaped folds of the wall of the gastrovascular sinus (Text-fig. 33). There is an endodermal core in the marginal tentacle (Maas, 1897, p. 39), but in poorly preserved specimens this is often disintegrated so that the tentacles appear to be hollow.

Stomach and gastrovascular system

The shape of the stomach in specimens about 20–40 mm. in diameter is conical, but its summit becomes more evenly rounded as the medusa increases in size. In many specimens, from about 30 mm. in diameter upwards, the apex of the stomach becomes compressed to form a narrow and pointed canal somewhat resembling the 'stiel-canal' or umbilical canal of Hydromedusae. All authors are now agreed that this is in no way homologous with a true 'stiel-canal', since it is not present in the smallest specimens (see e.g. Broch, 1913; Stiasny, 1934; and Bigelow, 1938), but Stiasny (1940) considers that it may have some special function since he found well-preserved specimens in which canals 12–15 mm. in length have broken through to the exterior. It seems to me that it must result from the constriction of the summit of the stomach of the small medusa as the jelly thickens when the medusa grows in height. It is not always present in large specimens, and occurs most commonly in medusae of intermediate size.

The four triangular gastric septa formed by the fusion of the stomach to the subumbrella become very elongated towards the apex as the umbrella grows in height. The simple gastric filaments, which are very numerous, are arranged along the edges of these septa and extend downwards along the margins of the ostia leading into the coronal gastrovascular sinus. These gastric ostia are long vertical slits which extend downwards nearly to the level of the coronal groove (Text-fig. 35).

The coronal sinus is continuous round the umbrella from the level of the bases of the gastric septa to the level of the proximal margin of the coronal muscle, except for interruption by four short septal nodes beneath the interradial deltoid muscles at the lower end of the gastric ostia. In some specimens there may be one or two small additional fusions above these septa. Evidently these septa are to anchor the strongly developed and elongate deltoid muscles (Text-fig. 35).

Below the level of the proximal margin of the coronal muscle there are sixteen radial septa which run into the middle of each marginal lappet. These begin about midway between the coronal groove and the bases of the marginal tentacles and extend to about the middle of each marginal lappet, ending a short distance beyond the level of the rhopalia. These septa divide the gastrovascular sinus into a peripheral festoon canal system. Those adjacent to the rhopalia are nearer one another than the others.

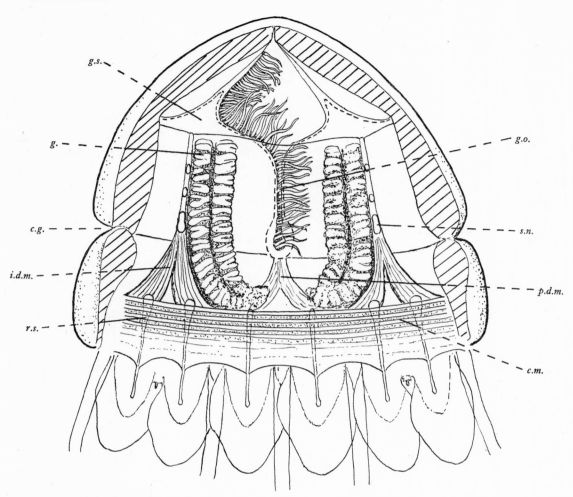

Text-fig. 35. *Periphylla periphylla.* 165 mm. in diameter: half-section viewed from subumbrellar side after removal of manubrium and subumbrellar wall of coronal sinus. The whole preparation is slightly flattened, and the gastric filaments, of which there are about 140 in each group, are shown on one side only of one gastric septum. *c.g.* coronal groove; *c.m.* coronal muscle; *g.* gonad; *g.o.* gastric ostium; *g.s.* gastric septum; *i.d.m.* interradial deltoid muscle; *p.d.m.* perradial deltoid muscle; *r.s.* radial septum; *s.n.* septal node.

Manubrium

The manubrium has four shield-shaped perradial lips with stiffened mesogloea separated from one another by interradial folded curtain-like areas with thick but softer mesogloea. These interradial areas form pillar-like bulges into the cavity of the manubrium (Text-fig. 36).

Gonads

The eight gonads (Text-fig. 37) first appear in the young medusae as straight ridges in the endoderm of the subumbrellar wall of the gastrovascular sinus, situated adradially about midway between the perradii and interradii. These extend vertically from a short distance above the coronal groove, sloping towards the perradii and reaching somewhat below the mid-point between the coronal groove and the proximal margin of the coronal muscle. The development of the gastric sinus leads to the formation of inward facing folds (Text-fig 37 B). These folds increase in length and their distal ends turn upwards so that hook-shaped folds are produced, described by some authors as horseshoes whose adradial arms are shorter than their interradial arms. The folds increase in width until the inner margins of the adradial and interradial arms become nearly

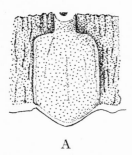 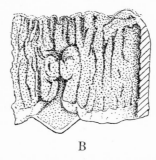

A   B

Text-fig. 36. *Periphylla periphylla*. Manubrium. A, external view of perradially thickened lip; B, internal view of same, with thick mesogloea shown in section on the right.

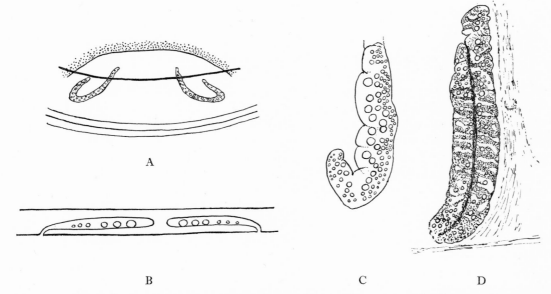

A

B   C   D

Text-fig. 37. *Periphylla periphylla*. Development of gonad. A, specimen 25 mm. in diameter; B, transverse section to show disposition of gonad; C, specimen 22 mm. high, subumbrellar view, largest egg 0·25 mm. in diameter; D, specimen 130 mm. high, subumbrellar view seen through wall of coronal sinus, eggs 1·2–1·3 mm. in diameter.

contiguous (Text-fig. 37 D). At the same time horizontal folds are developed across the arms. As the medusa grows the distal ends of the gonads become displaced towards the perradii along the margins of the deltoid muscles. In its fully developed state the gonad reaches from near the margin of the gastric septum to the proximal margin of the coronal muscle.

Owing to the horizontal folding of the gonad the subumbrellar wall of the gastrovascular sinus becomes puckered in conformity with the folds, leaving an apparent groove between the two arms of the gonad. When viewed through the subumbrellar wall of the gastrovascular sinus the horizontal folds of the gonads thus look as though they are directed outwards arising from this groove, but of course this is not so.

In the female gonad the young eggs first appear near the point of origin of the fold and the oldest eggs lie towards its distal margin. There is thus a tendency towards the arrangement of the eggs in series of increasing size. The largest eggs are 1·0–1·3 mm. in diameter.

In the male gonads the horizontal folds have secondary diverticula, so that the general appearance of folding is more complicated.

According to Stiasny (1934, 1940) gonads may already appear mature in specimens 8–10 mm. in diameter. He thought that this might indicate that there are two spawning periods, since in medusae of intermediate sizes the gonads may not be mature. Ranson (1945), however, thought that these might only be instances of individual variation. I have seen a small specimen with mature gonads. While the sex may be clearly indicated in quite young specimens, in the female the eggs are still very small. Presumably they are not mature until about 1 mm. in diameter (see Text-fig. 37C in which the largest are only 0·25 mm. diameter).

The minute structure and histology of the gonads has been described in detail by Maas (1897).

COLORATION

In life *P. periphylla* has an extremely dark red or brown coloration, but this changes to almost blackish brown on preservation. In preserved specimens the jelly of the exumbrella is usually quite colourless and transparent and the deep coloration beneath shows clearly through it. The colour is very dark brown to almost black and the bluish tinge imparted by the thick overlying jelly gives it a velvety appearance. Well-preserved specimens show patches of pigmented exumbrellar epithelium and it may well be that in life the whole surface of the umbrella is pigmented.

In general the distribution of the colour changes with age and size (Bigelow, 1938). In the smallest specimens the colour is usually restricted to the stomach and the peripheral zone of the umbrella is transparent and colourless. Soon, however, endodermal pigment appears in the peripheral zone, and eventually the whole medusa becomes deeply coloured. Nevertheless, there is considerable variation in the degree of coloration, and often the zones round the muscles at the bases of the tentacles are more heavily pigmented, imparting a pattern to the coloration of the periphery of the umbrella.

Bigelow (1938) has given some observations on specimens from Bermuda. Of thirty-eight specimens 5 to 12 mm. in diameter at the level of the coronal groove thirty-seven had so transparent a peripheral zone that the gonads were entirely visible, and one was so densely pigmented that the gonads were concealed. Of thirty-seven specimens 15 to 20 mm. in diameter at the level of the coronal groove the gonads were fully visible in twenty-one; the upper parts of the gonads were obscured by pigment in seven specimens, and they were wholly obscured in nine. Of twenty-seven specimens 25–40 mm. in diameter at the level of the coronal groove the gonads were completely obscured in eighteen; they were partially visible in eight specimens and only completely visible in one.

Unlike *Nausithoë* and *Paraphyllina* it seems that *Periphylla* may not lose its blackish coloration in preservative, for I have seen specimens over fifty years old still deeply pigmented. If kept in the light, however, all trace of coloration eventually disappears.

GENERAL OBSERVATIONS

Periphylla periphylla is a deep-water species of wide distribution and observations have been made on its vertical distribution based on expedition collections. The information is scattered over a number of publications, the more important being those of Broch (1913), Kramp (1913, 1924), Stiasny (1934), Bigelow (1938) and Kramp (1947, 1968). In the last-named reports Kramp summarizes and discusses these results in some detail.

It would appear that in the open ocean the depth of maximum occurrence is probably between 400 and 1,500 m. In some regions they may occur higher in the water and even at the surface in high latitudes. In the Mediterranean also they have been recorded at the surface. The maximum depth from which records are available is about 2,700 m. Pérès (1958), as a result of observations from the bathyscaphe, says that *Periphylla* was common in the Mediterranean between 1,000 and 1,300 m. and he had the impression that they kept within 200 m. of the bottom but avoided its immediate neighbourhood.

There seems to be a tendency for the older, larger and more deeply pigmented medusae to live deepest, but this is not a constant feature, Kramp (1947) having found a large number of specimens 4–10 mm. in diameter at 1,000 m. and 1,350 m. (see also Kramp, 1957).

Apparently the medusae do not grow large in the Mediterranean, Kramp (1924) recording no specimen over 35 mm. in diameter.

Nothing is known of the life history of *Periphylla*. Kramp (1947) regards the occurrence of small individuals far out in the ocean as an indication of direct development without a fixed stage. Kramp (1924) found that it lived throughout the year in the Mediterranean, and Bigelow (1938) thought that it probably did so off Bermuda.

Nicol (1958) recorded luminescence in this species...'When mechanically stimulated they emitted a blue glow around the margin, either in the tentacles or in the rim of the umbrella.' Krumbach (1925, p. 601) very briefly describes the swimming movements of the living medusa; when the pedalia contract strongly and the marginal lappets are raised the manubrium sticks out clearly. They are, however, rarely seen alive.

HISTORICAL

As with other deep-sea medusae *Periphylla periphylla* has received a number of different names, the different forms and growth stages each having been described as separate species. It is now universally regarded that these are all the same species which was first named *Charybdea periphylla* by Péron & Lesueur (1909). The synonymy of this species is fully discussed by Kramp (1947).

SEMAEOSTOMEAE

Scyphomedusae with umbrella without coronal groove and without pedalia; with or without hollow marginal tentacles; without interradial gastric septa; with radial septa in gastrovascular sinus or with gastrovascular canal system; with four simple oral arms with frilled or folded lips.

Semaeostome medusae are quite distinct from the coronate medusae, especially in the lack of a coronal groove and of pedalia. They also lack the four interradial crescent-shaped gastric septa which fuse the exumbrellar and subumbrellar endoderm round the margin of the stomach. Thus in those forms without a system of radial canals or gastrovascular network of canals there is unrestricted communication between the gastrovascular sinus and the central stomach all round its circumference. In these forms, the Pelagiidae and Cyaneidae, there are radial septa which divide the gastrovascular sinus into separate rhopalar and tentacular pouches. In the Ulmaridae the fusion of the exumbrellar and subumbrellar endoderm of the gastrovascular sinus forms a pattern so that distinctive radial and branching canal systems are produced.

In general form all the semaeostome medusae have somewhat flattened saucer-shaped umbrellas, and they do not show the extreme diversity of outline shown by coronate medusae.

There are three families, Pelagiidae, Cyaneidae, and Ulmaridae, distinguishable by the following characters.

1. Gastrovascular sinus divided by radial septa into separate rhopalar and tentacular pouches.
Pouches simple and unbranched . . . . **Pelagiidae**
Pouches branched . . . . **Cyaneidae**
2. Gastrovascular system in form of unbranched and branching radial canals, or with anastomosing radial canals . . . . **Ulmaridae**

In addition the Pelagiidae and Cyaneidae can also be distinguished by the position of the marginal tentacles. These are situated on the margin of the umbrella in the Pelagiidae, but in the Cyaneidae they arise from the subumbrellar surface at a distance from the umbrella margin.

Family PELAGIIDAE

Semaeostome Scyphomedusae with marginal tentacles arising from umbrella margin; with gastrovascular sinus divided by radial septa into separate unbranched pouches; without ring canal; with elongated oral arms with frilled lips.

The characteristic feature of the Pelagiidae is their apparent rather simple general form.

There are three genera in the family, *Pelagia* Péron & Lesueur, *Chrysaora* Péron & Lesueur, and *Sanderia* Goette. Of these only the first two are represented in British waters and each by a single species, *Pelagia noctiluca* (Forskål) and *Chrysaora hysoscella* (L.). These two medusae cannot be confused. *Pelagia noctiluca* is the simpler of the two, having eight marginal tentacles alternating with the eight marginal sense organs. *Chrysaora hysoscella* has twenty-four marginal tentacles arranged in eight groups of three alternating with the eight marginal sense organs. Their colour patterns are also completely distinctive. *Pelagia noctiluca* is an oceanic species and *Chrysaora hysoscella* is a coastal form.

Genus **Pelagia** Péron & Lesueur

Pelagiidae with eight marginal tentacles alternating with eight marginal sense organs.

Three species of *Pelagia* are at present recognized: *P. noctiluca* (Forskål), *P. flaveola* Eschscholtz and *P. colorata* Russell. Of these only *P. noctiluca* occurs in British waters.

Pelagia noctiluca (Forskål)

Plate II; Plate X; Text-figs. 38–46

Medusa pelagia Linné, 1766, *Syst. Nat.* Ed. 12, p. 1098.
Medusa noctiluca Forskål, 1775, *Descriptiones animalium*, p. 109.
Medusa perla Slabber, 1781, *Physikal. Bel.* p. 58, pl. XIII, figs. 1, 2.
   Modeer, 1790, *N. Abh. Schwed. Akad.* **12**, 245.
Medusa pelagica Swartz, 1791, *K. vet. Akad., nya Handl.* **12**, 185, pl. V.
Medusa panopyra Péron & Lesueur, 1807, *Voyages aux Terres Australes*, pl. XXXI, fig. 2.
Pelagia noctiluca, Péron & Lesueur, 1809, 'Tableau…de Méduses…', *Annls Mus. Hist. nat. Paris*, **14**, 350.
   Eschscholtz, 1829, *System der Acalephen*, p. 77.
   Wagner, 1841, *Ueber den Bau der* Pelagia…, pl. XXXIII, figs. 1–22.
   Hertwig, O. & R., 1878, *Das Nervensystem der Medusen*, p. 109, pl. IX, figs. 1, 3, 4, 6, 7, 8, 12; pl. X, figs. 15, 18.
   Haeckel, 1879, *System der Medusen*, p. 505.
   Metschnikoff, 1886, *Embryologische Studien an Medusen*, p. 100, pl. X, figs. 23–8.
   Vanhöffen, 1888, *Bibliotheca Zoologica*, **3**, 8, pl. VI, figs. 1–5.
   Mayer, 1910, *Medusae of the World*, **3**, 572, pl. 60, figs. 1–3.
   Krumbach, 1930, *Tierwelt der Nord und Ostsee*, III*d*, 1–88, figs. 15, 16, 17, 39, 40, 54.
Aurellia phosphorica Péron & Lesueur, 1809, *Annls Mus. Hist. nat. Paris*, **14**, 358.
Pelagia cyanella Péron & Lesueur, 1809, *Annls Mus. Hist. nat. Paris*, **14**, 349.
   Eschscholtz, 1829, *System der Acalephen*, p. 75, pl. VI, fig. 1.
   M'Andrew & Forbes, 1847, *Ann. Mag. Nat. Hist.* **19**, 390, pl. IX, fig. 5.
   Forbes, 1848, *Monogr. Brit. Medusae*, p. 76.
   L. Agassiz, 1860, *Contr. Nat. Hist. U.S.* **3**, pl. XII; 1862, **4**, 128.
   Haeckel, 1879, *System der Medusen*, p. 507.
   Mayer, 1910, *Medusae of the World*, **3**, 574, pl. 61, fig. 1.
   R. P. Bigelow, 1910, *J. exp. Zool.* **9**, 765, figs. 13–16.
Pelagia panopyra, Péron & Lesueur, 1809, *Ann. Mus. Hist. nat. Paris*, **14**, 349.
   Eschscholtz, 1829, *System der Acalephen*, p. 73, pl. VI, fig. 2.
   Brandt, 1838, *Zap. imp. Akad. Nauk*, sér. 6, sci. nat. **2**, 382, pl. XIV, fig. 1, pl. XIVA.
   Haeckel, 1879, *System der Medusen*, p. 509.
   Vanhöffen, 1888, *Bibliotheca Zoologica*, **3**, 14, pl. VI, fig. 21.
   Mayer, 1906, *U.S. Fish. Comm. Bull for 1903*, pt. III, p. 1139, pl. II, figs. 3, 4.
   Bigelow, 1909, *Mem. Mus. comp. Zool. Harv.* **37**, 43, pl. I, fig. 1.
   Mayer, 1910, *Medusae of the World*, **3**, 575.
Dianaea cyanella, Lamarck, 1816, *Animaux sans Vertèbres*, **2**, 507.
Rhizostoma perla, Eschscholtz, 1829, *System der Acalephen*, p. 53.
Pelagia discoidea Eschscholtz, 1829, *System der Acalephen*, p. 76, pl. 7, fig. 1.
   Haeckel, 1879, *System der Medusen*, p. 510.
Pelagia phosphorea Eschscholtz, 1829, *System der Acalephen*, p. 78.
Pelagia labiche Eschscholtz, 1829, *System der Acalephen*, p. 78.
Pelagia denticulata Brandt, 1838, *Zap. imp. Akad. Nauk*, sér. 6, sci. nat., **2**, 383, pl. XIV, fig. 2.
   Haeckel, 1879, *System der Medusen*, p. 508.

PELAGIIDAE

Pelagia phosphora, Haeckel, 1879, *System der Medusen*, p. 506.
   Vanhöffen, 1888, *Bibliotheca Zoologica*, **3**, 11, pl. VI, figs. 18, 19.
   Vanhöffen, 1902, *Wiss. Ergebn. dt. Tiefsee-Exped. 'Valdivia'*, **3**, 36.
Pelagia perla, Haeckel, 1879, *System der Medusen*, p. 506.
   Delap, 1907, *Fish. Ireland, Sci. Invest.* 1905, no. VII, p. 25, pl. I, figs. 2, 3, pl. II.
Pelagia papillata Haeckel, 1879, *System der Medusen*, p. 509.
Pelagia placenta Haeckel, 1879, *System der Medusen* p. 510.
   Vanhöffen, 1888, *Bibliotheca Zoologica*, **3**, 12, pl. VI, fig. 20.
Pelagia neglecta Vanhöffen, 1888, *Bibliotheca Zoologica*, **3**, 9, pl. I, figs. 1, 2, pl. VI, figs. 13–15.
Pelagia crassa Vanhöffen, 1888, *Bibliotheca Zoologica*, **3**, 10, pl. VI, figs. 6–12.
Pelagia minuta Vanhöffen, 1888, *Bibliotheca Zoologica*, **3**, 12, pl. VI, figs. 16, 17.
Pelagia noctiluca var. *neglecta*, Mayer, 1910, *Medusae of the World*, **3**, 574.
Pelagia panopyra var. *placenta*, Mayer, 1910, *Medusae of the World*, **3**, 575.
Pelagia purpuroviolacea Stiasny, 1914, *Zool. Anz.* **44**, 12, p. 529, figs. 1, 2.
Pelagia rosacea Stiasny, 1914, *Zool. Anz.* **44**, 12, p. 531, figs. 3, 4.
Pelagia curaçaoensis Stiasny, 1924, *Bijdr. Dierk.* **23**, 83, fig. 1.

SPECIFIC CHARACTERS

Exumbrella with medium-sized warts of various shapes; marginal tentacles with longitudinal muscle furrows embedded in mesogloea; up to 100 mm. in diameter.

DESCRIPTION OF ADULT

Umbrella hemispherical to somewhat flattened; with thick jelly; with surface covered with nematocyst warts of varying shapes, sizes and distribution. Sixteen marginal lappets, rectangular in shape with rounded corners and shallow median notches of varying extent. Well-developed coronal muscle on subumbrellar surface. Eight hollow adradial marginal tentacles, up to twice or even three times diameter of umbrella in length with thirty or more longitudinal muscle furrows. Four perradial and four interradial marginal sense organs; rhopalia consisting of statocyst and sensory bulb, but without ocellus, protected by exumbrellar extension of umbrella margin and sides of marginal lappets. Four interradial much-folded elongated gonads with central folds bulging downwards between thickened bases of manubrium. Stomach without interradial gastric septa with many, probably 400 or more, gastric filaments arranged in four interradial groups on subumbrellar wall where central stomach passes into gastrovascular sinus; gastrovascular sinus divided into compartments by sixteen radial septa extending from proximal edge of coronal muscle to its distal margin where they widen out to merge with wide central areas of fusion of exumbrellar and subumbrellar surfaces of marginal lappets, which divide the marginal extensions of gastrovascular sinus into tentacular and rhopalar pouches with no peripheral connecting canals. Manubrium arising from four perradial regions of much-thickened mesogloea to form continuous oral tube whose wall divides distally into four elongated arms with much-frilled edges; length of whole manubrium up to one and a half times diameter of umbrella, with oral tube forming a quarter to one-third of total manubrium length; covered with nematocyst warts. Colour rather variable, umbrella usually brownish yellow or with rose to purple tinge especially pronounced on tentacles and gonads; nematocyst warts also more strongly coloured; statocyst dark brown. Size up to 100 mm. in diameter.

DISTRIBUTION

Pelagia noctiluca is an oceanic species widely distributed in all warm and temperate waters. Off the British Isles it may be found anywhere over deep water off the west and north coasts. Kramp

(1947) says: 'It is very abundant in the Bay of Biscay, where it is frequently carried into the mouth of the Channel and northwards along the western coasts of Ireland and Scotland.'

But its occurrence in inshore waters of the English Channel is rare. Forbes (1847) recorded it from Mount's Bay, Cornwall; Russell (1938) recorded it occasionally off Plymouth. It did, however, occur in numbers in 1951 when Hunt (1952) recorded many specimens in the River Yealm estuary, South Devon. It was exceptionally abundant in 1966 when swarms invaded the south coast of Devon and north coasts of Devon and Cornwall and southern Irish Sea (Russell, 1967). Other west coast records are, north coast of Devon (Gosse, 1853, p. 378); Valencia, south-west Ireland (Browne, 1900; M. & C. Delap, 1907); and Port Erin (Cole, 1952).

It is carried round the Orkneys, Shetlands and Fair Isle into the North Sea (Fraser, 1954), and vast numbers were present off the Northumberland coast in November 1946 (letter from H. O. Bull).

It is apparently very rare in the southern North Sea. Slabber (1791) recorded it off the Dutch coast, but Maaden (1942a) had never seen it there. However, it was present off the Texel Lightship in mid-August to mid-September 1966 (Baan, 1967a, b). Stiasny (1930) recorded small specimens off Zeebrugge. A chart of records in the north Atlantic is given by Legendre (1940).

STRUCTURAL DETAILS OF ADULT

Pelagia noctiluca is subject to much variation in general form and colour and this has been fully discussed by Bigelow (1928), Broch (1913), Kramp (1924), and Stiasny (1934).

Umbrella warts

The shape, number and distribution of the nematocyst warts on the exumbrella are very variable, and this has given rise to the erection of a number of species. Three main types are found:

with round warts . . . . *perla* type
with narrow oval warts . . . . *panopyra* type
with marginal zone without warts . . . . *cyanella* type

In fact these distinctions are not clear-cut, and as more and more specimens have been examined it has been found that the shapes of warts differ according to the age of the medusa or their position on the exumbrella. The largest warts occur over the summit and central part of the umbrella and they are smaller on the periphery. There is a tendency for them to be arranged in radial rows; and they may sometimes be sparsely distributed or they may coalesce. They vary in shape from circular to elongated ridges, and all shapes may be found in the same specimen. While their bases may be fairly regular in outline their summits are often very irregular (Text-fig. 39).

The warts extend right up to the margins of the lappets on the exumbrellar surface, but they do not occur at all on their subumbrellar surfaces. They are absent on the lappets on the areas of fusion between the rhopalar and tentacular gastrovascular pouches.

Warts are numerous on the manubrium, both on its thick basal junction with the subumbrella and on the whole exterior surface of the oral arms and their folded lips (Text-figs. 38, 41 B). There is a complete circle of warts on a very narrow zone of the subumbrella just inside the proximal margin of the coronal muscle (Text-fig. 38).

Kramp (1956b) recorded two specimens from the Persian Gulf with no traces of warts and the exumbrella completely smooth.

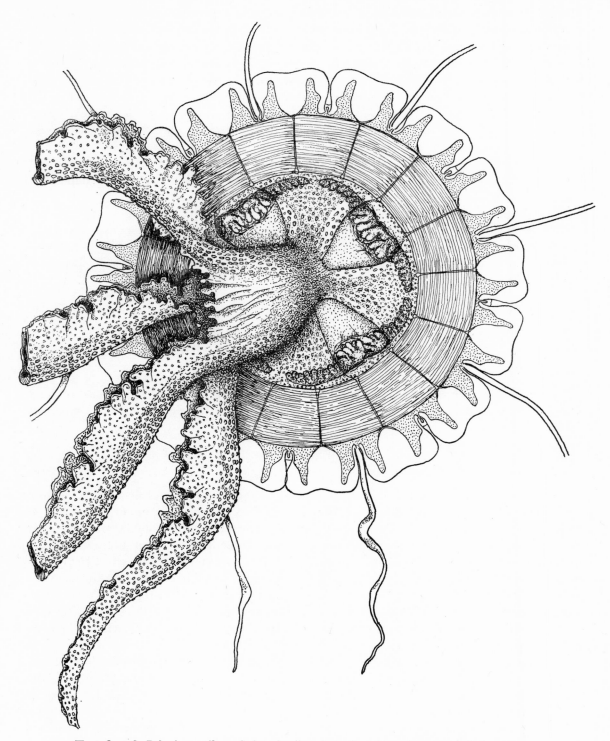

Text-fig. 38. *Pelagia noctiluca*. Subumbrellar view of adult to show general characters.

In the warts there are concentrations of nematocysts, and it is possible that they may be sites for luminescent cells according to early observations by Forbes (see p. 85). Vanhöffen (1888) gave measurements of the warts: in a specimen 65 mm. in diameter the average width of the wart was 2 mm., and they varied between 3 mm. long and 0·75–1·0 mm. wide. The warts on the oral arms were smaller than on the umbrella.

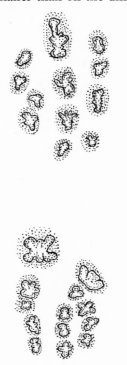

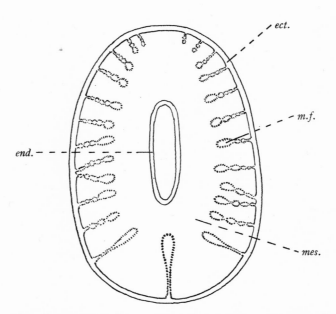

Text-fig. 39. *Pelagia noctiluca*. Shapes of nematocyst warts on exumbrella.

Text-fig. 40. *Pelagia noctiluca*. Cross-section of marginal tentacle (after Krasińska, 1914, text-fig. 3). *ect.* ectoderm; *end.* endoderm; *mes.* mesogloea; *m.f.* muscle fold.

The detailed structure of the mesogloea has been described by Bouillon & Vandermeerssche (1956, 1957) and G. Chapman (1959). There are no cells in the mesogloea. The fibrils run vertically between the subumbrellar and exumbrellar surfaces where they branch and run tangentially to form a plexus beneath each surface. This plexus is denser on the exumbrellar side.

Marginal lappets

The marginal lappets (Text-fig. 38) are rectangular with rounded corners. The substance of each lappet is very thin in the central area of fusion of exumbrellar and subumbrellar surfaces, and more solid on either side in the regions of the gastrovascular pouches. For this reason, in preserved specimens there often appears to be a deep cleft or furrow down the middle of each lappet owing to the folding inwards of the central area when the margins of the umbrella are curled inwards. Owing to its delicate nature also this central area is often frayed or torn so that there is quite a deep indentation on the lappet margin. In fact, in well-preserved specimens the marginal embayment is extremely shallow. Some authors (e.g. Kramp, 1924) have described the depth of this embayment as very variable, possibly for the above reason.

PELAGIIDAE

The width of the central area of fusion is over half that of the whole lappet. The extensions of the tentacular and rhopalar gastrovascular pouches into the marginal lappets are thus rather narrow and pointed. Their adjacent sides are more or less parallel until they suddenly converge towards the radial septa. The sides next to the rhopalium and the tentacle tend to be concave distally.

Manubrium

One of the characters which has in the past been regarded as having specific value is the length of the manubrium including the oral arms. The total length from the base of the oral tube to

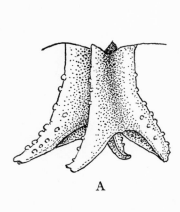

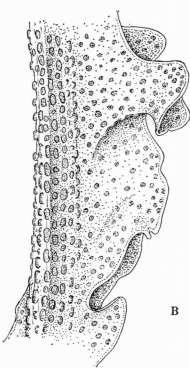

Text-fig. 41. *Pelagia noctiluca*. A, manubrium of young specimen 8 mm. in diameter;
B, portion of oral arm of adult.

the tips of the oral arms may equal the radius of the umbrella or exceed its length by two or three times. In preserved material this must of course be very dependent upon the degree of contraction. The proportional lengths of the oral tube and arms are also variable, and both characters vary according to the size of the medusa. Vanhöffen (1902), Broch (1913), Ranson (1945) and others have studied this feature, but the most comprehensive examination was that made by Kramp (1924), who studied 100 specimens from the Mediterranean and 59 from the North Atlantic. His results can be summarized as follows.

Umbrella diameter (mm). ...	10–	20–	30–	40–	50–	60–	70–	80–90
Length of oral tube (mm.)								
North Atlantic	3·0	6·2	7·0	8·2	10·0	14·5	17·0	18·0
Mediterranean	6·8	7·9	9·8	11·4	14·2	16·8	18·4	20·7
Length of oral arms (mm.)								
North Atlantic	7·0	17·1	21·0	32·9	43·6	55·5	68·2	67·3
Mediterranean	13·1	21·5	30·3	42·6	47·7	56·5	67·7	75·5

Kramp concluded that there was a constant though small difference between Mediterranean and North Atlantic specimens, both the oral tube and the oral arms being on the average slightly longer in Mediterranean specimens. As regards the nematocyst warts and marginal lappets Kramp could find no difference between Mediterranean and North Atlantic specimens. Kramp concluded that the differences between medusae from the two areas could at most only be regarded as racial. Ranson (1945) regarded the Mediterranean Indo-Pacific race with the longer oral tube and oral arms as the typical form and the Atlantic race as forma *perla*.

From Kramp's data given above it appears that at first the oral arms increase in length more rapidly than the oral tube.

Marginal tentacles

The marginal tentacles are flattened from side to side and oval in cross-section. Forbes (in M'Andrew & Forbes, 1847) and Browne (1900) noted that in living medusae the tentacles when contracted are about 4 in. long, but that they are capable of extending to several feet. Bozler (1926*b*) mentions that tentacles extended to a length of 30 to 40 cm. contracted quickly to a length of 4 or 5 cm.

A detailed study of the marginal tentacles was made by Krasińska (1914). The tentacle is hollow, but has a very thick mesogloea into which are sunk a number of longitudinal folds from the ectoderm along the walls of which the longitudinal muscles are arranged (Text-fig. 40). The deepest of these muscle folds of the ectoderm is on the adaxial side of the tentacle (personal observation); the others are symmetrically disposed on either side of this deep median fold, decreasing in depth the farther they are from it. Near the base of the tentacle Krasińska found twenty-one to twenty-five such folds, and these became fewer as the end of the tentacle was approached, but remained always in equal numbers on either side of the median fold. All the folds, except the median one and its two immediate neighbours, had longitudinal constrictions which could lead to complete separation into 'muscle-tubes' lying one above the other. The number of muscle folds evidently depends on the age and size of the tentacle as I have seen as many as thirty-three folds in a section of a large tentacle.

Marginal sense organs

The structure of the marginal sense organ and its development from the ephyra stage of the medusa were described in detail by O. & R. Hertwig (1878) and their description was slightly amended by R. P. Bigelow (1910). A description was also given by Eimer (1878).

The rhopalium itself consists only of statocyst and sensory bulb; there is no ocellus. The rhopalium is housed in a sensory niche formed by the umbrella margin (Text-fig. 42). The exumbrella tissue is extended between the bases of the adjacent marginal lappets to provide a hood over the rhopalium on the exumbrellar side. On the subumbrellar side the lateral edges of the marginal lappets extend inwards to form a partial floor beneath the rhopalium. According to the degree of contraction these edges may be seen to overlap so that the rhopalium is then housed in a tube. The statocyst is coloured completely with dark brown endodermal pigment.

On the exumbrella, immediately over the base of the rhopalium, there is a shallow sensory pit (Text-fig. 42).

PELAGIIDAE

According to R. P. Bigelow (1910) the statocyst in *Pelagia* has a smaller circumference than that of a young *Chrysaora* in the *Pelagia* stage, and the latter has not yet developed the exumbrellar sensory pits.

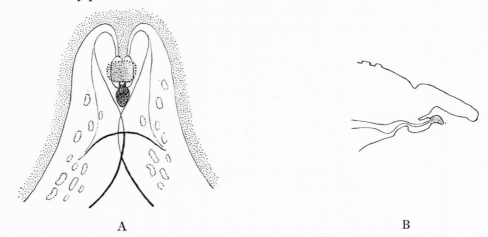

Text-fig. 42. *Pelagia noctiluca*. Marginal sense organ. A, exumbrellar view; B, radial sectional view to show position of exumbrellar sensory pit.

Gonads

An excellent detailed description of the gonads of *Pelagia noctiluca* was given by O. & R. Hertwig (1879, pp. 145–54; pl. IX, figs. 1–6; pl. X, figs. 6–8) (see also Claus, 1883, p. 40, pl. XI, figs. 80, 81), and the early development of the gonad was described by Hamann (1883, p. 423, pl. XXII, figs. 14–15).

The four gonads arise as elongated endodermal proliferations developing into ribbon-like folds in the interradial sectors of the stomach wall slightly distal to the rows of gastric filaments (Text-fig. 43). In these interradial sectors the subumbrellar wall of the stomach is thin compared with the much-thickened radial sectors at the base of the oral tube. Being bordered on each side by thick jelly these interradial areas thus appear as depressions which have been called the gonadial pits.

Each gonad has the form of a horizontal ribbon-like fold of the stomach wall whose free edge lies distally. The gonadial ribbon is closely apposed to the stomach wall leaving only a very narrow intervening space communicating with the main stomach cavity, the genital sinus, the two walls of which are connected at intervals by irregularly spaced trabeculae.

As the gonad develops it is thrown into lateral folds which are taken up by the stomach wall itself. The folds become very prominent in the middle region of the gonad where, when mature, they push the stomach wall outwards so that it protrudes into the subumbrellar subgenital area. (In preserved fully mature specimens the subumbrellar wall of the stomach is sometimes broken and the gonads and gastric filaments extrude.)

The side of the gonad immediately adjacent to the stomach cavity is covered by a continuation of the typical endoderm cells of the stomach wall, but both walls of the genital sinus are lined with more flattened cells. Between the two endodermal layers of the gonadial fold there is fairly thick mesogloea, but this is very attenuated at the base of the fold where it arises from the stomach wall.

78

The eggs are developed in the wall of the gonad on the genital sinus side and protrude in a single layer into the mesogloea on that side. The eggs first appear in a germinal zone along the proximal region of the gonad and they increase in size towards its distal margin. The Hertwigs drew attention to a small thin portion at the edge without eggs, but Claus considered that this was secondary, resulting from discharge of ripe eggs. Over each of the most fully developed eggs there is a cap of cylindrical vacuolated endoderm cells forming a crown protruding into the genital sinus.

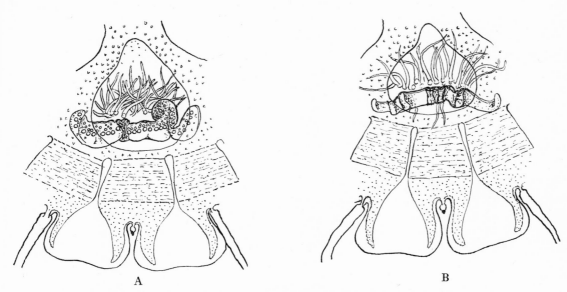

Text-fig. 43. *Pelagia noctiluca*. Sectors of umbrella of young specimens,
20 mm. in diameter. A, female; B, male.

The male gonads are essentially similar in general form to those of the female, and the sperm follicles extend right up to the distal margin. The follicles, which start as small hollow clumps of cells, branch into complicated diverticula which completely fill the gonad. The gonads are already beginning to appear in specimens as small as 10 mm. in diameter. The detailed structure of the egg was studied by Schaxel (1910, 1912).

Nematocysts

Weill (1934, p. 340) records three types of nematocyst in *Pelagia*.

Atrichous haplonemes in two sizes: (*a*) 7·0 × 3·0 μ; (*b*) 17·0 × 12·0 μ.
Both sizes occur on all parts of the medusa; intermediate forms are found in the nematocyst band which borders the mouth lips.
Holotrichous haplonemes: 30·0 × 29·0 μ.
Microbasic heterotrichous euryteles: 20·0 × 14·0 μ.

In her description of the histology of *Pelagia* Krasińska (1914) describes nematocysts, probably the euryteles, which have at their bases several muscular stalks with branching ends running down to the mesogloea. Each stalk consists of a single cell with a nucleus. After removal of these cells there still remains a protoplasmic process running from the nematoblast to the mesogloea (see also Weill, 1934, p. 119; Toppe, 1910).

PELAGIIDAE

Pelagia noctiluca was the first scyphomedusa whose nematocysts were observed (Wagner, 1841). Other references are Toppe (1910) and Alvarado (1923). Photomicrographs of nematocysts of *Pelagia* are given in Bennett (1967, pl. 29).

COLORATION

The coloration of the medusa is rather variable; the most usual colours mentioned are rose pink to purple and brownish yellow or orange. Coloration is deepest on the nematocyst warts of the umbrella and manubrium, on the marginal tentacles, and on the gonads. The membranous fringes of the oral arms are tinged with rose. The degree of coloration varies with age; for instance, Vanhöffen (1902) gives the following information about specimens collected on the 'Valdivia' Expedition.

Diameter of umbrella (mm.)	COLOUR			
	Umbrella	Stomach and manubrium	Gonads	Marginal tentacles
10	Colourless	Yellow ochre	—	White
20	Yellow ochre, brighter on margin	Yellow ochre	—	Wine red or rose
30	Bright rose	—	Rose	Yellowish or reddish

The largest specimens had a brownish violet tone and brownish coloured nematocyst warts. Browne (1900) gave the colour of specimens occurring in Valencia harbour as usually purplish brown or pale mauve. Griffiths & Platt (1895) isolated the violet pigment which they called Pelagein, $C_{20}H_{17}NO_7$. The pigment faded in light. (See also Fox & Millott, 1954, and Russell, 1964, *P. colorata*.)

DEVELOPMENT

Pelagia noctiluca has direct development, with no fixed scyphistoma stage. Krohn (1855) was the first to show this, and the early sequence from the egg to the ephyra was described by L. Agassiz (1860; 1862), Goette (1893), and Metschnikoff (1886) who recorded that the eggs, which were opaque brownish or violet coloured, were laid between 12 noon and 2 p.m. in December in Mediterranean specimens. He described the embryological development up to the 4-day-old larva.

Subsequently Delap (1907) reared *Pelagia* from the egg to ephyrae up to 3 mm. in diameter, using specimens taken in Valencia harbour (Text-fig. 44). After three days the eggs had developed to planulae which had a dark spot at either end but were colourless in the middle region. After five days the wider hindmost end had become broader and a ring of small knobs had appeared on it. The dome-shaped stomach extended upwards from this ring towards what had been the front end of the planula. By the sixth day the circle of knobs had developed into eight lobed lappets, each with a sense organ. The stomach had increased in height and the whole animal was now bell-shaped, but was still moved by ciliary action.

On the seventh day the mouth had broken through and the general shape was that of an ephyra with a pointed apex, but swimming was still only by cilia. Two of the ephyrae had nine arms and two had ten; the largest was 1·0 mm. in diameter.

A month after fertilization the largest ephyra was 3 mm. in diameter and had a very wide stomach cavity with four gastric filaments, but there were still no marginal tentacles.

The development from the ephyra to the young medusa was described by Uchida (1934a, as *P. panopyra*) from the Pacific and I have myself examined a number of small specimens (Text-figs. 45, 46) from the south-west of the mouth of the English Channel (Russell, 1967).

Uchida stated that the youngest ephyra, 2 mm. in diameter, had pointed marginal lappets, narrow rhopalial canals, and conspicuous nematocyst clusters sparsely distributed over the whole surface. Delap (1907) gave a drawing of an ephyra she had reared, 3 mm. in diameter. She remarked that it differed from the ephyrae of *Chrysaora, Cyanea* and *Aurelia* in the shortness of the marginal lobes in relation to the width of the disk, and in that their margins are more rounded.

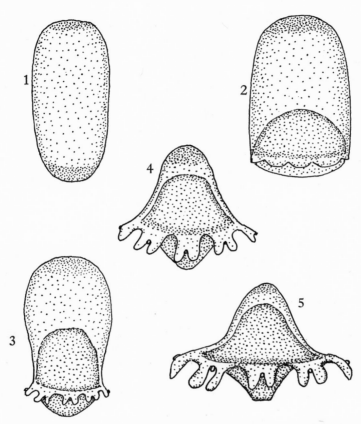

Text-fig. 44. *Pelagia noctiluca*. Development of ephyra from planula
(after Delap, 1907, pl. II, figs. 1–5; cilia omitted).

I think, however, that the shape of the marginal lobes and their lappets is likely to be variable, especially in preserved specimens, and that this is an unreliable character for identification. The presence of conspicuous nematocyst warts on the exumbrella is, however, a useful character for identification (Text-fig. 45). These are distributed in a regular pattern: there are two rows on each marginal lobe diverging from the base of each rhopalar canal and each running along either side of a canal and out to the ends of the marginal lappets. These warts are distributed also over the centre of the exumbrella, but mostly concentrated round the periphery of the central stomach.

There is one gastric filament in each interradius. The coronal muscle is apparent though weakly developed. The manubrium is cruciform. The colour is slightly pinkish or yellow.

At a diameter of 5 mm. the mesogloea at the umbrella centre is already thickening. The marginal lobes are shorter and wider in relation to the central disk; and the rhopalar canals are well bifurcated. The eight adradial canals have increased in length and started to bifurcate. The coronal muscle is more developed and radial muscles are appearing on either side of the perradial and interradial rhopalar canals. There are two to three gastric filaments in each group. The rudiments of four marginal tentacles are present, one at the end of each of four alternate adradial canals.

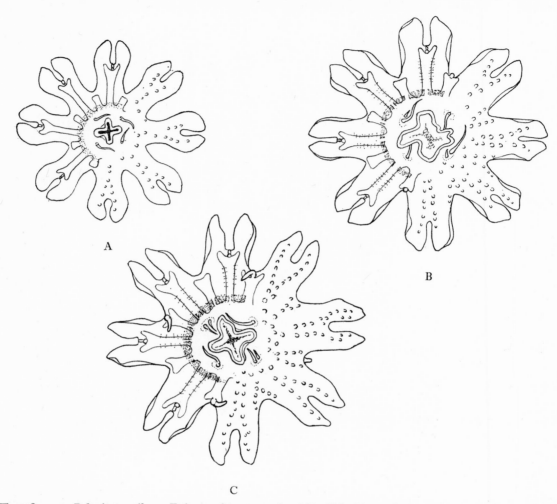

Text-fig. 45. *Pelagia noctiluca*. Ephyrae from mouth of English Channel. A, 3·5 mm. in diameter; B, 4·4 mm.; C, 4·8 mm. In each the subumbrellar view is on the left and the exumbrellar view on the right to show the arrangement of the nematocyst warts.

At a diameter of 8 mm. (Text-fig. 46) the four primary marginal tentacles are well developed, and the four secondary tentacles have appeared but are still quite small. The four oral lips are clearly developed and pointed. They have nematocyst warts on their outer surfaces (Text-fig. 41 A).

I have made an examination of the development of gastric filaments and marginal tentacles in thirty-two small specimens between 3·2 mm. and 8·0 mm. in diameter. There was a certain degree of variation.

Usually up to a size of about 4·5 mm. diameter there was only one gastric filament in each quadrant, and the marginal tentacles were either completely absent or, at about 4 mm. diameter, just appearing as two opposite or four alternate very small rudiments. Specimens up to 5·5 mm. in diameter might have two gastric filaments in each quadrant, but between 5 and 7 mm. there was some variation between two in each and three in each, or combinations of two and three. At 8 mm. diameter there were four or five gastric filaments in each quadrant.

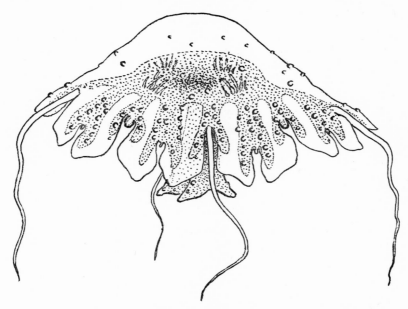

Text-fig. 46. *Pelagia noctiluca*. Young specimen 8 mm. in diameter.

The first four marginal tentacles were usually fully formed at about 5·0 mm. diameter when they were as long as the marginal lappets, and presumably were capable of further extension. The second series of four marginal tentacles started to develop also at a diameter of 5 mm., but their development was irregular, one 8 mm. in diameter still having them very short and stump-like, while another only 6 mm. in diameter had them fairly well developed.

The distribution of the nematocyst warts on the exumbrella is quite distinctive.

The rhopalium is already covered by a hood in specimens only 3·4 mm. in diameter. Notches are developing in the marginal lappets when the young medusa is about 8 mm. in diameter.

The medusa is now obviously a *Pelagia*, and in the Channel specimens is a bright golden yellow.

LIFE HISTORY

From its occurrence off the British coasts it appears that *Pelagia* breeds in summer and autumn and into the winter. M. & C. Delap (1907), referring to the shoal seen in Valencia harbour in August 1904, stated that early in September a number of ephyrae were caught and young stages were found up till the middle of October.

Kramp (1924) found that in the Atlantic, if the catches made in the months March, June, September and November were compared, the smaller specimens 3–10 mm. in diameter only occurred in November; those of a diameter of 20–50 mm. occurred in March and June; and the largest, 50–90 mm., occurred in March, June and September. He concluded that there was a well-marked breeding period in the autumn in the Bay of Biscay and that the entire development takes about a year. Observations in 1966 showed that in the western end of the English Channel *Pelagia* was breeding from September to December (Russell, 1967).

Lo Bianco (1888) stated that mature individuals were to be found in the Gulf of Naples throughout the year, especially during the winter, and that ephyrae were very abundant from November to March. Kramp (1924), however, recorded conflicting observations and concluded that in the Mediterranean *Pelagia* had a very extended breeding period.

HABITS AND PHYSIOLOGY

Alimentary system

The feeding reactions of the adult medusa were studied by Bozler (1926*b*). When a piece of food was given to a marginal tentacle extending to 30–40 cm. the tentacle contracted quickly to a length of 4–5 cm. There was then a slow contraction of the coronal muscle which brought the tentacle nearer to the mouth. The food was grasped by the lip of one of the oral arms and transported slowly along until it reached the stomach in about 10 min.

Metschnikoff (1880) recorded the taking of food particles by amoeboid processes of the endoderm cells.

Delap (1907) found that *Pelagia* fed on the salps *Thalia democratica* (as *Thalia mucronata*), and she fed ephyrae, which she had reared, on young *Obelia* and other small medusae.

Neuromuscular system

Experiments on the muscular control of movement in *Pelagia* were made by Bozler (1926*a*, *b*) and Bauer (1927). Bozler (1926*b*) found that removal of all marginal sense organs did not result in complete cessation of pulsation. Von Skramlik (1945) found that removal of more than four marginal sense organs induces periodic rhythm in pulsation. He said that medusae should be studied during the first two hours after capture.

Bozler (1926*a*) found that *Pelagia* were negatively geotactic in sunlight in an aquarium, but the next spring (1926*b*) he noted that they were positively geotactic and regarded this as escape reaction from water movement. Horridge (1959) observed at Naples that *Pelagia* swam upwards when left undisturbed and downwards when held by a clip for observation.

Heymans & Moore (1924) recorded the results of experiments on varying the ionic content of sea water on pulsation rate.

Komai (1942) described the distribution of nerve cells and nerve fibres of the marginal sense organ and its surrounding area of the ephyra of *Pelagia*.

G. Chapman (1959) discussed the relation between the musculature and the structure of the mesogloea and its fibrils in connection with the contractions of the umbrella when swimming.

Denton & Shaw (1962) studied the composition of the body fluid of *Pelagia*, and related buoyancy to low sulphate content.

Wagner (1841) recorded that the muscle is striated in this and other Scyphomedusae. He also described briefly the swimming behaviour. When the medusa pulsates, the marginal tentacles are contracted and straight and stiff.

Effects of temperature

Von Skramlik (1945) studied the effects of temperature on the pulsation rhythm. *Pelagia* has a continuously rhythmical pulsation between 8° and 26° C., and the rate of pulsation increases with increasing temperature. Below and above these limits pulsation becomes periodic and ceases altogether below 5° C. and above 35° C.

GENERAL OBSERVATIONS

Swarming

Pelagia generally occurs in large swarms in the open sea. Vanhöffen (1896; 1902) drew attention to the tendency for these swarms to occur in long rows; from observations of the occurrence of such rows on the cruise of the 'Valdivia' he showed that they usually occurred at the boundary between opposing oceanic currents.

When found off British shores they have usually occurred in shoals. Browne (1900) recorded shoals in Valencia harbour in 1896 and 1897. The medusae swam close to the surface and were brought in by the tide and stranded on the shore. Cole (1952) recalled astonishing numbers in Port Erin bay, Isle of Man, in October 1899, when 'the sea looked as if converted into a solid mass of jellyfish', and that on 10 August 1903 a large shoal appeared in Valencia harbour. The latter was the shoal from which Miss M. J. Delap (1907) obtained specimens for her rearing experiments. Hunt (1952) recorded the occurrence of about sixty specimens in the estuary of the river Yealm, South Devon, in December 1951.

Fraser (1954) recorded *Pelagia* in 1953 off Shetland, Fair Isle, west of Orkney, and in the North Sea east of 1° W. at times 'in sufficient numbers to cause inconvenience to fishermen'. These were accompanied by salps, as they were in 1903 in Valencia harbour.

Bourne (1890) recorded passing through a shoal 'miles in extent' in July in the mouth of the English Channel between Plymouth and Cork.

Luminescence

As its name implies *Pelagia noctiluca* is well known for its power of luminescing. Vanhöffen (1902) describes vividly the flashing-up and fading of the light given out by the medusae when stimulated by turbulence created by a ship's motion or by waves. The flashing is only of short duration and gradually fades.

Early observations by Forbes (in M'Andrew & Forbes, 1847) on the location of luminescence indicated that the intense light green is given out most vividly by the exumbrellar warts, especially on the marginal lobes. He states that if the undersurface of the umbrella is irritated the warts on the exumbrella opposite luminesce, but not the subumbrella itself.

Panceri (1872) and Dahlgren (1916) made further observations. A direct stimulus—mechanical, chemical, or electric—was necessary to make the medusa luminesce. The luminescence, which was light greenish in colour, covered the umbrella, manubrium and tentacles in rather an irregular pattern which might be blotched or spotted. Dahlgren gave a photograph of some swimming in an aquarium which had been struck with a glass rod. (This figure is reproduced in Krumbach, 1930, fig. 17.) If the medusa is touched lightly the luminescence, at first localized, spreads out in lines and streaks, and occasionally patches. Sufficiently violent contact could produce simultaneous luminescing of the whole animal.

85

PELAGIIDAE

Nicol (1955, p. 303) recorded that on stimulation with condenser shocks the animal gave a flash to each stimulus of a duration of 3 sec. He found no evidence of facilitation.

Effects on luminescence of altering the ion content of the sea water were given by Heymans & Moore (1923, 1924) and Hykes (1928), and of galvanic stimulation by Moore (1926).

Newton Harvey (1952, p. 167) records work on the biochemistry of luminescence.

Luminescent material sticks to the fingers (see for example Dubois, 1914, p. 42). Hunt (1952) recorded that the specimens he found in the Yealm estuary, South Devon, in December 1951 were faintly luminescent.

Vertical distribution

Kramp (1924) records that the bulk of specimens in the open ocean are found between 200 m. and the surface. Occasional specimens are taken at considerable depths. Stiasny (1937) suggests that they might live deep in tropical waters since sixty-nine specimens were caught in an Agassiz trawl at 549–640 m.

Pérès (1958) observed *Pelagia* in dives in the bathyscaphe and found them very abundant at a depth of 44 m. His paper includes a photograph of one taken at 500 m.

Bauer (1927) recorded that swarms of *Pelagia* appeared at the surface when the sky became overcast which were not to be seen previously.

Enemies

Legendre (1940) found only one specimen of *Pelagia* in his examination of the stomach contents of tuna.

Stinging

Wagner (1841) remarked that *Pelagia* could sting quite actively. Hunt (1952) recorded that handling *Pelagia* produced a stinging sensation on tender areas of skin between the fingers. Some people may be affected more than others, and Cleland & Southcott (1965) say that *Pelagia* does not sting very strongly, though repeated contact may produce severe results. During the invasion of this medusa on the north coast of Cornwall in 1966 there were reports of stinging in the Press.

Parasites and Commensals

Tattersall (1907) says that the commonest host for the amphipod *Hyperia galba* on the west coast of Ireland is *Pelagia*. Stiasny (1937) in his report on the collections of the John Murray Expedition to the Indian Ocean cites a note by R. B. Seymour-Sewell that small black amphipods seemed to be associated with *Pelagia*.

M'Andrew & Forbes (1847) also recorded the presence of a crustacean in the gastric cavity. Thomas (1963) recorded phyllosoma larvae of *Ibacus* attached to the umbrella surface in Sydney Harbour. They were unharmed by the marginal tentacles and quickly climbed back on to the exumbrella. I have found young horse mackerel, *Trachurus trachurus*, in association with *Pelagia*.

SEASONAL OCCURRENCE

On British coasts in years when shoals have been reported they generally appear in late summer and winter. Browne (1900) recorded *Pelagia* in Valencia harbour in July, August and September in 1896 and July, August and November in 1897. None were seen in 1895 or 1898. The specimens were mostly $2\frac{1}{2}$ to 4 in. (6·5–10 cm.) in diameter. Delap (1907) and Delap, M. & C. (1907) recorded shoals in Valencia harbour in December, 1902; August, September, October and November, 1903; and August 1904. Cole (1952) mentioned shoals in Port Erin bay in October 1899. Hunt (1952) recorded the shoal in the Yealm estuary in December 1951, the specimens being 1 to 4 in. in diameter, with the majority about 2 in. In 1966 *Pelagia* was unusually abundant off the coasts of Devon, Cornwall and Pembrokeshire (Russell, 1967) from January to October.

ABNORMALITY

There do not seem to be records of abnormality in *Pelagia*. I have seen a specimen 7·5 mm. in diameter collected for the Singapore Fishery Research Station on 9 March 1956 which had four fully developed marginal tentacles with ten others in course of development and fourteen marginal sense organs.

HISTORICAL

Pelagia noctiluca was first described by Forskål in 1775 as *Medusa noctiluca*. Subsequently many different species were described based on variable characters such as the size and shape of nematocyst warts, the shape of the marginal lappets, and the lengths of the oral tube and arms. A number of authors generally agreed that all these species are the same and Kramp (1955*b*) re-examined Haeckel's specimens of *perla*, *phosphora*, and *cyanella* and found them all to be *noctiluca*. Stiasny (1934) also agreed that there was only one species of *Pelagia* until he found in the collection of the 'Snellius' Expedition specimens which could be ascribed to *Pelagia flaveola* Eschscholtz, evidently synonymous with *P. tahitiana* A. Agassiz and Mayer.

Subsequently Russell (1964) added a third species *P. colorata* which is found off the coast of California, of which illustrations are given by Johnson & Snook (1927), MacGinitie & Mac-Ginitie (1949), and Fox & Millott (1954).

Genus **Chrysaora** Péron & Lesueur

Pelagiidae with groups of three or more marginal tentacles alternating with eight marginal sense organs.

Kramp (1961) lists eleven species of *Chrysaora* but the identity of some of these is still uncertain. There is, however, only one British species, *Chrysaora hysoscella* (L.) whose characters make identification easy. It has twenty-four marginal tentacles arranged in groups of three, each group lying between adjacent marginal sense organs, so that the eight groups of three tentacles alternate with the eight sense organs. The dark brown exumbrellar colour pattern of sixteen radiating triangles, although variable, is also quite distinctive.

Chrysaora hysoscella (L.)

Plate III; Plate XI; Text-figs. 47–55

'*Urtica marina*' Borlase, 1758, *Nat. Hist. Cornwall*, p. 256, pl. 25, figs. 7, 8. (Rejected for nomenclatorial purposes because the author did not apply the principles of binominal nomenclature; Hemming and Noakes, 1958, p. 6, opinion 332 (26).)
Medusa hysoscella Linnaeus, 1766, *Syst. Nat.* Ed. 12, p. 1097.
    Forbes, 1848, *Monogr. Brit. Medusae*, p. 77.

PELAGIIDAE

Chrysaora Lesueur Péron & Lesueur, 1809, 'Tableau...de Méduses...', *Annls Mus. Hist. nat. Paris,* **14,** 365, no. 110.

Chrysaora aspilonota Péron & Lesueur, 1809, *ibid.* p. 365, no. 111.

Chrysaora cyclonota* Péron & Lesueur, 1809, *ibid.* p. 365, no. 112.
 Gosse, 1853, Devon coast, pp. 363–77, pl. XXVII (frontispiece), figs. 1–4.

Chrysaora spilhemigona Péron & Lesueur, 1809, *ibid.* p. 365, no. 113.

Chrysaora spilogona Péron & Lesueur, 1809, *ibid.* p. 365, no. 114.

Chrysaora pleurophora Péron & Lesueur, 1809, *ibid.* p. 365, no. 115.

Chrysaora heptanema Péron & Lesueur, 1809, *ibid.* p. 366, no. 119.
 Lesson, 1843, *Acalèphes,* p. 399.

Chrysaora macrogona Péron & Lesueur, 1809, *ibid.* p. 366, no. 120.

Chrysaora mediterranea Péron & Lesueur, 1809, *ibid.* p. 366, no. 116.
 Eschscholtz, 1829, *System der Acalephen,* p. 82.
 Haeckel, 1879, *System der Medusen,* p. 511, pl. XXXI.

Medusa fusca Pennant, 1812, *Brit. Zool.* (1812 ed.), **4,** 121.

Medusa tuberculata Pennant, 1812, *Brit. Zool.* (1812 ed.), **4,** 122.

Aurelia crenata Chamisso, 1820, *Nova Acta Phys. Med.* **10,** 359, pl. 29.

Chrysaora hysoscella, Eschscholtz, 1829, *System der Acalephen,* p. 79, pl. 7, fig. 2.
 Lesson, 1843, *Acalèphes,* p. 396.
 Claus, 1877, *Denkschr. Akad. Wiss. Wien,* **38,** 33, pl. I, figs. 1–10 (development of egg to scyphistoma); pl. VI, figs. 29–31; pl. VII; pl. IX, figs. 40 (2, 3); pl. XI, figs. 50, 51.
 Wright, 1861, *Ann. Mag. nat. Hist.* n.s. 3, **7,** pl. XVIII, figs. 1, 2, 4.
 Mayer, 1910, *Medusae of the World,* **3,** 579.
 Stiasny, 1927, *Zool. Meded. Leiden,* **10,** 73, pl. I–III.
 Kramp, 1961, *Synopsis of Medusae of the World,* p. 325.

Chrysaora lutea Quoy & Gaimard, *Mém. Ann. des Sci. Nat.* **10,** pl. 4B, fig. 1.
 Blainville, 1834, *Actinologie,* p. 299, 'La Chrysaore jaune', pl. XLIII.

Chrysaora oculata Lesson, 1843, *Acalèphes,* p. 402, pl. 4, fig. 1.

Medusa stella Dalyell, 1847, *Remarkable Animals of Scotland,* p. 106, pl. XVII. (Pl. XV, XVI, planulae and young polyps; pp. 102–10, describes hydratuba and ephyrae as *Medusa bifida.*)

Cyanea chrysaora Milne Edwards, 1849, in Cuvier's *Règne animal illustré. Zoophytes,* pl. 47.

Chrysaora isosceles Haeckel, 1879, *System der Medusen,* p. 513.

Taeniolhydra roscoffensis Hérouard, 1908, *C. r. hebd. Séanc. Acad. Sci. Paris,* **147,** 1336.

SPECIFIC CHARACTERS

Exumbrella typically with sixteen brown V-shaped radial markings with varying degrees of pigmentation between, with dark apical circle or spot, with brown marginal lappets; twenty-four marginal tentacles in groups of three alternating with eight marginal sense organs.

DESCRIPTION OF ADULT

Umbrella saucer-shaped, with thick jelly; surface smooth. Thirty-two marginal lappets, semi-circular in outline; rhopalar lappets shorter than tentacular lappets. Moderately well-developed coronal muscle on subumbrellar surface. Twenty-four hollow marginal tentacles in eight groups of three. Four perradial and four interradial marginal sense organs each consisting of rhopalium with statocyst and sensory bulb but without ocellus, protected by exumbrellar extension of umbrella margin to form a hood and by approximation of free edges of marginal lappets on subumbrellar side; deep exumbrellar sensory pit present immediately over rhopalium. Four interradial much-folded gonads bulging downwards into cavity of subgenital pit. Hermaphrodite; embryos developing to planulae in gastric cavity and on mouth lips. Stomach without interradial gastric septa; with many gastric filaments arranged in four interradial groups on subumbrellar

* This and the following species of Péron and Lesueur were placed in the genus *Cyanea* by Lamarck, 1817, *Anim. sans Vert.* **2,** 519–21.

wall where central stomach passes into gastrovascular sinus; gastrovascular sinus divided into sixteen pouches by sixteen radial septa extending from proximal edge of coronal muscle to its distal margin where they diverge to merge with the fused edges of the marginal lappets adjacent to the rhopalia on the sides nearest the neighbouring marginal tentacles; thus the tentacular gastrovascular pouches dilate at the periphery while the rhopalial pouches contract; without peripheral connecting canals. Manubrium arising from four perradial regions of much-thickened jelly to form a short continuous oral tube whose wall divides distally into four elongated oral arms with frilled edges; length of whole manubrium up to several times diameter of umbrella; thickened subumbrellar basal portions of manubrium fused to form four gonadial pouches with circular or oval orifices. Colour very variable, typically in British waters with sixteen V-shaped brown markings with varying amounts of pigmentation between, with brown apical circle or spot, and with brown patches on marginal lappets. Size up to 200–300 mm. in diameter.

DISTRIBUTION

Chrysaora hysoscella occurs in coastal waters all round the British Isles; it may possibly be more prevalent off the south and western shores but in 1968 I found it to be very abundant in Loch Ewe, north-west Scotland, in September.

It is common in the southern North Sea, being frequently stranded in numbers along the Dutch coast. It possibly does not occur so frequently farther north, as Kramp (1934) drew attention to unusual numbers in Danish waters in 1933 when it reached Frederikshaven in the Kattegat. He quotes some unpublished notes by D. Damas recording it off the west coast of Norway without precise details. Papenfuss (1936) stated that it may be common at Kristineberg, Sweden, during the autumn months. An unusual occurrence in the Baltic was recorded by Rathke and Zaddach (1849) when a swarm appeared in the Bay of Danzig in 1848, but Kramp (1934) was inclined to doubt this.

Southwards it extends along the Bay of Biscay into the Mediterranean and Adriatic. Kramp (1955 a) recorded eight specimens from Monrovia, Liberia, on the West African coast; and Vannucci & Tundisi (1962) recorded two specimens, 17 cm. and 20 cm. in diameter, as far south as Puerto Melchior, Argentina, and Ushuaia, Tierra del Fuego.

STRUCTURAL DETAILS OF ADULT

Umbrella

The saucer-shaped umbrella, with its thick jelly and smooth surface, thickens somewhat at the bases of the marginal tentacles and rhopalia so that the umbrella margin appears slightly fluted. G. Chapman (1953) found no cells in the mesogloea. Fibres run parallel with the surface of the exumbrella, and large fibres run at right angles to the surface where they branch.

Marginal Lappets

The marginal lappets are semicircular in outline, but the rhopalar lappets are slightly broader and less rounded than the tentacular lappets. As a result the tentacular lappets project a little beyond the rhopalar lappets (Text-fig. 47).

Manubrium

The manubrium, the four oral arms of which are fused for a short distance at its base, is characteristically long and slender. The frilled lips of the oral arms are almost uniformly narrow

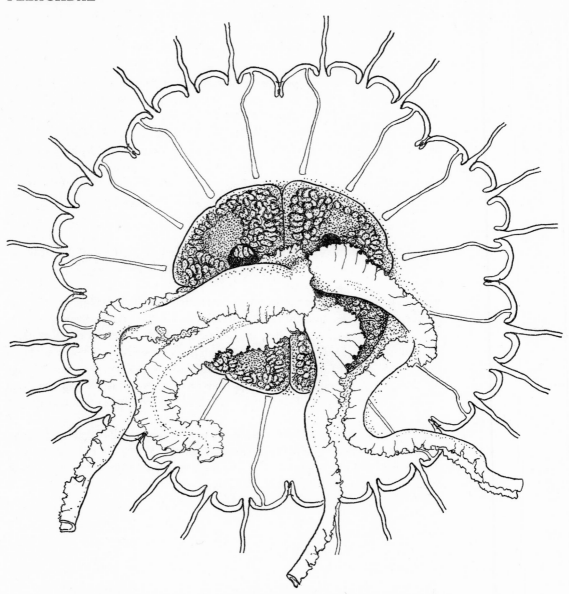

Text-fig. 47. *Chrysaora hysoscella*. Subumbrellar view of adult to show
general characters, Plymouth.

over the greater part of their length and their inner surface may be covered with the small blister-like sperm sacs (Text-fig. 51 A). In life the oral arms may coil spirally at their tips, like tentacles (Text-fig. 49).

At the base of the manubrium the thickened portion of each of the basal pillars of the four oral arms continues over the surface of the subumbrella to form two diverging branches. The ends of these branches are fused with those of their neighbours to surround oval interradial openings, or subgenital ostia, leading into cavities into which the gonads, enclosed in their gastro-genital membranes, may protrude (Text-fig. 47). These are the four subgenital cavities. In each perradius there is a rather deep embayment into this thickened jelly and interradially the margin is slightly concave. Thus the whole thickening has a margin somewhat like that of a four-leaved clover.

Marginal tentacles

The marginal tentacles, which are capable of great elongation, have conical, somewhat laterally flattened, bases. Beyond the basal bulb the tentacle remains laterally flattened and is covered with small nematocyst warts (Text-fig. 48).

Text-fig. 48. *Chrysaora hysoscella*. Marginal tentacle.

Text-fig. 49. *Chrysaora hysoscella*. End of oral arm showing spiral coiling.

The tentacle is hollow and on the subumbrellar or adaxial side of the basal bulb there is a longitudinal furrow.

Gosse (1853) remarked that the tips of the tentacles are very liable to be torn off. He saw them frequently stretched to a length of 1 ft. and noted that they were never drawn up in spiral coils. They could contract to about 1 in. in length.

Haeckel (1879) records finding in a large swarm at Granville, Normandy, a specimen 160 mm. in diameter with forty marginal tentacles and forty-eight marginal lappets.

Marginal sense organs

The structure of the marginal sense organ was described by Claus (1877), but in greatest detail by R. P. Bigelow (1910) for the *Chrysaora*-like form from Chesapeake Bay, regarded as being the *Chrysaora* stage of *Dactylometra quinquecirrha* L. Agassiz (see L. G. Cowles, 1930, p. 331).

PELAGIIDAE

The rhopalium has the form of a thick bent finger. It is set in a deep tube-like sensory niche formed by the umbrella hood above and the approximation of the free edges of the marginal lappets below (Text-fig. 50A). The rhopalium is attached at its base to a ridge running to the proximal wall of the sensory niche. Immediately above the rhopalium, on the exumbrellar surface of the hood, there is a well-developed exumbrellar sensory pit which is deep and cone-shaped (Text-fig. 50B). The whole of the surface of the subumbrellar sensory niche is covered with thickened ectoderm. This layer is thickest on the proximal wall of the niche and two thickened

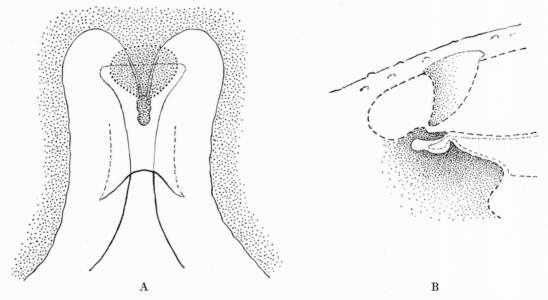

A   B

Text-fig. 50. *Chrysaora hysoscella, ca.* 150 mm. in diameter. Marginal sense organ. A, exumbrellar view; B, radial sectional view to show position of exumbrellar sensory pit.

areas extend outwards for a short distance along the sides of the marginal lappets. Similar columnar ectoderm cells line the exumbrellar sensory pits (although Schäfer (1878) said that these cells did not differ from those of the exumbrellar ectoderm). A folded and pitted epithelium covers a small part of the lower surface of the rhopalial ridge and extends along lateral grooves into pockets at the side of the mouth of the rhopalial canal. Beneath this epithelium is a felted mass of extremely fine nerve fibres, similar to that of the nerve layer of the rhopalium itself.

In the ephyra the rhopalium lies between marginal lappets, but is not enclosed in a sensory niche. When the medusa is 6 mm. in diameter, in the *Pelagia* stage, the niche and hood are already developed. At the beginning of the *Chrysaora* stage, 10 mm. in diameter, the dorsal sensory pit begins to appear as a shallow depression in the surface of the exumbrella above the base of the rhopalium, but it is still lined with flat epithelial cells like those of the rest of the exumbrellar surface.

Gonads

The development of the gonads is similar to that for *Pelagia*, but the degree of folding is much greater, so that when mature the convolutions of the gonad are extremely complicated (Text-fig. 47). These folded gonads are, however, only female.

The occurrence of hermaphroditism in *C. hysoscella* was first discovered by Derbès (1850) in specimens in the Mediterranean near Marseilles. It was next reported by Wright (1861) for *Chrysaora* in British waters.

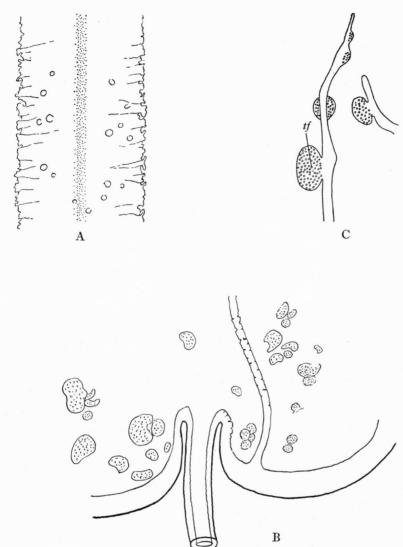

Text-fig. 51. *Chrysaora hysoscella*. Sperm sacs. A, on oral lips; B, in margins of gastric pouches; C, on genital filaments (C from Widersten, 1965, text-fig. 9; *tf*, follicle).

The medusa is a protandrous hermaphrodite and Claus (1877) found that in the Adriatic individuals of $1\frac{1}{2}$ to 2 in. in diameter already had ripe sperm. Wright stated that small specimens 4 in. in diameter had only male elements. Claus found that specimens with a diameter of at least 10 in. were only female; smaller specimens had male and female elements present together. Claus thought that in the Adriatic the occurrence and ripening of the male organs was limited to the period February to April.

93

The sperm are developed in outgrowths which arise from the endoderm in the gonadial region of the stomach wall. Sperm sacs are also developed in small blister-like structures on the gastric filaments, the walls of the stomach, in the gastrovascular pouches and on the inner surfaces of the oral arms (Text-fig. 51). Widersten (1965) states that the outgrowths from the gonadial region, or genital filaments, were always bigger than the gastric filaments in his specimens.

Claus (1883) noted that the developing female gonads first appeared in specimens from the Adriatic of about 25–30 mm. in diameter.

The most detailed account of the female gonad has been given by Widersten (1965). The genital epithelium is strongly flattened in contrast to the prominent nurse cells. The oocytes migrate into the mesogloea, which has tangential and transverse strands of fibre but no mesogloeal cells. Fertilization takes place in the mesogloea in which the embryos develop to the blastula and gastrula stages.

Nematocysts

The nematocysts (Text-fig. 5, p. 9) have been described by Weill (1934) at Wimereux and Papenfuss (1936) at Kristineberg; and Tchéou-Tai-Chuin (1930b) studied their development.

There are three kinds, namely, atrichous haplonemes, holotrichous haplonemes, and heterotrichous microbasic euryteles. The atrichous haplonemes are of two size groups.

The measurements given by Weill and Papenfuss respectively show a slight difference in the large atriches: Weill gave only the range of lengths of nematocysts and Papenfuss the average lengths and widths, as follows.

| | Atriches | | | |
	large	small	Holotriches	Euryteles
Weill (length)	11–14 μ	7–9 μ	15–17 μ	10–13 μ
Papenfuss (averages)	16·2 × 9·2 μ	8·0 × 3·8 μ	13·8 × 12·0 μ	12·0 × 7·0 μ

Papenfuss paid particular attention also to the arrangement of the coils of the thread within the capsule. She noted that the large atriches differed from those of *Cyanea capillata*, *C. lamarckii* (as *C. palmstruchii*) and the American *Dactylometra quinquecirrha* in having two loose and slanting coils at the stomal end. She also noted differences in the small atriches.

She found that differences in size and structure of the holotriches separated *Chrysaora* from *Dactylometra*, and also that there were considerable differences between the euryteles of these species.

Weill recorded that both sizes of atriches are present in all parts of the medusa, and that they were particularly abundant in the marginal tentacles and oral lips. He said that intermediate forms are localized in the band of nematocysts which fringes the oral lips, where there is a continuous size series from large to small. The holotriches and euryteles occur in the marginal tentacles and oral lips, the holotriches being much the rarer of the two.

Weill also recorded that in planulae 180 μ long there were atriches (7 × 5 μ) and microbasic euryteles (10 × 7 μ). They were distributed irregularly in the ectoderm, being more abundant at the posterior end. He also gave measurements of nematocysts in the scyphistoma and ephyrula, which had been reared by Lambert, as follows:

Scyphistoma: atriches (6–9 μ long) and euryteles (11–14 μ long), in all parts but particularly abundant in the tentacles.

Ephyrula: atriches (6–9 μ long) abundant in ectoderm of lobes and irregularly dispersed in other organs; in the peristome they reached 13 μ in length. Holotriches (12 × 10 μ–14 × 13 μ) sparsely distributed on the surface of the exumbrella and abundant in lobes. Not found in tentacles. Euryteles (10 × 8 μ–13 × 9 μ) occurring everywhere.

COLORATION

Chrysaora hysoscella shows much variation in the pattern of colour of the exumbrella. This gave rise at first to the erection of a large number of species now known to be the same.

The colour patterns have been described by many authors, including, for example, Eschscholtz (1829, pp. 79–81), but the most comprehensive study of the subject was made by Stiasny (1927) who examined several hundreds of specimens collected from the Dutch shores, especially between Katwijk and Noordwijk. Stiasny listed in all eight major colour varieties which can be summarized as follows.

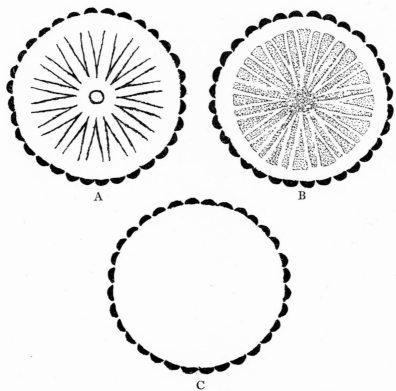

Text-fig. 52. *Chrysaora hysoscella*. Three variations of exumbrellar pigment pattern.
A, type 2 *b*; B, type 3 *a*; C, type 1 *a*.

Type 1   Without a pattern
 (*a*) Almost colourless, white, with or without a dark apical spot; with dark marginal lappets (Text-fig. 52 C).
 (*b*) As for (*a*) but with colourless marginal lappets.

Type 2   With dark pattern on light ground
 (*a*) Heavy strongly marked pattern on light umbrella (pl. XI, fig. 3).
 (*b*) Fine delicate pattern on light umbrella (Text-fig. 52 A; pl. XI, fig. 2).
 (*c*) Strongly marked pattern on more or less heavily pigmented umbrella (pl. XI, fig. 1).

Type 3   With light pattern on dark ground
 (*a*) Weak coloration, umbrella fairly evenly pigmented apically, both within the triangles and between them (Text-fig. 52 B).

(*b*) With more pigment within the triangles than between them, perhaps with the spot formation standing out more strongly than the pattern.

(*c*) With less pigment within the triangles than between them so that the triangles stand out whitish on a dark ground.

Of these variations Type 3 is characteristic of the mediterranean form *C. mediterranea*, and Stiasny suggested that this type of coloration might be that of warmer waters. He saw no specimen of Type 1 with colourless marginal lappets, but found five specimens in the collections in the Rijksmuseum, so they are evidently very rare. It should also be remembered that the colour can disappear entirely after long periods in preservative, although the areas where the pigmentation was remain obvious.

As regards the form of the pattern there are typically sixteen triangles. Stiasny said that the number varied between twelve and sixteen; I have myself seen specimens with fourteen and seventeen.

The triangular markings may be in the shape of an elongated inverted V or Y, and they may be short or long, thick or thin, straight, bent, or irregular (pl. XI, fig. 4).

The apical spot, when present, may be circular, oval or polygonal; it may be evenly light or dark, or with a darker or lighter circumference.

In well-pigmented specimens the small pigment spots on the exumbrella are continuous to the marginal lappets; but there is a tendency for these spots to be absent or few in areas over the marginal sense organs and round the umbrella margin opposite the ends of each line of the triangles. The marginal lappets may be dark brown, or lighter brown with dark peripheries.

The thickened axes of the oral arms are usually darkly coloured and the fringes of the lips themselves are covered with pigment spots. This latter pigmentation is continued on to the thickened bases of the mouth pillars which coalesce to form the subumbrellar walls of the genital pits. These areas are pigmented irrespective of the degree of pigmentation of the exumbrella. But the remainder of the subumbrellar surface is unpigmented. The colour of the pigment varies from light brown through red-brown and dark brown to a blackish brown. The pigment spots are on small warts containing nematocysts.

When I first examined specimens with different degrees of pigmentation I wondered whether the less heavy pigmentation in some specimens might have been due to loss of epidermis. Examination by staining showed, however, that although much epidermis might be missing the pigmented areas remained attached to the mesogloea, indicating that colour variations could not be due to loss of epidermis. Section of the epidermis shows that the pigment appears to be slightly sunk into the mesogloea and thus anchored to it. In fresh specimens it is not possible to rub the V-shaped pigmentation off.

In specimens which have been long in preservative and in which there is no coloration left the original pigmented areas show clearly opaque whitish against the background.

SEASONAL OCCURRENCE

In British waters young *Chrysaora* first make their appearance in May. These have grown to large adults by July. The medusae are still common in August and September but only occasional specimens occur in October. Probably the peak period for obtaining fair-sized specimens is mid-June to September.

On the east coasts the season may be slightly later, for van der Maaden (1942*a*) gave records of strandings on the Dutch coast showing maximum abundance from mid-August to mid-

October, specimens being last seen in December. Hartlaub (1913) also recorded them as especially abundant in September off Helgoland.

Graeffe (1884) and Claus (1877) recorded that in the Adriatic ephyrae are produced in September and October and that the medusae are mature in March and April, as Derbès (1850) stated for Marseilles.

DEVELOPMENT

Development from the egg was first described by Claus (1877, 1883) and later in fuller detail by Hadži (1909), Okada (1927) and Teissier (1929). The egg is small, about 47 μ in diameter. Segmentation is total and irregular giving rise to ciliated planulae varying much in size; Teissier gives, for instance, extreme measurements of $85 \times 65 \mu$ to $750 \times 400 \mu$, while Widersten (1968) gives dimensions of 300×110–$360 \times 165 \mu$. Gastrulation according to Okada is by true invagination, but owing to the unusually large volume of the blastula cavity the archenteron is reduced to a mere slit.

Development to the planula stage takes place in the parent. A fragment of gonad will have in it at the same time oocytes, ova, segmenting ova, and planulae in various stages of development. Teissier made analyses of the organic contents of planulae which showed that the planulae contained more material than the egg and had presumably derived nourishment from the parent medusa. He states that the free life of the planula can last four or five days (see also Delap (1901) cited, below, p. 98).

The planula has nematocysts which, according to Claus (1877), Hadži (1909) and Weill (1934), are distributed irregularly in the ectoderm, being more abundant at the posterior end.

Hadži (1909) made a careful histological study of the development of the planula to the scyphistoma. It was this work probably more than any which led to the final separation of the Scyphozoa from the Anthozoa on the grounds that the ectoderm played no part in the formation of the mouth tube, contrary to the opinions of Goette, Hein and Hyde.*

Hadži showed that the planula attaches to the substratum by its broader end which was anterior when swimming. After settlement this broader end narrows and becomes stalk-like, while the free oral end becomes broad and flattened. The mouth is formed by the breakthrough of the epithelial layers in the centre of the oral disk. At the stage when the four primary tentacles have developed the four endodermal folds of the gastric septa are developing.

Heric (1909) gave a detailed account of the histology of the scyphistoma during strobilation. In this he again confirmed the endodermal origin of the pharynx. Heric also showed that the four subgenital cavities in the ephyra do not originate from the four peristomial pits of the scyphistoma, but are new formations.

Hadži (1909) and Hérouard (1912) described the formation of podocysts by the scyphistoma, as did T.-T.-Chuin(1930b). Hérouard (1908) at first thought that the podocysts were statoblasts and, not having seen the scyphistoma strobilate, he described it under the name *Taeniolhydra roscoffensis*. However, after intensive feeding with the ovaries of the echinoderm *Strongylocentrotus* the scyphistoma produced ephyrae in December 1909. Hérouard (1911c) stated that the podocysts could exist for at least three years.

Detailed studies on the scyphistoma were made by Tchéou-Tai-Chuin(1930b) who incorporated in his paper earlier brief reports (T.-T.-Chuin, 1928, 1929a–e, 1930a). He studied the general

* A recent paper by Gohar and Eisawy (1961) on the development of *Cassiopea andromeda* would, however, appear to reopen this controversy.

histology, digestion, stolonization and budding, and strobilation. His observations on strobilation confirmed those of Heric.

T.-T.-Chuin found that in the laboratory tanks at Roscoff from June to October multiplication by budding proceeded actively giving rise to individuals of all sizes. In June and July long slender stolons were produced, but as the season advanced these were more and more reduced until there were none in October. Any budding produced under artificial conditions in winter was without the formation of a stolon; he recorded such observations made in the laboratory at temperatures of 15–20° C. (Tchéou-Tai-Chuin, 1928).

He also observed that an isolated piece of stolon would attach and develop directly into a scyphistoma, but this took 8–10 days instead of the 3 days taken by a normal scyphistoma bud.

Rearing the scyphistoma in the laboratory

T.-T.-Chuin (1930b) kept scyphistomas in small vessels to which he added small pieces of the calcareous alga *Lithothamnion* to prevent the water from becoming too acid. He fed them on *Tubularia* gonophores, pieces of sea-urchin ovary, *Ciona* intestine, the liver of molluscs and crabs, and especially veal liver. Hérouard (1911b) noted that scyphistomas would not take free eggs of the echinoderm *Strongylocentrotus*, but only the ovary.

Under natural conditions strobilating scyphistomas refuse food, but individuals reared in the laboratory in winter and, as a result, not yet strobilating, would feed (T.-T.-Chuin, 1928).

Claus (1891, p. 21, footnote) recorded that he had *Chrysaora* scyphistomas in an aquarium which had already been living there for fourteen years. Each year they strobilated at the end fo October to beginning of November, and sometimes also in the spring, and produced masses of ephyrae. The basal stock then regenerated sixteen tentacles and produced stolons and buds. Hagmeier (1933) stated that at strobilation the polyp assumed a yellowish colour.

Chrysaora has been reared right through from the planula to the adult medusa at Valencia, south Ireland, by Delap (1901). Five days after their liberation from the parent medusa on 22 June 1899 the planulae were attaching to the glass and two days later tentacles were beginning to develop, at first four in number, but according to Claus (1877) two develop first. Three weeks after liberation of the planulae some of the resulting scyphistomas had their maximum number of sixteen tentacles.

Verwey (1942) considered that the upper limit for strobilation might be between 7 and 10° C.

The scyphistomas were kept through the winter and fed on small medusae and ctenophores which they preferred to copepods. On 3 April 1900 the first ephyrae were observed. Observation on scyphistomas showed that ephyrae were liberated three days after the beginning of strobilation. The ephyrae were about 2 mm. in diameter with eight marginal lobes each with a marginal sense organ. At first pinkish in colour, they soon changed to a translucent white.

In just under three weeks they were $\frac{1}{2}$ in. in diameter and had four marginal tentacles which could extend to 4 in. (At a comparable size, 10 mm., in the Adriatic Claus (1877) records that they have eight tentacles.) The circular mouth now had four oral arms. After six weeks the largest specimen, having eaten all the others, was $1\frac{1}{2}$ in. in diameter; it had eight marginal tentacles, with rudiments of others. The oral arms now had frilled margins and were $2\frac{1}{2}$ in. long.

After seven weeks, on 22 May, this medusa was $2\frac{1}{2}$ in. in diameter, had twenty-four marginal tentacles and its oral arms were 5 in. long. A fortnight later brown markings were beginning to appear on the umbrella, the medusa being $3\frac{3}{4}$ in. in diameter with oral arms 9 in. long.

On 21 June the medusa, now nearly eleven weeks old, was $6\frac{1}{2}$ in. in diameter and 3 in. thick;

it had rich dark brown coloration on the marginal lobes and on the exumbrella, as bright as in a specimen taken from the sea. When thirteen weeks old the medusa was 9 in. in diameter and the gonads were visible, showing yellow through the umbrella. Further growth ceased owing to lack of food, but the medusa must have been nearly full grown.

During these observations Delap noted that the temperature in her aquarium in February was 3 to 5° F. below that of the sea (sea minimum 47·5° F.; aquarium 42° F.), but during May and June the aquarium was usually 3 to 4° above that of the sea (aquarium 59–62° F.) In July the aquarium temperature reached 66° F., that of the sea being 64° F.

A specimen reared at Helgoland grew to a diameter of 35 mm. after forty-nine days (Hagmeier, 1933); and a photograph of a living young specimen 10 mm. in diameter is reproduced in Kühl (1967).

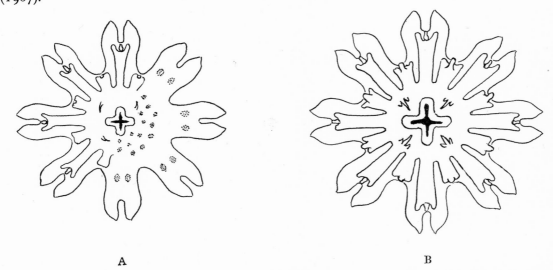

A   B

Text-fig. 53. *Chrysaora hysoscella*. Ephyra. A, 2·7 mm. in diameter, Solway Firth, 22 July 1965; B, 4·0 mm., Solway Firth, 14 May 1966 (coll. T. G. Skinner). In A the exumbrella is shown on the right with nematocyst clusters.

The ephyra

The newly liberated ephyra of *Chrysaora* has been described by Claus (1883) from the Adriatic and by Delap (1901) from south-west Ireland. Claus noted the following characters which distinguished it from the ephyra of *Aurelia*:

(1) Greater size and robust form; (2) the more conspicuous protrusion of the developing adradial gastric pouches, which in *Aurelia* are so small that they can be overlooked; (3) the presence of a pair of nematocyst clusters at the base of each marginal lappet and a single cluster nearer the centre of the disk opposite each lappet; (4) the retarded development of the gastric filaments which are as yet not so elongated as in *Aurelia*; (5) the slenderness of the sense organs compared with those in *Aurelia*.

The ephyra drawn by Delap is possibly a little older than that drawn by Claus, for Claus remarks that as they grow the marginal lappets become less pointed and more rounded and broad. The ephyra in Delap's drawing also has more advanced development of the adradial gastric pouches than that drawn by Claus. The two drawings differ in that while Claus shows the single nematocyst clusters at the bases of the perradial and interradial gastric pouches, Delap places them at the distal ends of the adradial canals. Whether this constitutes a regular difference between the Mediterranean and northern forms needs to be examined.

T.-T.-Chuin (1930*b*) gave the numbers of gastric filaments in each group in newly liberated ephyrae as four; the smallest specimen I have drawn (Text-fig. 53) had only one filament in each group.

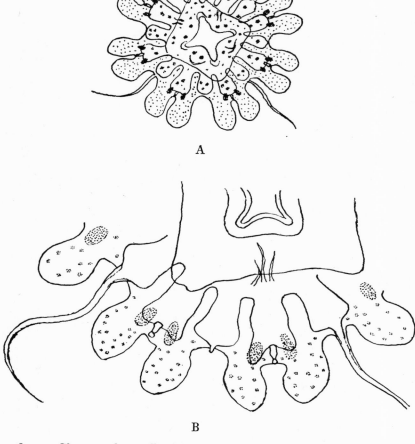

A

B

Text-fig. 54. *Chrysaora hysoscella*. A, specimen 6–7 mm. in diameter; B, quadrant of same enlarged (after Claus, 1883, pl. V, figs. 38, 39).

Claus also comments on the high degree of development of the ring muscle and the sixteen radial muscles in the ephyra of *Chrysaora*. I have seen specimens of ephyrae from the Solway Firth kindly sent me by T. G. Skinner (Text-fig. 53). These agreed in the distribution of nematocyst clusters with Delap's drawing.

While size of ephyra and shape of the marginal lappets are not good diagnostic characters (see p. 23), a further character seems possible in that the protrusions at the corners of the distal ends of the rhopalar canals appear to be more prominent in *Chrysaora* than in *Aurelia*.

Delap states that when $\frac{1}{2}$ in. (*ca.* 12 mm.) in diameter the first four marginal tentacles are developed; and that at $1\frac{1}{2}$ in. (*ca.* 40 mm.) diameter there were eight tentacles with rudiments of others. Probably development in the warmer Mediterranean waters is more rapid, for Claus (1883)

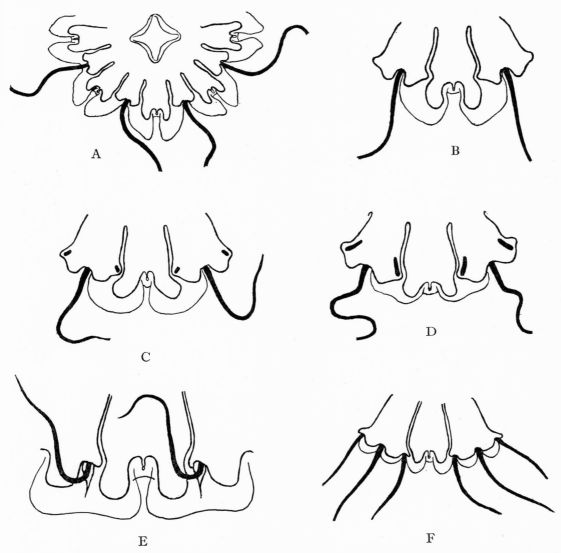

Text-fig. 55. *Chrysaora hysoscella*. Development. A, B, 10 mm. in diameter, *Pelagia* stage; C, 20 mm. in diameter; D, slightly later stage; E, showing splitting of marginal lappet at its base; F, all marginal tentacles developed. (Modified after Claus, 1877, pl. VI, figs. 29–31; pl. VII, figs. 32–4.)

gives a drawing of a specimen with four marginal tentacles already developed when the medusa was 6–7 mm. in diameter (Text-fig. 54), and he recorded a specimen 10 mm. in diameter which already had eight tentacles (Claus, 1877).

According to Claus (1883) when the medusa is 25–30 mm. in diameter, with sixteen adradial tentacles, the gonads are beginning to appear.

PELAGIIDAE

The development of the sixteen adradial tentacles (Text-fig. 55) was described and illustrated by Claus (1877). The rudiments of these tentacles appear in lateral outgrowths in the marginal corners of the tentacular gastric pouches. At the same time slight embayments appear in the margins of the rhopalar lappets opposite the sites of the rudimentary tentacles. Eventually, when the tentacles are developed, splits appear in these marginal lappets which finally become parted to form the new tentacular lappets (Text-fig. 55 E).

When the medusa has only eight marginal tentacles it closely resembles *Pelagia noctiluca* of the same size in general appearance. It differs, however, in that the tentacular pouches in *Pelagia* do not have the lateral outgrowths in which the sixteen adradial tentacles of *Chrysaora* will develop, nor is there the same pattern of exumbrellar nematocyst warts.

HABITS AND PHYSIOLOGY

Food and feeding

Delap (1901), when she reared *Chrysaora*, found that at all stages the medusa showed a preference for hydromedusae, siphonophores and ctenophores. It would eat *Tomopteris* and *Sagitta*, but did not eat copepods and other crustacea nor, apparently, young fish. It was notable, however, that it would not eat the anthomedusa *Leuckartiara*, nor the ctenophore *Beröe*, even after a few days of starvation.

Later, Lebour (1923) confirmed Delap's observations, stating that *Chrysaora* was an omnivorous feeder with an evident preference for cnidarians and *Sagitta*. She also noticed that *Leuckartiara* (as *Turris pileata*) was refused. She saw two *Aurelia*, about 55 mm. in diameter, eaten by a larger *Chrysaora*. As regards young fish Lebour found that they were eaten by *Chrysaora* only before they had reached the scaled stage. This perhaps explains the findings of Gosse (1853) and Delap that young fish were not eaten.

The method of feeding was recorded by Gosse (1853), Delap (1901) and especially by Lebour (1923). When hungry the medusa extends its marginal tentacles to many times the diameter of the umbrella, and the oral lips are also extended. When prey is caught by the tentacles these contract and quickly transfer it to the oral lips. The lips either pass the food up into the stomach, or enclose it in a temporary bag in which it is wholly or partly digested outside the stomach—a statement which requires confirmation. While digestion proceeds, and when the medusa is satisfied with its meal, the marginal tentacles and oral arms remain contracted. Delap said that 'a good meal consisted of several dozen medusae and ctenophores'; this referred to the specimen she reared when 9 in. in diameter.

Fritz Müller (1859) was the first to show that the gastric filaments are solid and that they have digestive properties. Using crustacean flesh and gastric filaments of the Cubomedusan *Tamoya* he showed their powers of digestion. After treating gastric filaments of *Chrysaora* with chromic acid he removed the epithelium leaving the solid core.

Water content

Krükenberg (1880) gave the water content for specimens from Trieste as 95·75 % with 4·25 % solids, and 96·3 % and 3·7 % respectively. Teissier (1926) investigated the effect of maturity on water content. He found that a *Chrysaora* weighing 1 kg. would liberate in 2 or 3 days more than 10 g. of planulae. Thus in some individuals the genital elements might represent at least 2–3 %

of the wet weight. A fully mature *Chrysaora* could therefore have more than twice as much organic substance as an animal of the same size when not in the reproductive period.

Teissier (1932) found that planulas 1 or 2 days old were 66% water, 31% organic substance, and 3% mineral substance. He estimated the protein content as 59–63% of the dry weight, and the total lipids as 20% of the dry weight.

Mesogloea

G. Chapman (1953) found the following amino-acids in hydrolysates of the mesogloea: alanine, glutamic acid, glycine, leucine, lysine, proline, serine, valine and (?) histidine.

Bateman (1932) investigated its osmotic properties and concluded that there was no bound water. Alexander (1964) studied the visco-elastic properties of the mesogloea.

Neuromuscular system

Lehmann (1923) studied experimentally the function of the marginal sense organs in *Chrysaora* and *Cyanea capillata* (see also p. 122). He concluded that they were not statocysts, but that the medusae were orientated by the specific weight of the subumbrella and manubrium as opposed to that of the jelly of the exumbrella; that the sense organs have the function of nervous regulation, having a great influence on the frequency and strength of the contractions of the umbrella; and that the excitement produced by the sense organs gives the stimulus for rhythmic contraction.

Woollard & Harpman (1939) made observations on the nerve fibres in *Chrysaora*.

Effects of temperature

Schaefer (1921) studied the effects of temperature and drugs on pulsation rate. He also noted that *Chrysaora* from the North Sea was positively thermotactic, its optimum in autumn being about 17° C. Movement ceased at upper temperatures of 28–32° C. and lower temperatures of 4–6° C.

Stinging

Gosse (1863) recorded that a lady felt the sting of *Chrysaora* but that he himself could not feel it.

Vertical distribution and behaviour

Verwey (1966) found that *Chrysaora* were more abundant in the water column at times of ebb and flow when currents were strong than during slack water of high and low tides (see p. 195).

I have watched *Chrysaora* swimming several feet below the surface in late August in Loch Ewe. One swam upwards and as soon as its umbrella broke the water surface it turned over and swam down.

Commensals

Tattersall (1907) records having once seen the amphipod *Hyperia galba* with *Chrysaora*.

HISTORICAL

Chrysaora hysoscella received its specific name from Linnaeus in 1766, who placed it in the genus *Medusa*. The species had been previously recorded and satisfactorily figured as early as 1758 by Borlase in his *Natural History of Cornwall*. Borlase called it ' *Urtica marina* ', but his name has had to be rejected because he did not apply the principles of binominal nomenclature (Hemming, 1958, p. 6).

Owing to the great diversity in individual coloration the medusa has been given many specific names, especially by Péron & Lesueur (1809) who made as many as nine distinct species. Haeckel (1880), however, regarded eight of Péron and Lesueur's species as synonymous, but kept their *C. mediterranea* separate. Mayer (1910) agreed with Haeckel but regarded *C. mediterranea* as probably also synonymous.

Haeckel preferred the specific name *isosceles* of Eschscholtz (1829) on grounds of orthography. The species was well figured in colour for the first time by Dalyell (1847, as *Medusa stella*) and by Milne Edwards (1849, as *Cyanea chrysaora*) in Cuvier's *Règne animal illustré*.

As regards the question of specific identity of the many species described in the past the most important contribution was supplied by Stiasny (1927), who made a thorough analysis of the variations in colour and pattern. His findings were in agreement with the opinions of Haeckel and Mayer, but he definitely regarded *C. hysoscella* and *C. mediterranea* as the same species, while regarding the latter as a warmer-water form with a tendency for its own distinctive colour pattern.

Family CYANEIDAE

Semaeostome Scyphomedusae with marginal tentacles arising from subumbrellar surface at a distance from umbrella margin; with gastrovascular sinus divided by radial septa into separate branched pouches, which may or may not be connected by anastomoses; without ring canal; with four oral arms with much-folded lips.

There are three genera in the family, *Cyanea* Péron & Lesueur, *Desmonema* L. Agassiz, and *Drymonema* Haeckel. Of these only *Cyanea* is represented in British waters.

Genus **Cyanea** Péron & Lesueur

Cyaneidae with radial and circular muscles on subumbrella; with eight adradial groups of marginal tentacles arranged in more than one row; with eight marginal sense organs.

The most striking features of medusae in the genus *Cyanea* are the arrangement of the marginal tentacles into eight adradial crescent-shaped groups on the subumbrella some distance from the umbrella margin, and the pattern of the well-developed muscle folds of the subumbrella. The coronal muscle is divided into sixteen areas or fields by the sixteen radial septa, the tentacular fields being wider than the rhopalar fields. Between the inner ends of the radial septa and the stomach there are a few muscle folds, each of which forms a continuous uninterrupted circle. Over the regions of the radial septa the muscle folds are also continuous (but see p. 131). There are well-developed radial muscle folds lying on the rhopalar and tentacular sides of each radial septum arising from the distal margin of the coronal muscle.

The genus *Cyanea* contains a number of forms whose specific identity is uncertain. In British waters there are two species, *C. capillata* (L.) and *C. lamarckii* (Péron & Lesueur).

C. capillata, which is mainly a northern species, occurs on both sides of the Atlantic, but *C. lamarckii* is restricted to European waters. It is possible that *C. lamarckii* has its counterpart on the eastern coasts of North America in *C. fulva* L. Agassiz or *C. versicolor* L. Agassiz, but a more detailed comparison of these species is required.

C. capillata occurs in the Pacific where a number of species have been described whose identity will remain uncertain pending further research. These are: *C. annaskala* von Lendenfeld, *C. buitendijki* Stiasny & Maaden, *C. ferruginea* Esch., *C. mjobergi* Stiasny & Maaden, *C. muellerianthe* Haacke, *C. nozakii* Kishinouye, and *C. purpurea* Kishinouye, together with a number of varieties.

The genus was reviewed in detail by Stiasny and Maaden (1943). They divided the genus into two main groups, the *capillata* group and the *nozakii* group. The most important distinguishing character was the presence in the *nozakii* group of anastomosing connections between the branches of the tentacular and rhopalar gastrovascular pouches, and their absence in the *capillata* group. There is, however, a certain degree of anastomosis across the radial septa in the region of the coronal muscle in *C. capillata* (Text-fig. 61). Much more detailed examination of many specimens is needed, however, before final conclusions can be drawn, although Kramp (1965) feels sure that *C. annaskala* and *C. muellerianthe* are *C. capillata*. It is evident, however, that *C. annaskala* comes nearer to *C. lamarckii* in that its exumbrella is papillose and there are no pit-like intrusions into the coronal muscle folds. But it differs from both *C. capillata* and *C. lamarckii* in having cells in the mesogloea.

At any rate in European waters recent careful counting of tentacles by M. E. Thiel (1960, 1962b) has confirmed the specific identity of *C. capillata* and *C. lamarckii*. The identity of these species has been much discussed over the years in the literature and an excellent account of these discussions has been given by M. E. Thiel (1962b). The two species can be distinguished by the following characters (see Text-figs. 58 and 68).

	C. capillata	*C. lamarckii*
Average number of marginal tentacles per group	70–150 or more	40–60
Pit-like intrusions from gastrovascular sinus into coronal and radial muscle folds	present	absent or only few
Number of coronal muscle folds between radial septa	13–15	16–20
Colour	yellow ochre to reddish brown	pale yellow to blue
Maximum diameter of umbrella	1000 mm. or more	300 mm.
Distribution	boreal	southern boreal

In addition, *C. lamarckii* matures at a much smaller size than does *C. capillata*.

In the synonymy of these two species which follows I have only included those names known for certain to belong to the species in question. I have no doubt that in time other names will be added to these lists, for instance I think it most probable that *C. ferruginea* Esch. is *C. capillata*.

The most detailed anatomical description of a *Cyanea* is that of Lendenfeld (1882) on *C. annaskala*, and much of the histological detail described by him will also be found to be similar in the British species.

Cyanea capillata (L.)

Plate IV; Plate XII; Plate XIII (above); Text-figs. 56–9, 61–7

Medusa capillata Linnaeus, 1746, *Fauna Suecica*, Ed. 1, p. 368, nr. 1286.
  Linnaeus, 1747, *Westgöta Resa*, pl. III, fig. 3.
  Linnaeus, 1758, *Syst. Nat.* Ed. 10, **1**, 660.
  Baster, 1765, *Natuurkund. Uitspann.* **2**, 63 (in part), pl. V, fig. 1.
  Fabricius, 1780, *Fauna Groenlandica*, p. 364.
  Gaede, 1816, *Beitr. Anat. Phys. Medus.* p. 21, pl. II.
Cyanea capillata, Eschscholtz, 1829, *System Acaleph.* p. 68.
  Forbes, 1848, *British naked-eyed Med.* p. 77.
  Agassiz, L., 1862, *Contr. Nat. Hist. U.S.* **4**, 161.
  Haeckel, 1880, *System der Medusen*, p. 529.
  Vanhöffen, 1906, *Nord. Plankt.* **11**, Acrasp. p. 52, fig. 15.
  Mayer, 1910, *Medusae of the World*, **3**, 596, text-fig. 380, pl. 65, figs. 3–4.
  Stiasny, 1930, *Mém. Mus. r. Hist. nat. Belg.* no. 42, p. 1 (in part), probably pl. I, figs. 5, 6, 7.
  Neppi, 1931, *Boll. Zool.* **2**, 143, figs. 1–4.
  Kramp, 1937, *Danm. Fauna*, **43**, 176, figs. 76, 78, 79.
  Kramp, 1942, *Meddr Grønland*, **81**, nr. 1, p. 128, figs. 34–6.
  Stiasny & Maaden, 1943, *Zool. Jb.* (Syst.), **76**, p. 244 (in part), figs. 7, 8, 9.
  Ranson, 1945, *Résult. Camp. scient. Prince Albert I*, p. 53, pl. I, fig. 8.
  Thiel, M. E., 1960, *Abh. Verh. naturw. Ver. Hamburg*, n.f., **4** (1959), 89, figs. 1–16.
  Kramp, 1961, *J. mar. biol. Ass. U.K.* **40**, 332.
  Naumov, 1961, *Fauna of U.S.S.R.* no. 75, p. 67, figs. 44–6.
  Thiel, M. E., 1962*b*, *Abh. Verh. naturw. Ver. Hamburg*, n.f., **6** (1961), 277, figs. 1, 2.
Cyanea arctica Péron & Lesueur, 1809, 'Tableau...de Méduses', *Annls Mus. Hist. nat. Paris*, p. 363, nr. 105.
  Agassiz, L., 1862, *Contr. Nat. Hist. U.S.* **3**, pl. III–V*a*, pl. X, figs. 1–17, 19–21, 23–30, 33–5 (development), pl, X*a*, figs. 1–12*a*, 14, 15 (Scyphistoma); **4**, 17, 87.
  Agassiz, A., 1865, *N. Amer. Acaleph.* p. 44, fig. 67.
  Haeckel, 1880, *System der Medusen*, p. 530.
  Vanhöffen, 1906, *Nord. Plankt.* **11**, Acrasp. p. 53, figs. 16–19.
  Stiasny & Maaden, 1943, *Zool. Jb.* (Syst.), **76**, 243 (syn. of *C. capillata*).
  Ranson, 1945, *Résult. Camp. scient. Prince Albert I*, p. 52, pl. I, fig. 7.
Cyanea lamarckii, Forbes, 1848, *British naked-eyed Med.* p. 78.
Stenoptycha dactylometra, Haeckel, 1880, *System der Medusen*, p. 526 (*vide* Kramp, 1942, p. 128).
Cyanea capillata var. *arctica*, Mayer, 1910, *Medusae of the World*, p. 597.
  Bigelow, 1926, *Bull. Bur. Fish. Wash.* **40**, 1924, pt. II, doc. no. 968, p. 357.
Cyanea capillata var. *capillata*, Bigelow, 1913, *Proc. U.S. Nat. Mus.* **44**, 93, pl. 4, figs. 8, 9.

SPECIFIC CHARACTERS

Coronal and radial muscle folds with pit-like intrusions from gastrovascular sinus; thirteen to fifteen coronal folds between radial septa; average number of marginal tentacles in each group 70 to 150 or more; terminal ramifications of gastrovascular sinus pouches without anastomoses; colour usually yellowish brown or reddish; size usually up to 300–500 mm. in diameter in British waters, extreme size up to 2,000 mm.

DESCRIPTION OF ADULT

Umbrella saucer-shaped with uniformly fairly thick jelly thinning suddenly at bases of marginal lappets; central exumbrellar surface smooth, without prominent nematocyst clusters; umbrella margin with eight deep adradial tentacular clefts, and eight less deep perradial and interradial rhopalar clefts. Sixteen marginal lappets broad and rounded, sloping asymmetrically outwards from tentacular clefts towards rhopalar clefts, sometimes with slight median indentation. Coronal muscle well developed, consisting of four or five continuous circular muscle folds, and eight wide tentacular and eight narrower rhopalar fields divided by radial septa each with thirteen to fifteen muscle folds with rows of pit-like intrusions from gastrovascular sinus into mesogloea of each fold. Well-developed radial muscles running from outer circumference of coronal muscle along either side of radial septa towards umbrella margin, also with pit-like intrusions from gastrovascular sinus; rhopalar radial muscles slightly narrower and longer than tentacular radial muscles. Marginal tentacles hollow, arising from subumbrellar surface and arranged in eight somewhat horseshoe- or rectangular-shaped groups each with 70 to 150 or more tentacles in up to four complete serial rows. Four perradial and four interradial marginal sense organs each consisting of rhopalium with statocyst and sensory bulb, but without ocellus, protected by long exumbrellar extension to form a hood and by approximation of free edges of marginal lappets on subumbrellar side; exumbrellar sensory pit present immediately over each rhopalium. Four interradial very much folded gonads hanging freely downwards from subumbrellar surface. Stomach without interradial gastric septa; with many gastric filaments arranged in four interradial groups on subumbrellar wall where central stomach cavity passes into gastrovascular sinus; gastrovascular sinus divided into sixteen pouches by sixteen radial septa extending from short distance from proximal border of coronal muscle to distal margin of umbrella where they merge into fused areas of marginal lappets; near distal ends of radial muscles pouches of gastrovascular sinus start to branch and ramify towards periphery, but without anastomosis of ramifications. Manubrium with very short thickened basal pillars merging into four wide membranous and much folded curtain-like lips, united interradially for about half their length; oral arms about as long as diameter of umbrella. Colour typically yellowish brown or reddish, sometimes nearly colourless. Size in British waters usually less than 500 mm. in diameter, extreme size in northern waters up to 2,000 mm. in diameter; largest recorded British specimen 910 mm. (3 ft.) in diameter (M'Intosh, 1927).

DISTRIBUTION

Cyanea capillata is a northern boreal species with a circumpolar distribution; further research may show it to be cosmopolitan in its distribution. Around British coasts it occurs normally all along the east coast of England and Scotland in the North Sea. It is found off the north coast of Scotland and down the west coast and is common in the Irish Sea. It occurs off the north coast of Ireland. It has never been recorded for certain from the English Channel nor southwest of Ireland, though Schoddyn (1926) recorded it in the Pas de Calais area in the Bay of Ambleteuse.

Elsewhere it occurs in North Atlantic and Pacific waters and along the north-east seaboard of America. It is common in the southern North Sea and in the Skagerak and Kattegat, and of regular occurrence in the western half of the Baltic. Its presence in the Baltic is dependent on the inflow of high-salinity water, and it was recorded as far east as off the Gulf of Riga in the eastern Baltic after the increase in salinity in the Baltic which occurred in the 1940s, culminating in 1951 (Segerstråle, 1965, fig. 1 after Lindquist, 1960). For records of its occurrence in the Gulf of Finland see also Haahtela and Lassig (1967).

Verwey (1965) states that as a rule *C. capillata* is much less numerous than *C. lamarckii* on the Dutch coast, but that in 1963 in early August between Scheveningen and Den Helder it outnumbered the latter by 130 to 1; this may have been a reflection of the cold winter.

The records of Kramp (1959*a*) off the west coast of Africa, and Kramp (1965) in Australian waters indicate its possible cosmopolitan distribution.

In the Pacific Uchida (1954) associates it with the cold current—'Oyashio'—and he says that it often occurs in abundance off the coasts of Hokkaido.

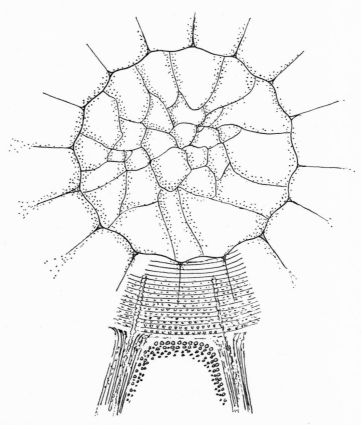

Text-fig. 56. *Cyanea capillata*. 68 mm. in diameter, Millport; centre of umbrella disk viewed from subumbrellar side to show pattern of furrows.

STRUCTURAL DETAILS OF ADULT

Umbrella

The jelly of the umbrella is fairly thick and firm but it thins out suddenly near the bases of the marginal lappets. The mesogloea is very transparent and contains no mesogloeal cells (Eimer, 1878). The exumbrellar surface is faintly papillose in the peripheral region due to clusters of nematocysts, but the central part of the disk is smooth; this was specially commented on by Östergren (1909). In very small specimens the nematocyst warts on the marginal lappets may be quite prominent.

When viewed from above the central stomach region appears to be outlined by a heavily pigmented circle whose circumference is wavy (Text-fig. 56). The troughs of the waves, which are sixteen in number, are opposite the radial septa, while from their more peripheral crests lines can be seen radiating outwards. These lines run along the mid-lines of the rhopalar and tentacular gastrovascular pouches. The radiating lines are in fact furrows in the aboral walls of the gastrovascular pouches which are deep proximally and become progressively shallower and wider towards the margin. The wavy circumference of the central circle is also a furrow where the rounded centripetal ends of the segments of jelly lying between the radial furrows meet the edges of the thickened central disk of jelly. This pattern becomes clearly visible because where the two layers of pigmented endodermal epithelium are pressed together in the furrows their colour is intensified. Sometimes however these furrows appear pale and give the impression of false radial septa (see p. 116 and Text-fig. 63).

The central circular disk of jelly forming the upper wall of the stomach has an irregular honey-comb-like pattern of shallow furrows. The spaces between these furrows are slightly elevated; they are large and radially elongated around the periphery (Text-fig. 56). Nathorst (1881) repro-duced photographs of plaster impressions and casts of *Cyanea* showing these patterns.

I have made a few measurements of the diameter of the central disk for comparison with that of the whole umbrella. On an average the over-all diameter of the umbrella was about three times the diameter of the central disk.*

Marginal lappets

In perfect specimens it may be that the margin of each lappet is entire (Text-fig. 58) so that there are only sixteen lappets. Starting at the V-shaped notch opposite the centre of the tentacle group the margin slopes outwards in a gentle curve to its most distal point opposite the termina-tion of a radial septum. The margin then bends rapidly inwards to the rhopalium which is situated on about the same circumference level as that of the tentacular cleft. Each lappet is, however, joined on its exumbrellar side to its neighbour by the rhopalial hood which extends distally rather far beyond the rhopalium itself. This gives the impression that the tentacular notch is much deeper than the rhopalar notch. In some specimens there may be a slight embayment along the tentacular side of the lappet near the end of the radial septum. The mesogloea is always very thin in this region, where the radial septum merges into the broad area of fusion of the exumbrella with the subumbrella. This thin area is rather wide and lies between the main branches of the rhopalar and tentacular gastrovascular pouches. As a result the margin often tears at this point of weakness so that the medusa has sometimes been described as having a division into rhopalar and tentacular marginal lappets making thirty-two lappets in all.

The outline of the marginal lappet was originally used as a specific character, but it is of such a delicate and contractile nature that it is quite unsuitable for this purpose. It was, for instance, used as a character to distinguish *C. arctica* from *C. capillata*, but Kramp (1942) showed quite conclusively, by comparing marginal outlines of Greenland specimens with others from Danish waters, that they were in fact indistinguishable on this character, as indeed he found them to be on all other characters.

Subumbrellar musculature

A general account of the subumbrellar musculature has been given under the generic characters (p. 104). The following characters are, however, of specific value, at any rate for distinguishing *C. capillata* from *C. lamarckii*.

The number of folds in the circular muscle between the radial septa is fewer than in *C. lamarckii*,

* In the drawing of fig. 1 on pl. V of L. Agassiz (1862) the central disk appears to be only about a quarter of the umbrella diameter.

being usually thirteen to fifteen. In small medusae a number of radial furrows can be seen running at right angles across the muscle folds, on the subumbrellar surface of the gastrovascular sinus. After the medusa has reached a size of about 30 mm. in diameter small pit-like intrusions of the

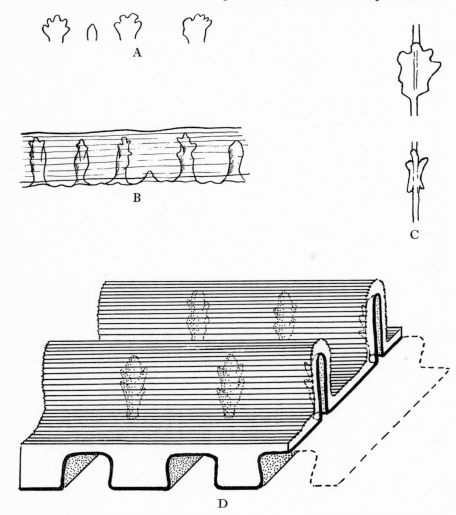

Text-fig. 57. *Cyanea capillata*. Illustrations of pit-like intrusions from gastrovascular cavity into coronal muscle folds. A, various shapes of pits in specimen 100 mm. in diameter, Millport, and B, 150 mm. in diameter Gruinard Bay; C, pit in relaxed and in contracted state of muscle; D, diagram of coronal muscle folds and gastrovascular pits arising from radial grooves. The gastrovascular endoderm is shown in thick black; the clear white area is mesogloea.

gastrovascular sinus into the muscle folds begin to develop in these radial furrows. These appear first in the distal folds, but they are soon present in all folds. As they arise in the radial furrows these intrusions are arranged in radial rows running at right angles to the muscle folds (Text-fig. 58). There may be up to about fifteen or sixteen of these rows between adjacent radial septa in the tentacular muscle fields in larger specimens, and slightly fewer in the narrower rhopalar muscle fields.

At first the intrusions are simple conical or finger-like pits, but as the medusa grows the pits develop small diverticula (Text-fig. 57). They assume different appearances according to whether the muscles are relaxed or contracted (Text-fig. 57C).

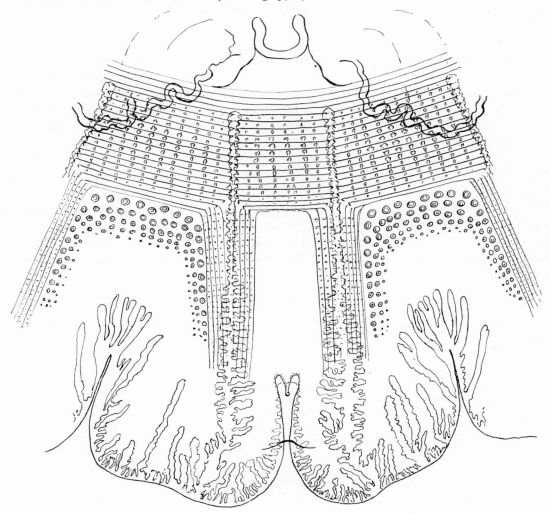

Text-fig. 58. *Cyanea capillata*. Portion of umbrella margin of specimen *ca.* 100 mm. in diameter, Millport, to show diagnostic characters, namely muscle folds with pit-like intrusions from gastrovascular cavity, number of marginal tentacles, and extent of gonad.

The number of radial muscle folds also seems to be usually less than in *C. lamarckii*, but these are generally difficult to count for certain. A few I have examined had three or four folds in the radial muscle on the rhopalar side of the radial septum and four to six in that on the tentacular side. These radial muscle folds also have pit-like intrusions from the gastrovascular sinus.

In a specimen about 70 mm. in diameter the height of a circular muscle fold from its outer edge to the subumbrellar surface of the gastrovascular sinus was 0·53 mm., and a typical intrusion was 0·33 mm. deep and 0·25 mm. wide. In a specimen 160 mm. in diameter a corresponding measurement of the muscle fold was 2·0 mm. and the depth of an intrusion 1·8 mm.

In larger specimens the continuity of the circular muscle folds appears to be broken at the radial septa, and they seem to differ in this respect from *C. lamarckii*.

Manubrium

Surrounding the base of the manubrium there is a ring of thickened subumbrellar mesogloea from which the four arms of the mouth arise. There is a very short gelatinous root or pillar at the base of each arm which soon thins out to form very wide and much-folded membranous lips on either side of a slightly thickened median furrow. Between adjacent oral arms there is a membranous connecting region which hangs down like a much-folded curtain between them. While the oral arms are about as long as the diameter of the umbrella, the connecting curtain between the arms is only about half the radius of the umbrella. The membranous frill of the lip is about as wide as the umbrella radius at its base and it gradually tapers off in spiral whorls to a point at its end (see Text-fig. 71 A for *C. lamarckii*).

The margin of the lip is smooth and even, apart from the folds, and has a narrow thickened edge filled with nematocysts. The membranous frill has nematocysts widely scattered over it, but a short distance from the margin itself there is a band of more closely spaced nematocyst clusters, each consisting of several nematocysts in a globular cluster raised above the surface (as in *C. lamarckii*, Text-fig. 71).

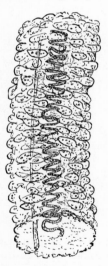

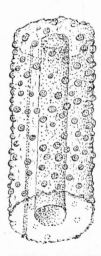

Text-fig. 59. *Cyanea capillata*. Portion of contracted marginal tentacle.

Text-fig. 60. *Cyanea lamarckii*. Portion of relaxed marginal tentacle.

Marginal tentacles

The marginal tentacles are bilaterally flattened at their bases. Apart from a short smooth basal portion the tentacles are covered throughout their lengths with fairly evenly distributed clusters of nematocysts. The mesogloea is thick, and running through the length of a tentacle is a tube lined with endoderm.

The whole tentacle has bilateral symmetry in that there is a small longitudinal thickening of ectoderm cells penetrating the mesogloea along its adaxial side. This can be clearly seen in cross-section.

When the tentacle is relaxed the central endodermal tube is round and spacious (as in *C. lamarckii*, Text-fig. 60). On contraction of the tentacle this tube becomes compressed and at first takes up a screw-like appearance. On further contraction of the tentacle the cavity of the tube becomes completely reduced and the endodermal lining assumes the form of a tightly coiled spiral spring (Text-fig. 59).

The detailed histology of the tentacles was described by von Lendenfeld (1882) for *C. annaskala*, but this description will probably be found to hold for other species of *Cyanea*.

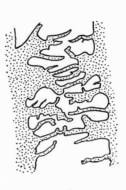

A   B

Text-fig. 61. *Cyanea capillata*. A, portion of radial septum to show branching and anastomosis of gastrovascular system; B, enlarged portion of A.

In the development of a group of tentacles, one appears first, successive tentacles growing on either side of it until a crescentic row of primary tentacles is formed. Before the first row is complete tentacles of a second row are already beginning to develop in the middle region. Eventually there may be as many as four rows of tentacles composed of old large tentacles and smaller ones in all degrees of development. On the whole the tentacles are slenderer than those of *C. lamarckii* in medusae of comparable size. In medium-sized specimens the oldest tentacle in each group is often dark red.

As originally pointed out by von Lendenfeld the cross-section of the tentacle is rather small at its point of attachment to the subumbrella. They often break off at this weak point, when their original positions can be clearly seen as circular or oblong openings into the gastrovascular sinus.

In full-grown medusae there may be 150 or more tentacles in each group, approximately twice as many as in *C. lamarckii*. For instance, Östergren (1909) recorded 120 to 150 tentacles in each group for *C. capillata* 80 to 150 mm. in diameter. The range of variation in numbers of tentacles in each group in medusae of different sizes has been a subject of careful investigation by M. E. Thiel (1960). The table on p. 114 summarizes his results for specimens whose diameter was measured between opposite marginal sense organs.

From his results Thiel produced an approximate formula to fit the number of tentacles per group to the size of the medusa: $y = \frac{3}{2}x$, where y equals the number of tentacles and x the diameter of the medusa between marginal sense organs in mm.

Diameter (mm.)	Average number of tentacles per group	Extremes of averages	Number examined
1–4	1·9	1–3	15
5–9	7·7	1–11	58
10–19	14·9	9·8–23·7	37
20–29	29·8	16·0–38·3	48
30–39	41·5	28·6–60·0	30
40–49	63·0	45·0–84·3	28
50–59	73·7	60·5–92·8	23
60–69	92·4	72·0–113·0	19
70–79	103·3	84·5–121·5	13
80–89	115·6	106·0–136·7	9
90–99	132·0	121·0–155·1	4
100–109	140·8	116·0–166·0	3
120–179	179·5	—	4

Counts that I have made myself were somewhat higher than those of Thiel.

The tentacles are capable in life of great extension, for instance L. Agassiz (1862) said that he could see the tentacles of medusae about 3 ft. in diameter stretching to 20 or 30 ft. He described graphically their behaviour. When fully extended their thickness is uniform throughout, but they may contract at isolated points along their length causing local swellings. The extremity is generally the most swollen part. He remarked especially on their independence of movement, one tentacle shortening suddenly in a jerky movement without disturbing its neighbours. He never saw the majority of the tentacles all contract together at one time. The relaxation of the tentacles proceeds irregularly, sometimes a swollen tip suddenly descending as though it were falling through the water.

Marginal sense organs

The marginal sense organ was described in detail by Eimer (1878) and Fewkes (1881), although it was previously figured by L. Agassiz (1862, pl. IV, fig. 3; pl. V a, fig. 8). More recent descriptions were given by Wu (1927) and Horstmann (1934a). The rhopalium is short and straight, and directed slightly downwards. It is attached at its base to a ridge which runs to the proximal edge of the sensory niche and through which the rhopalar canal runs from the gastrovascular sinus. This ridge may be smooth or show varying degrees of papillosity (Text-fig. 62). The rhopalium is covered on its exumbrellar side by a hood formed by the junction of the exumbrellar surface of the neighbouring marginal lappets, while the lateral edges of these lappets overlap on the subumbrellar side to form a deep tube. The rhopalium is set well back in this tunnel, the distal margin of the hood being about half-way between the rhopalium itself and the periphery of the umbrella (see Text-fig. 72 for *C. lamarckii*).

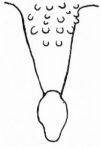

Text-fig. 62. *Cyanea capillata*. Rhopalium to show small warts on its base.

On the surface of the exumbrella above the rhopalium there is an elongated sensory pit.

Fewkes (1881) stated that the crystals or otoliths in the rhopalium are bright orange. He also mentioned a cluster of small otoliths on the underside of the rhopalium near its junction with the ridge.

Gonads

In the interradial areas of the subumbrella there is no thickening of the mesogloea to form genital pits. As a result the fully developed gonads with their many folds receive no support and hang freely down beneath the umbrella. The genital products are situated in a genital band which appears like a folded ribbon. An indication of the degree of folding of the gonad is given by the observation of von Lendenfeld (1882, p. 528) on *Cyanea annaskala* that a specimen 90 mm. in diameter had a gonadial ribbon which when straightened out was 300 mm. long.

The development of the gonads has been described by Widersten (1965). In a position in the subumbrellar wall of the stomach just distal to the gastric filaments endoderm cells immigrate into the mesogloea. Here they form aggregations in which isolated cavities appear lined with epithelium supplied by the immigrating cells, some of which are primary germ cells. The epithelial walls of these cavities break down to form larger chambers which often fuse to form a large genital sinus interrupted by the remaining lamellae.

The spermatozoa develop in follicles which are also formed by the immigration of cells from the genital epithelium into the mesogloea. The follicles are never quite disconnected from the genital epithelium, and there is a connection between the cavity of the follicle and the genital sinus through which the sperm can escape.

The eggs have been described by Wagner (1835), L. Agassiz (1862) and Harting (1875), while Monné (1957) studied their staining properties.

The spermatozoa have been described by Retzius (1904); its head is very long, narrow and pointed. Franzen (1956) referred to aberrations mentioned also by Retzius.

The gonads develop much later in *C. capillata* than they do in *C. lamarckii*. In specimens as large as 60 mm. in diameter they only protrude very slightly and when viewed on the subumbrellar surface do not extend over the coronal muscle (Pl. XII). C. Edwards tells me that he had two ripe specimens 100 and 160 mm. in diameter off Millport in November 1966; he thought that they might have been stunted by the cold spring.

Nematocysts

The nematocysts of *C. capillata* have been studied by Weill (1934) in specimens from Wimereux and by Papenfuss (1936) at Kristineberg (Text-figs. 5 and 6, on pp. 9 and 10).

There are three kinds: atrichous haplonemes in two sizes, holotrichous haplonemes, and heterotrichous microbasic euryteles.

Measurements of the undischarged capsules in adult medusae given by the two authors are as follows:

	Atriches		Holotriches	Euryteles
	Large	Small		
Weill	$35 \times 25\,\mu$	$10 \times 6\,\mu$	15–$20\,\mu$ diam.	$15 \times 8\,\mu$–$22 \times 12\,\mu$
Papenfuss (averages)	$20.8 \times 14.2\,\mu$	$9.8 \times 5.6\,\mu$	$17 \times 14.4\,\mu$	$17.6 \times 10.6\,\mu$

It is to be noted that there is a considerable difference between the measurements given by the two authors for the large atriches.

Papenfuss noted differences in the coiling of the thread within the capsule between *C. capillata* and *C. lamarckii* (as *C. palmstruchii*). In the small atriches there was a smaller number of transverse coils in *C. capillata* and the capsule was pyriform, while in *C. lamarckii* it was broadly ovoid. But the holotriches and euryteles each had larger numbers of coils in *C. capillata* than in *C. lamarckii*.

Weill also studied the nematocysts in scyphistomas reared at Wimereux, and in others reared by Lambert at Leigh-on-Sea, Essex. There were atriches 7–9 μ long, and euryteles $11 \times 9\,\mu$ and

$12 \times 10\,\mu$. Both types of nematocyst were present in all parts of the polyp, but the euryteles were often the more abundant. A nematocyst from the scyphistoma of *C. capillata* was figured by Möbius (1866, pl. II, figs. 1, 2).

COLORATION

The coloration of *C. capillata* shows some variation; most authors have described the medusae as usually yellowish brown or reddish brown, but they may be pale yellow or almost colourless.

Text-fig. 63. *Cyanea capillata* and *C. lamarckii*. Pattern of coloration leaving radial septa and false septa colourless.

Haeckel (1880) stated that the stomach and gastrovascular system were mostly dark red-brown, and described the oral arms as pale yellow, dark red-yellow, rust brown or chestnut brown; the gonads he described as whitish or flesh-red to brick-red. Östergren (1909) remarked that the tentacles might be almost crimson, while Agassiz (1862) said that the endoderm, which gives the colour to the tentacles, might be yellow, orange, purple or brown. Verwey (1965) said that all shades between dark brown and white are found, but that in thousands of specimens he never saw one with any shade of blue.

The many *C. capillata* that I have seen round the coasts of Scotland and north-east England have almost all been of the same golden brown colour with dark red manubria. They have varied slightly in depth of colour, but I have only seen a few very pale ones. The pigmentation is both ecto-dermal and endodermal, but the coloration of the stomach and gastrovascular pouches is so much darker than the rest that when the medusa is viewed from above swimming in the water the pattern of the stomach and pouches shows up strongly, the umbrella margin and groups of marginal tentacles generally appearing whitish in contrast to the deep red of the manubrium. A pattern is often created by the radial septa and 'false septa' being only weakly pigmented (Text-fig. 63).

The manubrium darkens in colour with age. In small specimens it is yellow ochre. The solid basal stems of the oral arms are whitish. In young specimens one marginal tentacle in each group is often much darker in colour than the others.

DEVELOPMENT

The embryology of *C. capillata* has been studied by Hamann (1890), McMurrich (1891), Hyde (1894), C. W. & G. T. Hargitt (1910), Okada (1927), Widersten (1965, 1968) and Korn (1966). Segmentation is total and unequal. The conclusion reached by Okada is that invagination is typical and that after invagination many free cells appear in the cavity. These free cells, however, disintegrate and finally disappear without taking part in the formation of the endoderm, which would account for the divergent opinions of the earlier workers that invagination was not typical.

The oocytes, which lie in the mesogloea, are nourished by nurse cells which protrude into the mesogloea. While it was earlier thought that fertilization takes place in the gastrovascular cavity or in the sea, the Hargitts suspected that it takes place in the ovary or genital sinus. Widersten (1965) confirmed that fertilization takes place in situ, the spermatozoa entering through a funnel-shaped pit surrounded by elongated nurse cells. The ova are white, yellow or orange in colour. According to Hyde the fertilized eggs are surrounded by a membrane which might persist until the formation of the gastrula. The fertilized eggs are expelled into the genital sinus at the first cleavage and may develop there to the blastula or gastrula stages. According to Widersten they then move into distal distensions of the genital sinus where they congregate into small groups held together by a sticky substance which is often granular and fibrous. The packages of embryos are carried by ciliary action and muscular contraction into the gastrovascular cavity, and development to the planula takes place among the folds of the oral lips. Korn (1966) gave the dimensions of the planula as 225–400 μ, and Widersten (1968) as 175 × 125–240 × 170 μ.

McMurrich and the Hargitts record the occurrence of encystment of the planulae in the laboratory, but it is not known whether this happens in nature. The cysts were plano-convex and the developing polyp emerged through the convex surface. Early descriptions of the scyphistoma and its strobilation were given by Dalyell (1836), van Beneden* (1860) and L. Agassiz (1862). The subsequent embryological work was concerned rather with the part played by endoderm and ectoderm in the formation of the mouth (see p. 16). According to the Hargitts the polyps are relatively small compared with those of *Aurelia*, and whitish in colour. They remark that stolonization is less common than in *Aurelia*. Stolons may arise from the body or base of the polyp, and they may branch to give rise to a colony of polyps. They did not observe budding directly from the body. Two opposite primary tentacles are often the first to develop and in the fully grown polyp the average number of tentacles is sixteen, though there may be twenty or more. Ephyrae are often produced by mono-strobilation, but there may be as many as five annulae. Lambert (1936*a*) had many with four annulae, but never with more than eight.

Hagmeier (1933) recorded the rearing of *C. capillata* in the laboratory at Helgoland, where he noted the spawning period as August to the beginning of September. Planulae might remain free-swimming for as long as 4 weeks. He also recorded monodisk and polydisk strobilation. For instance in 1924 scyphistomas strobilated in December of the same year and were monodisk, each polyp retaining its tentacles. The scyphistomas produced in 1927, however, did not strobilate until February of the following year; they were polydisk and remarkable for their reddish colour.

* Van Beneden's scyphistomas had as many as eleven strobilations; it is possible that he may have wrongly ascribed them to *C. capillata*.

Kaufman (1957) stated that the scyphistomas have narrow basal stalks and that these contain zoochlorellae. Lambert (1936*a*, *b*) said that the scyphistoma might be scarlet in colour, though usually delicate bronze. Widersten (1967) studied the development of the periderm.

The description of a scyphistoma and its strobilation given by Berrill (1949*b*) on preserved material must remain a doubtful identification of that of *C. capillata*.

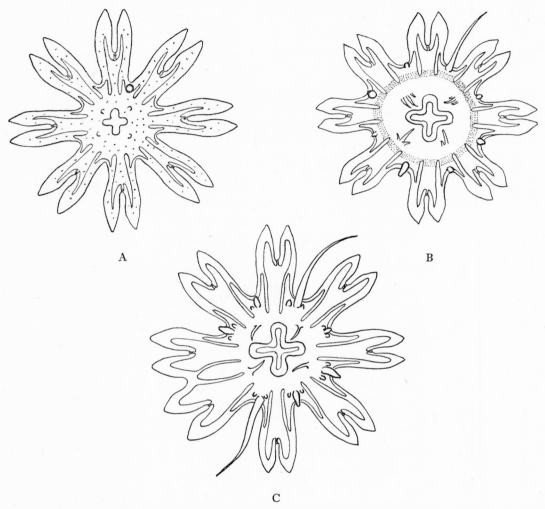

Text-fig. 64. *Cyanea capillata*. Ephyra. A, *ca*. 1·7 mm. in diameter, Millport, 24 March 1967; B, *ca*. 3·5 mm. Millport, 18 April 1967; C, abnormal specimen, *ca*. 5·0 mm. Millport, 3 April 1967.

The Hargitts stated that at Woods Hole the period from liberation of planulae to the free-swimming ephyra might be as short as 15–20 days in the aquarium, though the majority would take 30–40 days and some showed no sign of strobilation after 2 months. This period must obviously be affected by conditions of food and temperature.

Verwey (1942) considered that the upper temperature limit for strobilation was somewhat above 8° C. Loomis (1961, as *C. arctica*) stated that the scyphistoma would bud indefinitely if fed daily with brine shrimps and changed into clean water. But when left at 12° C. for one month, it was found to have strobilated and given off ephyrae. Korn (1966) observed that at 16° C. the planulae developed to the eight-tentacled schyphistoma in 6 days.

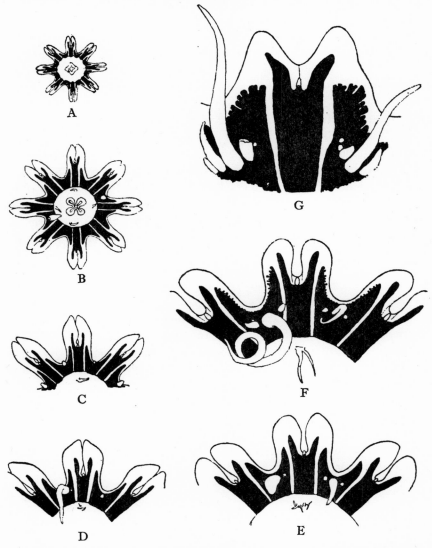

Text-fig. 65. *Cyanea capillata*. Stages of development (from Kramp, 1937, fig. 76). Diameters: **A**, 2·0 mm.; B, 3·5 mm.; C, 5·0 mm.; D, 6·0 mm.; E, 7·0 mm.; F, 8·5 mm.; G, 14·0 mm.

The ephyra (Text-fig. 64) is immediately distinguishable as that of a *Cyanea* by the long horn-like diverticula at the distal corners of the radial canals. In life *C. capillata* ephyrae are yellow ochre in colour at Millport; Mielck and Künne (1935) give the colour as reddish brown for specimens in the North Sea.

Hagmeier, who reared the medusa, stated that on liberation the ephyra was 3·5 mm. in diameter, but I have seen specimens as small as 1·5 mm. in diameter at Millport. At a size of 1·7 mm.

one marginal tentacle was developing. This first tentacle appears always to develop immediately to the right of a perradial marginal lobe, viewed from the subumbrellar side. This is followed by a tentacle diametrically opposite. The two tentacles in the remaining similar positions develop next, while the final four other tentacles develop last (Text-fig. 64).

Hagmeier's ephyra, 3·5 mm. in diameter, had two long and six short marginal tentacles when it was 8 mm. in diameter. The diameter of a 70-day-old medusa was 30 mm.; at an age of 3 months it was 50 mm. in diameter. There were no prominent nematocyst warts on the exumbrella.

A useful illustration of the development of the medusa is given by Kramp (1937) (see Text-fig. 65).

Thiel (1960) found that the average number of tentacles per group in medusae 1–4 mm. in diameter between marginal sense organs was 1·9, with extremes of averages of one to three; in specimens 5–9 mm. in diameter corresponding figures were 7·7 and one to eleven; and in specimens 10–19 mm. in diameter they were 14·9 and 9·75–23·7. In medusae of the latter size range I found rather higher numbers of tentacles in specimens from the Firth of Clyde.

Fewkes (1881) mentioned the occurrence on the exumbrella of small specimens of flexible, transparent, tapering, filamentous processes of brownish colour. These processes were solid and devoid of nematocysts. Similar processes are recorded by Mayer (1910) in specimens about 7 mm. in diameter.

SIZE

While very large specimens may occur in British waters, e.g. 910 mm. in diameter off St Andrews (M'Intosh, 1885), most specimens that one sees are under 500 mm. in diameter. In an investigation I made in September 1967 round the coasts of Scotland and north-east England (Text-fig. 67) most of the specimens I saw were between 200 and 500 mm. in diameter, and I saw an impression in the sand on the beach at Newton-by-the-Sea, Northumberland, of a specimen which must have been at least 900 mm. in diameter.

M. E. Thiel (1960), gives for Kiel Bay, 14 April, 4–19 mm.; 16 June, 10–179 mm.; 22 August, 40–169 mm. Maaden (1942a) measured large numbers of specimens stranded on the Dutch coast; up to the end of May the largest was 170 mm. in diameter, and by mid-June or later they had reached 300–350 mm. in diameter (Verwey, 1942). Kramp (1937) records that they rarely exceed 300 mm. in diameter in Danish waters.

REARING AND FEEDING

C. W. & G. T. Hargitt (1910) recorded that scyphistomas fed on echinoderm larvae and copepods, and on their own planulae. Gaede (1816) records medusae as feeding on fish and nereids; while Hargitt (1902) mentions fish as food. Hagmeier (1933) said that ephyrae were fed on *Mytilus* flesh, *Carcinus* ovaries, and small hydromedusae. Young medusae greater than 20 mm. in diameter preferred *Bolina* and *Eutonina*, and did not thrive if they had no medusae.

SEASONAL OCCURRENCE

The ephyrae of *C. capillata* may appear as early as February in British waters and small medusae will be present in April and May. The larger specimens are, however, not normally to be seen until June, and their main period of abundance is July to September. They become scarce in October, but a few large specimens may linger on over the winter in deep water.

Published records for different localities are as follows: M'Intosh (1885, 1927) St Andrews, August, many cast ashore in September, one specimen 26 in. in diameter at beginning of October, occasional huge specimens in 15–20 fathoms in mid-winter; Browne (1905 a) Clyde Sea area, July to October; Bruce, Colman & Jones (1963) Irish Sea off Isle of Man, late ephyrae in April, young in May, adults June to early September.

In the southern North Sea, while small specimens first appear in April, the peak season for mature specimens is June to September. Records of occurrence are given by the following authors: Hagmeier (1933) Helgoland; Maaden (1942 a) and Verwey (1942) Dutch coast; Künne (1952) German Bight; Kühl (1964) Elbe estuary.

In the northern North Sea and Skagerak Damas (1909 a, b) recorded that specimens up to 750 mm. in diameter may occur in May, but the period of greatest abundance is August and September. Observations in the Baltic are given among others by Thiel (1960) for Kiel Bay, and Kändler (1961) who states that ephyrae appear during January to April. Aurivillius (1898) recorded off the Swedish coast that ephyrae occurred from the end of February to the end of April, and that adults occasionally occurred in January and even from February to April.

Kramp (1937) stated that ephyrae may appear in Danish waters throughout the summer and that there is a second maximum of occurrence in the autumn. He presumed that the medusae arising from the autumn strobilation were rarely seen as they might seek the bottom to avoid the winter storms.

Verwey (1942) reviewed our knowledge of the early life history of *C. capillata* and other scyphomedusae in relation to temperature and geographical distribution.

The seasonal occurrence of this medusa on the eastern seaboard of North America is very similar to that in European waters (see Fish, 1925; Bigelow, 1926). Off west Greenland Kramp (1942) records that they are small in June, and full grown in August and September, and that he only had two specimens in October.

PHYSIOLOGY

Alimentary system

The passage of fluid through the digestive system was studied by Widmark (1911) in *C. capillata* at Kristineberg. The stomach fluid is driven outwards along the exumbrellar walls of the gastrovascular sinus to the marginal pouches by contractions of the genital sacs and of the subumbrellar musculature. Here the fluid is mixed with mucus and returned to the stomach along the subumbrellar side by ciliary action. The ciliary currents from the four perradial marginal pouches run direct to the oral arm bases. Those from the interradial and adradial pouches turn aside past the proximal ends of the septa and join the currents from the perradial pouches at the beginning of the troughs at the base of the oral arms; the currents from the interradial pouches usually divide to send branches to the two neighbouring oral arms on either side.

Henschel (1935), working with *C. capillata* from Kiel Bay, studied the reactions of the oral lips to the presence of food. He cut small pieces from the oral folds and used the enlargement of the area of each piece as an index of stimulation. Areas were enlarged in the presence of peptone, asparagine, leucine, tyrosine and skatole. No such reaction was shown with soluble starch or glycogen.

CYANEIDAE

Nervous and muscular systems

The first to study the nervous system of *C. capillata* were Eimer (1878), who experimented with specimens from Travemunde in the Baltic, and Romanes (1876) who worked with specimens from the Cromarty Firth.

Morse (1910) found that *C. capillata* (as *C. arctica*) did not regain pulsation after removal of the marginal sense organs. This was contrary to Romanes' finding, but agreed with that of Eimer, who Morse thought might have used *C. lamarckii*. Horstmann (1934*a*) found that after extirpation of all marginal sense organs specimens less than 5 cm. in diameter continued to pulsate while larger specimens became motionless. He also made observations on righting reactions in specimens with only two opposite sense organs left.

Schaefer (1921) and Horstmann (1934*a, b*) did neuromuscular experiments on specimens from Kiel Bay, and Bullock (1943) studied facilitation in *Cyanea* at Woods Hole. More recently Horridge (1954*a*; 1956*a*) made detailed observations on specimens usually 200–300 mm. in diameter at Millport. He distinguished the two systems: giant fibres controlling the rhythmic contractions of the umbrella, and a diffuse nerve net concerned with local muscular movements of the bell, and made a detailed study of their distribution and histology.

He noticed, using specimens 50–100 mm. in diameter, that the individual folds of the circular muscle contract independently to produce a wave of contraction starting at the proximal edge of the muscle and spreading outwards. Intervals between successive waves were $\frac{1}{2}$–2 min. The radial muscle contracts largely as a single unit.

Passano (1965) recorded that the time delays before a pace-maker triggers off its neighbouring centre are of the order of several hundred msec.

Lehmann (1923) concluded that the marginal sense organ was not a statocyst, but that it was mainly concerned with the rhythmic contractions (see also p. 103). He thought that the medusa was orientated by the difference between the specific weight of the subumbrella and manubrium and that of the exumbrella. Horstmann (1934*a*) found no reaction to light, but observed that the righting reaction was neuromuscular.

Horridge (1959) says that *Cyanea* usually swims upwards in the laboratory, but that when disturbed it can be induced to turn over and swim down.

Barnes & Horridge (1965) experimented on the effects of extracts from the rhopalar region. They also found that extracts of the tentacles of *Cyanea* killed the leptomedusan *Phialidium*. Welsh (1955) tested the effects of extracts of tentacles of *C. capillata* from Friday Harbour on autotomy in crabs and on heart activity in molluscs. Barnes (1964) records some original observations on the effects of different concentrations of various ions on *C. capillata* at St Andrews.

Kaufman (1957) did regeneration experiments on the scyphistoma. He found that a critical stage was when the scyphistoma had four tentacles; 35 % died and regeneration rate was very slow. Fixed planulae without tentacles when cut transversely regenerated as quickly, or even slightly quicker, than the development of the controls. Scyphistomas with eight and twelve tentacles also regenerated quickly.

CHEMICAL COMPOSITION

Stipos and Ackman (1968) state that *C. capillata* contains only low amounts of lipids in proportion to its wet weight, but a lipid system which is essentially similar to those of lower marine animals and phytoplankton feeders in general.

HABITS AND GENERAL OBSERVATIONS

Association with young fish

The young of the whiting (*Gadus merlangus*) and of the horse-mackerel (*Trachurus trachurus*) are commonly found in association with *C. capillata*. Earlier literature on the subject was reviewed by Scheuring (1915) and a full review and bibliography on the subject as a whole is given by Mansueti (1963). Scheuring found that young whiting selected small pieces of *Cyanea* from copepods and *Hyperia*, and suggested that they fed on the jellyfish. Dahl (1961), however, could not confirm this observation. He noticed that when jellyfish fragments were taken they were often quickly rejected. He found few nematocysts on whiting skin after contact with *Cyanea* marginal tentacles, but the skin of *Gobius flavescens* had whole batteries of nematocysts discharged into it.

Damas (1909*a*, *b*) drew attention to the part played by *C. capillata* in the dissemination of fish fry. It is notable that the young of cod (*Gadus morhua*), haddock (*G. aeglefinus*) and whiting (*G. merlangus*) associate with this medusa and take shelter under its umbrella and among the voluminous folds of the oral arms. The medusae, which originate in coastal areas, may be carried more than 250 miles off-shore and young fish probably go with them. While Damas thought that their influence in this was important in the western North Sea, Skagerak and Norwegian Sea, he did not think that their effect was so considerable off the east coast of Scotland where the medusae are generally smaller. Gosse (1856) gives an account of Peach's observations on the association of young whiting with *C. capillata* at Peterhead.

Vertical distribution

Information on vertical migration is reviewed by Verwey (1942) who quotes the vivid description given by Damas (1909, p. 42) on the appearance of the medusae at the sea surface at dusk. This habit is, however, not invariable for C. W. & G. T. Hargitt (1910) stated, among others, that there is no difference in their presence at the surface in full daylight, or at dusk or at dawn, and that they may be so common on the surface in full sunlight that the fishermen call them 'sun scald'. They stated that they come to the surface in calm weather and that a rough surface drives them down. Horstmann (1934*a*) records a similar observation in Kiel Bay, saying that when there are waves to disturb them they turn over and swim downwards. His experiments, however, showed no reactions to light stimuli. M'Intosh (1885), however, stated that in St Andrews Bay they could be seen swimming at a fathom or deeper by day and rose to the calm surface as evening approached. In late August and September I have often watched large *C. capillata* in Loch Ewe swimming slowly to the surface and remaining with their umbrellas awash, even when there was a considerable wind. Their rate of pulsation varied between about 12 and 16 per min., with periodic pauses.

The very young stages are said not to be common in catches. In view of this it was suggested first by L. Agassiz (1862) that they live on the bottom. The possibility that there might be some truth in this was given by Mayer (1910) who commented on the habit of young medusae about 7 mm. in diameter of not coming to the surface in an aquarium but remaining clinging to the bottom or sides of the tank by their oral arm folds. M. E. Thiel (1962*a*, p. 1075) states that he has often found *C. capillata* in shallow water in the Baltic resting with its oral arms spread out on the bottom. However, A. Agassiz (1865) recorded large numbers of young *Cyanea* on the surface on one occasion in early morning.

CYANEIDAE

Size

The largest specimen recorded in the literature was seen by A. Agassiz (1865) in Massachusetts Bay; it was $7\frac{1}{2}$ ft. in diameter and its tentacles stretched for 120 ft.

Swarming and stranding

It is common for *C. capillata*, as with other jellyfish, to be thrown ashore at times (see e.g. Maaden, 1942*a*). Their occasional occurrence in large swarms was first thought by L. Agassiz (1862) to be for breeding purposes, but it is generally realized by later authors that the swarms are the effects of wind and tide. Specimens which are thrown ashore in the adult state no doubt enable the planulae to settle in coastal waters. *C. capillata* is a less coastal species than *Aurelia*, and it is thought, but without direct evidence, that the scyphistomas of *C. capillata* might occur in water up to 30 to 70 m. in depth (Bigelow, 1926). (See also p. 12, and Text-fig. 67.)

Stinging

In contrast to that of *C. lamarckii*, the sting of *C. capillata* is relatively severe. This jellyfish is especially avoided by fishermen (see, e.g. Evans & Ashworth, 1909). Bigelow (1926) recorded an instance on 29 July 1921 when hundreds of bathers at Nantucket Beach near the mouth of Boston Harbour suffered more or less irritation. Kramp (1937) states that in Danish waters the jellyfish when very abundant may burst the nets of fishermen and cause nuisance to bathers from their stinging. Bad stings may give rise to blisters, lassitude, irritation of mucous membranes and muscular cramp, and may affect heart and respiratory activities. In these days of human underwater activities there is a danger that tentacles of jellyfish may touch the eye, and Mitchell (1962) records instances of lesions to the cornea inflicted by *Cyanea* in Australian waters. Barnes (1967) used the membranes from human afterbirths for extracting the poison from the nematocysts of Australian *Cyanea*.

Möbius (1866) sampled the sting of the ephyra on his tongue.

The stinging powers of *C. capillata* are retained long after the medusa has been stranded on the shore.

J. G. Wood (1882), who was evidently susceptible to the sting of *Cyanea*, gave a detailed account of the effects of a bad sting by this medusa. The tentacles had wrapped round one leg to just above the knee and round the greater part of his right arm; there were a few also on his face. Where the tentacles had been in contact his skin showed a light scarlet line of minute dots or pustules. On the walk back to his lodgings, apart from suffering severe pain, he recorded the following symptoms.

'Both the respiration and the action of the heart became affected, while at short intervals sharp pangs shot through the chest, as if a bullet had passed through the heart and lungs, causing me to fall as if struck by a leaden missile. Then the pulsation of the heart would cease for a time that seemed an age, and then it would give six or seven leaps, as if it would force its way through the chest. Then the lungs would refuse to act, and I stood gasping in vain for breath, as if the arm of a garotter were round my neck. Then the sharp pang would shoot through my chest, and so *da capo*.

'Several days elapsed before I could walk with any degree of comfort, and for more than three months afterwards the shooting pang would occasionally dart through the chest.'

It was this account of Wood's that inspired Conan Doyle (1927) to write his *Adventure of the Lion's Mane*. '"*Cyanea!*" I cried. "*Cyanea!* Behold the Lion's Mane."' A modern Sherlock

Holmes might well have solved his problem by examination of the nematocysts on the dead man's skin without having to find the jellyfish itself!

Högberg *et al.* (1956, 1957) and Uvnäs (1960) recorded the presence of a histamine-releasing principle whose action was blocked by a compound from hip seeds.

Chemical composition

It is of interest to record an early crude experiment by L. Agassiz (1862, **4**, 98) to estimate water content of this medusa. Using a specimen weighing 35 lb., he allowed it to dry and after washing the residual salt away with fresh water found its weight to be less than 1 oz. This is equivalent to a water content of 99·8%.

The inorganic composition of the medusa was studied by Macallum (1903) and Koizumi & Hosoi (1936) who confirmed Macallum's findings (but see p. 165). The osmotic properties of *Cyanea* were investigated by Bateman (1932); and the visco-elastic properties of the mesogloea by Alexander (1964).

Commensals

Künne (1948) quotes an observation by Carlgren on the attachment of the larva of the actinian *Peachia* to *C. capillata* (as *C. arctica*). Dahl (1959 *a, b*) found nematocysts in the stomach of the amphipod *Hyperia galba* (Montagu) collected from *C. capillata* which agreed with those of this medusa; this observation together with the presence of mesogloea among the mouth parts of one specimen led Dahl to conclude that the amphipod is a true ectoparasite. Bowman *et al.* (1963) thought that, owing to its inverted position on the jellyfish, if it was a true parasite it would probably feed on the marginal tentacles and the oral arms; they found both *Hyperia galba* and *Hyperoche medusarum* on *Cyanea capillata*.

Effects of temperature and pressure

Mayer (1914) states that this medusa (as *C. arctica*) dies at 27° C.

Digby (1967) recorded that *C. capillata* is sensitive to pressure changes.

Effects on commercial fishing

A north-east coast fisherman told me that he had heard that *Cyanea* preyed on planktonic lobster larvae and that a poor lobster fishery might result four years after a year of great abundance of these jellyfish. This seems to me quite possible. (See also Fraser, 1969; and p. 263 below.)

ABNORMALITIES

M. E. Thiel (1960) recorded a number of abnormalities in *C. capillata* from Kiel Bay and harbour. Out of 406 animals examined 10·5% had abnormalities. These consisted chiefly of variations above or below the usual number of eight peripheral organs, i.e. rhopalia, marginal tentacle groups, and coronal muscle fields. The lowest number was five and the highest twelve. The number of gonads and mouth lips however did not show corresponding abnormality in their number, being almost always four. Only four abnormalities were seen in these organs, namely 2, 5, 6 and 6. He found one instance of asymmetrical abnormality in which five octants were normal and three abnormal.

Thiel also found other types of abnormality such as subdivision of the marginal lappet, branching of a marginal tentacle, and twinning of the rhopalium.

Abnormal specimens that I have seen are shown in Text-figs. 64 and 66, and Pl. XII.

CYANEIDAE

Cyanea capillata was first named *Medusa capillata* by Linnaeus (1746). Péron & Lesueur (1809) described a number of new species in their genus *Cyanea*, including *C. arctica* and *C. lamarckii*. The name *C. arctica* was kept by later authors for medusae on the north-western side of the Atlantic, but confusion arose subsequently over the identity of *C. capillata* and *C. lamarckii*. While

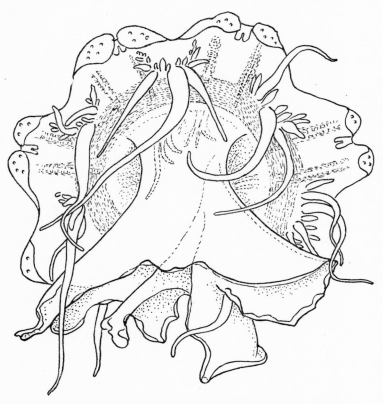

Text-fig. 66. *Cyanea capillata*. Abnormal specimen with seven rays;
13 mm. in diameter, Millport, June 1967.

Eschscholtz (1829), L. Agassiz (1862), Haeckel (1880) and Vanhöffen (1906) kept the two species distinct, Mayer (1910) regarded *C. lamarckii* as a variety of *C. capillata*. He also regarded *C. arctica* as a variety of *C. capillata*, and subsequent authors, especially Kramp (1942), have had no doubt that these two are the same species; Ranson (1945), however, still regarded them as separate.

Forbes (1848) was in error over *C. lamarckii*. He described this species as ferruginous in colour, predominating in the Irish Sea, and stinging as fiercely as *C. capillata*. His description is indeed that of *C. capillata*.

The further history of *C. capillata* and *C. lamarckii* is given on p. 138.

One cannot conclude without drawing attention to the remarkable drawing of this medusa, by A. Sonrel reproduced in L. Agassiz (1862, **3**, pl. III as *C. arctica*).

Cyanea lamarckii Péron & Lesueur

Plate V; Plate VI; Plate XIII (below); Text-figs. 60, 67–73

Cyanea Lamarck Péron & Lesueur, 1809, 'Tableau...de Méduses...', *Annls Mus. Hist. nat. Paris*, **14**, 363.
Cyanea britannica Péron & Lesueur, 1809, *ibid.* p. 364.
Cyanea lamarckii, Eschscholtz, 1829, System Acaleph. p. 71, pl. 5, fig. 2.
    L. Agassiz, 1862, *Contr. Nat. Hist. U.S.* **4**, 161.
    Gosse, 1863, *Intellectual Observer*, **4**, 149, 1 pl.
    Haeckel, 1880, *System der Medusen*, p. 530.
    Vanhöffen, 1906, *Nord. Plankt.* **11**, Acrasp. p. 53.
    Kramp, 1937, *Danm. Fauna*, **43**, 183.
    Kramp, 1961, Synopsis of Medusae of the World, p. 334.
    M. E. Thiel, 1962*b*, *Abh. Verh. naturw. Ver. Hamburg*, N.F. **6** (1961), 277.
Medusa capillata, Barbut, 1783, *Genera Vermium*, p. 79, pl. 9, fig. 3.
    Dalyell, 1848, *Rare and remarkable animals of Scotland*, **2**, 248, pl. LI, figs. 5, 6.
Cyanea imporcata, Norman, 1867, *Nat. Hist. Trans. Northumb.* **1**, 58, pl. XI.
Cyanea palmstruchii Östergren, 1909, *Zool. Anz.* **34**, 464.
Cyanea capillata var. *lamarcki*, Mayer, 1910, *Medusae of the World*, **3**, 596.
Cyanea capillata, Stiasny, 1930, *Mém. Mus. r. Hist. nat. Belg.* no. 42, p. 1 (in part), probably pl. I, figs. 5, 6, 7.
    Stiasny & Maaden, 1943, *Zool. Jb.* (Syst.), **76**, 244.

SPECIFIC CHARACTERS

Coronal and radial muscle folds without, or occasionally with only few, pit-like intrusions from gastrovascular sinus, sixteen to twenty coronal folds between radial septa; average number of marginal tentacles in each group forty to sixty; terminal ramifications of gastrovascular sinus pouches without anastomoses; colour pale yellow to blue; size 60 to 150 mm. in diameter, extreme size up to 300 mm.

DESCRIPTION OF ADULT

Umbrella saucer-shaped with uniformly fairly thick jelly thinning suddenly at bases of marginal lappets; central exumbrellar surface with prominent nematocyst clusters; umbrella margin with eight deep adradial tentacular clefts, and eight less deep perradial and interradial rhopalar clefts. Sixteen marginal lappets broad and rounded, sloping asymmetrically outwards from tentacular clefts towards rhopalar clefts, sometimes with slight median indentation. Coronal muscle well developed, consisting of up to four or five continuous circular muscle folds, and eight wide tentacular and eight narrower rhopalar fields divided by radial septa each with sixteen to twenty muscle folds without, or in large specimens with only few, rows of pit-like intrusions from gastrovascular sinus into mesogloea of each fold. Well-developed radial muscles running from outer circumference of coronal muscle along either side of radial septa towards umbrella margin, also without pit-like intrusions from gastrovascular sinus; rhopalar radial muscles slightly narrower and longer than tentacular radial muscles. Marginal tentacles hollow, arising from subumbrellar surface and arranged in eight somewhat horseshoe- or rectangular-shaped groups each with forty to sixty tentacles in up to four incomplete serial rows. Four perradial and four interradial marginal sense organs each consisting of rhopalium with statocyst and sensory bulb, but without ocellus, protected by long exumbrellar extension to form a hood and by approximation of free edges of marginal lappets on subumbrellar side; dorsal sensory pit present immediately over each rhopalium. Four interradial very much folded gonads hanging freely downwards from subumbrella surface. Stomach without interradial gastric septa; with many gastric filaments arranged in four interradial groups on subumbrellar wall where central stomach cavity passes into gastrovascular sinus; gastrovascular sinus divided into sixteen pouches by sixteen radial septa extending from short distance from proximal border of coronal muscle to distal margin of umbrella where they merge into fused areas of marginal lappets; near the distal ends of the radial muscles the pouches of the gastrovascular sinus start to branch and ramify towards the periphery, but without anastomosis of the ramifications. Manubrium with very short thickened basal pillars merging into four wide membranous and much folded curtain-like lips, united inter-radially for about half their length; oral arms slightly shorter than diameter of umbrella. Colour pale yellow to deep violet blue. Size usually 60 to 150 mm. in diameter, extreme size up to 300 mm.

DISTRIBUTION

Cyanea lamarckii is a more southern species than *C. capillata*, but nevertheless it occurs right round the British Isles.

There are only a few published records from British waters that can be attributed to this species for certain, e.g. Delap (1905) Valencia, South-west Ireland; Russell (1931, as *C. capillata*) Plymouth area; Lambert (1936*a*) Essex coast; and M'Intosh (1885) St Andrews, regarded by him as the young of *C. capillata*. Hillis (1967) recorded *C. lamarckii* in trawl catches off the coast of County Dublin in the Irish Sea in July 1966. This was the first time that it had been seen since these trawling observations were started in 1960. I can, however, now add considerably to these localities.

I have seen specimens sent me from the Irish Sea off Port Erin, from Whitstable, and from Robin Hood's Bay. On an investigation round the coasts of Scotland and north-east England (Text-fig. 67) in 1967 I either saw myself or had reported to me by fishermen and others the occurrence of blue *Cyanea* at the following points on the coast: north-west Scotland at Loch Ewe, Gruinard Bay, Kylesku, and Loch Clash; north and east Scotland at Lybster, Helmsdale, Stone-haven, St Andrews, and Pittenweem; and north-east England at Seahouses and Beadnell. H. T. Powell sent me specimens collected on 31 May and 1 June 1968 from Loch na Keal, on the west side of Mull; Hardy (1956, pl. 7) figures a specimen from the Norfolk coast, where it occurs commonly (R. Hamond personal communication); and F. Evans has sent me pure blue specimens taken off Blyth on the Northumberland coast on 30 August 1968.

It is then very evident that *C. lamarckii* occurs all round the British coasts. It is, however, noteworthy that it has never been seen at Millport.

In the North Sea it is common in the German Bight (Hartlaub, 1894; Künne, 1952) and in the Elbe Estuary (Kühl, 1964). The record by Leloup & Miller (1940) of a specimen in Ostend *bassin de chasse* with warts on its exumbrella indicates that *C. lamarckii* occurs on the Belgian coast. It is commoner than *C. capillata* along the Dutch coast (Maaden, 1942*b*), and J. Verwey tells me that it probably inhabits a different water mass from that of *C. capillata*. Kramp (1934) records that its distribution extends along the Jutland coast into the Skagerak and Kattegat. On the Swedish coast it was described by Östergren (1909) as *C. palmstruchii*. It has not been recorded from the Baltic. It has been recorded on the west coast of Norway a little north of Bergen (Kramp, 1937) and from Iceland and the Faeroes (Kramp, 1939), but these must be the extreme northern limits of its occurrence.

Little is known about its southerly limits of distribution; Le Danois (1913) recorded it from the Bay of Biscay south-west of Les Baleines on the Île de Ré off La Rochelle at the end of May.

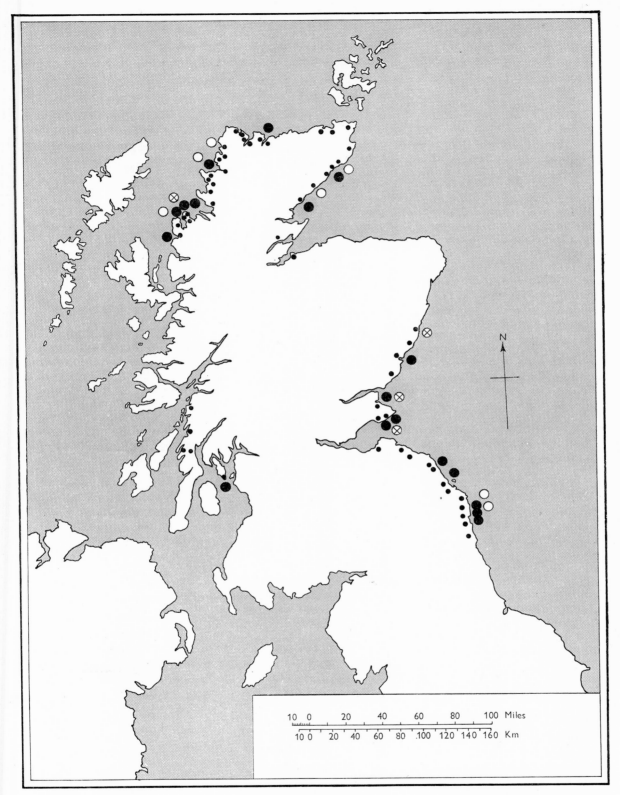

Text-fig. 67. Distribution of *Cyanea capillata* and *C. lamarckii* round coast of Scotland and north-east England, September 1967. • places visited; ● *C. capillata* seen; ⊗ *C. lamarckii* seen; ○ *C. lamarckii* reported.

CYANEIDAE

In general appearance many of the major structural details of *C. lamarckii* are similar to those of *C. capillata*, to which species the reader should refer. Mention will here only be made of those characters in which *C. lamarckii* differs from *C. capillata*.

Umbrella

The surface of the exumbrella is slightly papillose, especially in its central region where the scattered nematocyst clusters are particularly prominent. These can easily be seen by removing

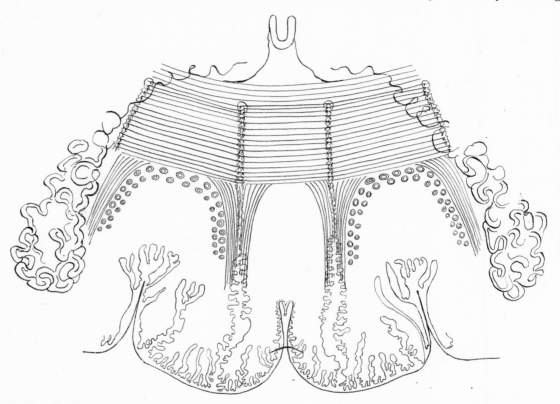

Text-fig. 68. *Cyanea lamarckii*. Portion of umbrella margin of specimen *ca.* 100 mm. in diameter to show diagnostic characters, namely muscle folds without pit-like intrusions from gastrovascular cavity, number of marginal tentacles, and extent of gonad (mouth of English Channel).

the medusa from water and viewing the exumbrella at an angle to the incident light. In *C. capillata* it is smooth.

G. Chapman (1953) could find no cells in the mesogloea; there are very fine crisscross straight fibres.

Subumbrellar Musculature

The number of folds in the coronal muscle between the radial septa (Text-fig. 68) is greater than in *C. capillata*, being usually sixteen to twenty. Radial furrows, sometimes branching (Text-fig. 70) are present on the subumbrellar wall of the gastrovascular sinus running approximately at right angles across the direction of the muscle folds, as in *C. capillata*. But there are no pit-like intrusions from the gastrovascular sinus along these furrows into the mesogloea of the muscle folds in specimens less than 100 mm. in diameter. They may begin to appear in larger medusae but are not developed to the same degree as in *C. capillata*. The muscle-fold areas thus appear more transparent in *C. lamarckii* and the folds more clearly defined when viewed with transmitted light.

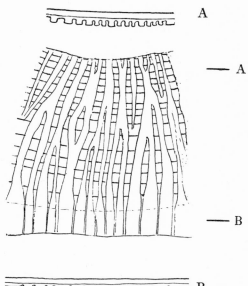

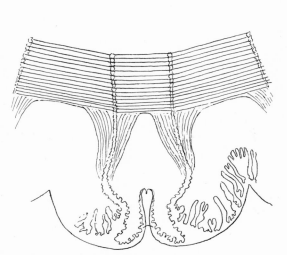

Text-fig. 69. *Cyanea lamarckii*. Portion of umbrella margin of a specimen *ca.* 50 mm. in diameter to show bowing of radial septa due to contraction.

Text-fig. 70. *Cyanea lamarckii*. Radial grooves across inner side of muscle folds. Above is a section at A contracted; below is a section at B relaxed.

The absence of the pit-like intrusions and greater number of folds constitute a quick method of identification of preserved specimens which have lost their coloration. The circular muscle folds also are not interrupted at the radial septa (Pl. XIII; see also p. 104).

The number of folds in the radial muscles of *C. lamarckii* also appears to be slightly higher than in *C. capillata*, there being about five to eight folds in the radial muscle on the rhopalar side of the radial septum and seven to ten on the tentacular side. Pit-like intrusions are also absent in these muscle folds.

Manubrium

The general appearance of the manubrium (Text-fig. 71) in *C. lamarckii* is the same as that of *C. capillata* but the oral arms are slightly shorter, being a little less than the diameter of the umbrella in length.

Marginal tentacles

The structure of the marginal tentacle is similar to that of *C. capillata* (Text-fig. 60).

There are on average about half as many marginal tentacles in each group in *C. lamarckii* as in *C. capillata*. For instance, Östergren (1909, as *C. palmstruchii*), counting the smallest tentacles

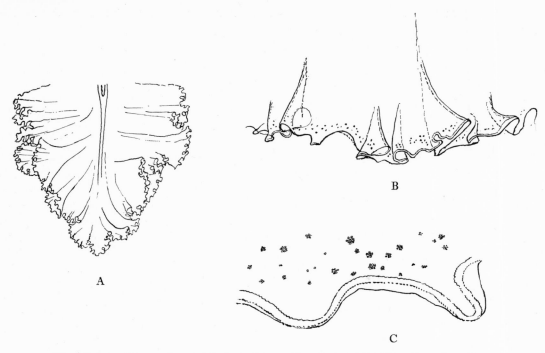

Text-fig. 71. *Cyanea lamarckii*. A, oral arm; B, margin of lip; C, enlarged to show nematocyst clusters.

which could be seen with the naked eye, gave fifty to sixty tentacles in each group for specimens 80–150 mm. in diameter. This agrees fairly well with the results of M. E. Thiel (1962) who made a careful investigation of this character. Thiel's results are summarized in the following table, in which diameter measurements of the umbrella are between rhopalia (10 or more mm. should be added to diameters of 50 or more mm. to convert to over-all diameter of umbrella).

Diameter (mm.)	Average number of tentacles per group	Extremes of averages	Number examined
30–39	25·2	22·8–29·0	2
40–49	29·3	20·1–39·4	12
50–59	31·8	19·4–46·5	28
60–69	37·0	25·1–49·6	21
70–79	40·0	27·1–60·0	12
80–89	45·0	28·6–59·0	12
90–99	52·0	44·6–60·8	4
100–109	49·0	42·6–57·1	2
110–119	60·0	—	1
120–129	54·5	—	1

From his results Thiel deduced a formula for the number of tentacles per group and size of medusa for specimens over 20 mm. in diameter, namely: $y = 2/5x + 12$, where y equals the number of tentacles and x the diameter of the medusa between rhopalia in mm.

I have counted the tentacles on a few specimens from the English Channel and the North Sea. In general I have had higher figures in the smaller size groups than Thiel; I have, however, included in my counts the minutest rudiments of developing tentacles, which may account for the higher figures. My averages were:

30–39 mm. 46	50–59 mm. 54
40–49 mm. 44	60–69 mm. 59

All the indications are that the final definitive number of tentacles does not exceed an average per group of sixty to seventy. That this is a final number receives support from the fact that in large *C. lamarckii* there are very few rudiments of developing tentacles to be seen, whereas in *C. capillata* there are always many small and rudimentary tentacles.

In specimens of comparable size in the lower size groups, since there are fewer tentacles in a group in *C. lamarckii* than in *C. capillata*, there is more space available for the tentacles and it is noticeable that the tentacle bases in *C. lamarckii* are considerably thicker than those in *C. capillata*, as was pointed out for instance by Östergren (1909). Östergren also noted, with other authors, that whereas in *C. capillata* there might be four complete rows of tentacles in a group, in *C. lamarckii*, although there might be as many as four rows in the older central area, there were never four complete rows.

Marginal sense organs

As far as I am aware the marginal sense organ of *C. lamarckii* has not been described previously. I can, however, find no obvious macroscopic difference between it (Text-fig. 72) and that of *C. capillata*. As in the latter species the basal ridge to the rhopalium may be smooth or papillose.

Gonads

The gonads are similar in general form to those of *C. capillata* but they develop much earlier; for instance, in specimens 40 mm. in diameter their folds already hang down as far beneath the subumbrella as do the mouth lips (Text-fig. 68).

Nematocysts

The nematocysts (Text-fig. 5, p. 9) of *C. lamarckii* have been studied by Weill (1934) from Wimereux and Papenfuss (1936, as *C. palmstruchii*) from Kristineberg.

There are three kinds: atrichous haplonemes in two sizes, holotrichous haplonemes, and heterotrichous microbasic euryteles.

Measurements of the undischarged capsules in adult medusae given by the two authors are as follows:

	Atriches		Holotriches	Euryteles
	large	small		
Weill	$15 \times 11 \mu$	$9 \times 5 \mu$	11–15 μ diam.	12–17 μ long
Papenfuss	$15 \cdot 6 \times 9 \cdot 8 \mu$	$9 \cdot 0 \times 5 \cdot 6 \mu$	$12 \cdot 6 \times 10 \cdot 4 \mu$	$14 \cdot 6 \times 8 \cdot 2 \mu$

Mention has been made under *C. capillata* (p. 115) of differences in the coiling of the thread between nematocysts of that species and those of *C. lamarckii* observed by Papenfuss. She drew attention also to the facts that the holotriches were smaller in *C. lamarckii*, as were the euryteles which also had a longer axial body than those of *C. capillata*.

Weill gave some observations on the distribution of the nematocysts in different parts of the medusa. The atriches were found in the marginal tentacles and in the nematocyst clusters of the oral lips, with intermediary stages along the mouth fringe borders. The holotriches were present in the marginal tentacles and on the oral lips but were not found in the borders of the fringe.

CYANEIDAE

The euryteles were abundant in the marginal tentacles and the borders of the fringe, but absent or rare in the nematocyst clusters on the oral lips.

Weill also studied the nematocysts in scyphistomas reared by Lambert at Leigh-on-Sea, Essex. There were atriches 8–10 μ long and euryteles 11–14 μ long.

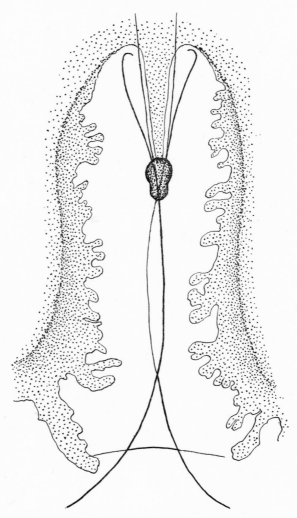

Text-fig. 72. *Cyanea lamarckii*. Marginal sense organ, exumbrellar view.

COLORATION

The degree of coloration in *C. lamarckii* varies from almost complete lack of colour, through pale yellow, pale brown and grey, to light blue and deep violet-blue. But it has none of the deep reddish brown colour of *C. capillata*. In faintly coloured specimens the blue tends to outline the pattern of the internal sculpturing of the umbrella, such as the wavy outlines of the central disk and the radiating furrows. The blue is very beautiful, varying between a cornflower blue and the deepest violet. Sometimes the pale yellowish brown specimens have blue marginal patterns and blue-violet tentacles. I have seen yellowish specimens from Whitstable, Robin Hood's Bay and St Andrews. C. I. D. Moriarty has kindly allowed me to reproduce in Pl. VI a photograph showing the colour range of specimens from the Irish Sea. As with *C. capillata* there is often the distinctive colour pattern shown in Text-fig. 63.

Very young specimens are generally colourless, but they tend to take on colour, which deepens as the season advances. Östergren (1909) thought that the increase in intensity of the blue might be due to the rising temperature, and it is significant in this respect that Verwey (1965) noticed that there was no blue colour in 1963 following the very cold winter, until the summer was well advanced.

Detailed descriptions of coloration have been given by Delap (1905), Östergren (1909) and Maaden (1942b). While the central part of the umbrella may be dark blue, its margin is more transparent and often slightly yellowish. All the darker coloration is in the gastrovascular endoderm, the jelly above it being clear and transparent. The marginal tentacles may range between pink, grey and dark blue. The gonads may be whitish, bluish or rose coloured. The oral lips may be yellowish, fading off to whitish at their extremities.

DEVELOPMENT

The early development of *C. lamarckii* has not been much studied, but it seems unlikely that it should differ markedly from that of *C. capillata*. Widersten (1965, as *C. palmstruchii*) noted that the eggs, which are white in colour, are smaller than those of *C. capillata*, being 150μ in diameter. The spermatozoa are similar to those of *C. capillata*, as is their mode of entry to the eggs. Widersten (1968) gave the dimensions of the planula as 125×75 to $225 \times 115 \mu$. He stated that after a planktonic stage of varying duration it enters an encysted stage.

The development of the longitudinal muscles, supporting fibrils, and periderm of the scyphistoma was studied by Widersten (1967).

Cyanea lamarckii has been reared from the planula to the adult medusa by Delap (1905) at Valencia, south-west Ireland. Eggs were obtained from specimens in September 1900, which in three days had developed to planulae. These began to settle after seven days, and ten days later the largest scyphistoma had eight tentacles and was 1·0 mm. high (including tentacles).

The rate of development from the eggs showed considerable variation. Some grew rapidly during the winter and strobilated in the following spring, while the majority remained dormant and only produced four or eight tentacles when several months old, remaining in this condition for a year or longer. The maximum number of tentacles was twenty-eight.

On 24 February 1901 the largest scyphistoma began to strobilate. It was 4 mm. high and 2 mm. wide, and its tentacles could extend to about 20 mm. The first ephyral constriction appeared just below the tentacles, and a second constriction developed the next day. The number of constrictions increased regularly until on 5 March there were nine, the seven uppermost being already lobed. By 8 March there were eleven ephyra segments. On 10 March the upper five were ready for liberation, the scyphistoma then being 6 mm. in height and 3 mm. in diameter, with pulsating ephyrae. Two ephyrae were liberated on 12 March and three more on the next day.

Lambert (1936a) found the number of ephyra segments to be four to twelve. Verwey (1942) considered that the upper temperature limit for strobilation might be 8° C.

Delap noted that at the time of liberation the ephyra was 4 mm. in diameter and white. It was about twice the size of the ephyra of *Chrysaora* reared by her, which was by contrast quite pinkish in colour. Lambert (1936a) stated that the ephyrae were blue-grey.

By 15 March ten ephyrae had been set free and there were two more nearly ready to come off.

The polyp now began to grow a new set of tentacles just below the twelfth segment, the last two ephyrae being liberated on 18 March.

During the development of an apical ephyra sixteen of the scyphopolyp tentacles remained on the margin of the ephyra itself, one just above each rhopalium and one in each cleft between the main lobes of the ephyra. The tentacles on this ephyra were gradually absorbed so that when the ephyra was liberated they had been reduced to tiny knobs which disappeared when the ephyra was nine days old. The remaining scyphopolyp tentacles 'were pushed in towards the mouth and formed isolated groups'.

By July the scyphistoma had grown thirty-three new tentacles and it started to strobilate again on 17 January 1902, producing eight ephyrae early in February.

On 21 March 1901 the largest ephyra, from which all traces of the original scyphopolyp tentacles had now gone, had developed in each of two opposite clefts a new bulb; three days later a tentacle started to grow from one of these bulbs.

By 5 April each ephyra had one large marginal tentacle and one large bulb opposite it, with minute bulbs in the remaining six clefts. The largest ephyra was 7 mm. in diameter.

At this stage the ephyra is now definitely recognizable as a *Cyanea*. Text-fig. 73 shows an ephyra 4·2 mm. in diameter which I saw at Plymouth.

Text-fig. 73. *Cyanea lamarckii*. Ephyra, 4·2 mm. in diameter, Plymouth.

By 15 April the largest young medusa was 10 mm. in diameter. It had two long opposite marginal tentacles and the remaining tentacle bulbs were well developed. The first signs of tentacle grouping were indicated by the presence of one minute bulb on each side of the largest tentacle. On 24 April this medusa had four long perradial marginal tentacles and four large interradial bulbs. By 30 April it was 20 mm. in diameter and had eight long tentacles and the oral arms were well developed. These oral arms were shorter and broader than those of *Chrysaora* of a corresponding size and they were less frilled along their margins.

By 9 May the medusa was 30 mm. in diameter with several tentacles of varying length in each of the eight groups. This specimen died on 4 June when it was 50 mm. in diameter, being deep blue in colour and having reached its adult form. Another grew more rapidly, being 50 mm. in diameter on 29 May and 80 mm. on 10 June; it was then very deep blue, with pink tentacles and yellowish white oral arms.

SIZE

The largest specimen in the shoal from which Delap (1905) obtained specimens for the above account of development was about 230 mm. in diameter.

Off Plymouth, Russell (1931) found that in April most specimens were 10 mm. or less in diameter, the largest being 75 mm. Later in the season the maximum sizes were generally between 130 and 150 mm. in diameter and the largest specimen caught was 190 mm. in diameter. M'Intosh (1885) recorded specimens in June off St Andrews about 50–180 mm. in diameter. Östergren (1909) off the Swedish coast found that they were mostly 60–120 mm. in diameter and rarely

150 mm. Similarly Hartlaub (1894) recorded that off Helgoland the average size did not exceed 120 mm. in diameter, and that he had seen none over 150 mm. Maaden (1942 a) recorded the largest specimen on the Dutch coast in May 1935 as 280 mm. in diameter. Kühl (1964) gave figures of 60–100 mm. in diameter for the Elbe estuary with an average of 70 mm.

It seems evident therefore that it is unlikely that specimens will be found more than 300 mm. in diameter.

REARING AND FEEDING

Delap (1905) fed the first liberated ephyrae on small copepods, small medusae, and especially fish eggs. They also fed on small *Sarsia tubulosa* medusae liberated from hydroids in the aquarium.

She found later on that *C. lamarckii* lived entirely on other medusae, especially *Phialidium*, *Cosmetira*, *Laodicea*, *Obelia* and *Steenstrupia*, the latter being greedily devoured. They sometimes ate the ctenophores *Pleurobrachia* and small *Beröe*, but preferred *Bolina*. *Neoturris pileata*, large *Sarsia tubulosa* and large *Beröe* were never eaten.

Lambert (1936 b, p. 83) says that the food of most of the *Cyanea* he caught, presumably both species, appeared to consist almost entirely of ctenophores.

Both Delap and Hagmeier (1933) mention that *Cyanea* does not keep very well in captivity.

SEASONAL OCCURRENCE

Being a more southern species it is probable that *C. lamarckii* tends to appear in the plankton a little earlier in the season than *C. capillata* in areas round the British coasts in which their distribution overlaps.

There are rather few published records for British waters. Off Plymouth (Russell, 1931), I made collections over a period of six years and only found specimens from April to July, but I have since found a late ephyra as early as January.

In April most specimens were very small, many being about 10 mm. or less in diameter and the largest 75 mm.

Delap (1905) and Browne (1896) recorded *C. lamarckii* at Valencia, south-west Ireland, in August and September.

Schoddyn (1926) records a tendency for the medusae to appear in May and June in the Pas de Calais area especially with strong westerly winds.

In the southern North Sea *C. lamarckii* occurs off Helgoland from mid-May to mid-September (Kramp, 1934); along the Dutch coast very small specimens have been recorded in February, but the season of greatest abundance with the largest specimens is mid-May to the end of June, and it is not usually found later than August (Maaden, 1942). Verwey (1964, 1965) records that ephyrae may appear as early as in November; he gave the following observations on their first and last appearances.

Earliest records		Latest records	
2 November 1961	9·8° C.	1 March 1961	7·0° C.
20 November 1962	8·9°	mid May 1962	8·1°
16 November 1963	11·4°	mid May 1963	7·1°

He attributed their continuation to as late as May in 1962 and 1963 to the cold spring of 1962 and the severe winter of 1962/3.

Künne (1952) states that in the German Bight the medusae predominate in spring and early summer, but that none is seen in October, while Kühl (1964) says that in the Elbe estuary it does not usually appear before June and disappears in August and September. According to Kramp (1934) it is found in the Skaw in July and August.

It would seem therefore that *C. lamarckii* tends to decrease in numbers about a month earlier than *C. capillata* and that its period of greatest abundance is also somewhat earlier.

PHYSIOLOGY

Little use seems to have been made of *C. lamarckii* for physiological studies. Woollard & Harpman (1939) made observations on the nerve fibres in this species.

G. Chapman (1953) found the following amino-acids to be present in the mesogloea: alanine, glutamic acid, glycine, ?histidine, leucine, lysine, proline, ?serine, and valine.

Ebbecke (1935) found that increase of pressure up to over 200 atmospheres caused more frequent and stronger pulsations of the umbrella using pieces of the umbrella margin. At 300 atmospheres and over the muscle went into tonic contraction. (He called it *C. capillata* but referred to its blue colour.)

Horridge & Mackay (1962) examined sections of the marginal ganglion of the 'blue North Sea' *Cyanea*, about 15 mm. in diameter, with the electron microscope.

HABITS AND GENERAL OBSERVATIONS

Damas (1909 *a*, *b*) noted that *C. lamarckii* in the North Sea often had the young of pollack (*Gadus pollachius*), poor-cod (*G. minutus*) and pouting (*G. luscus*) in association with it. Russell (1928, as *C. capillata*) showed also a correlation between the vertical distribution of this species and that of young whiting (*G. merlangus*). In the English Channel off Plymouth whiting between 12 and 22 mm. in length were caught with medusae of all sizes, but whiting over 23 mm. long were mostly only associated with medusae of 100 mm. or more in diameter. Whiting less than 12 mm. long showed little vertical migration to the surface at dusk, but those associated with *C. lamarckii* showed marked migration, presumably accompanying the medusae to the surface. It seems probable also that in the Plymouth area young horse-mackerel (*Trachurus trachurus*) *ca.* 11–15 mm. in length may associate with *Cyanea lamarckii* (see Russell, 1930, as *C. capillata*).

Rees (1966) records underwater observations in which young whiting were seen swimming and manoeuvring above the dome of the umbrella of *C. lamarckii* and apparently resting upon it.

Off Plymouth medusae 12–90 mm. in diameter were usually caught below 15 m. in the daytime (Russell, 1927, as *C. capillata*). Verwey (1966) states that off the Dutch coast Miss van der Baan had found very young stages to be nearly three times more numerous in collections made during the night than in those made by day.

By comparison with that of *C. capillata* the stinging powers of *C. lamarckii* are negligible (see e.g. Gosse, 1863). Lambert (1936) stated that *Cyanea*, presumably both species, are much infested by the crustacean amphipod *Hyperia*.

Fishermen on the north-east coast have remarked to me that the blue *Cyanea* foul their nets.

HISTORICAL

Cyanea lamarckii was first named *C. Lamarck* by Péron & Lesueur (1809) and its specific name was changed to its modern spelling by Eschscholtz (1829).

As already mentioned on p. 126 this species became confused with *C. capillata*. For instance M'Intosh (1885) and Stiasny (1930) thought that *C. lamarckii* medusae were the young of

C. capillata. It is evident that Forbes' (1848) description of *C. lamarckii* was really that of *C. capillata*, and Mayer (1910) regarded *C. lamarckii* as a variety of *C. capillata*.

M. E. Thiel (1960, 1962*b*) was the first to make a thorough investigation on the numbers of marginal tentacles in the two species. This fully bore out statements by previous authors that *C. capillata* had on average twice as many marginal tentacles as *C. lamarckii*. Thus a discrete character was established for distinguishing the two species, the other characters such as size, colour, and size of maturity being more of a nature that could be regarded as varietal only. Two more discrete characters can now be added, namely presence or absence of pit-like intrusions into the muscle folds from the gastrovascular sinus and the numbers of muscle folds in the coronal muscle between radial septa.

The *Cyanea imporcata* of Norman (1867) from the Northumberland coast must have been *C. lamarckii* on account of its blue colour.

Family ULMARIDAE

Semaeostome Scyphomedusae with gastrovascular system of unbranched and branched radial canals with varying degrees of anastomosis.

There are four subfamilies, Aureliinae, Sthenoniinae, Ulmarinae, and Stygiomedusinae. Of these only the first is represented in British waters by the common *Aurelia aurita*.

There are, however, representatives of the Sthenoniinae and Stygiomedusinae which live in deep water, namely *Poralia rufescens* Vanhöffen and *Stygiomedusa fabulosa* Russell respectively. These have not been taken in waters west of the British Isles but farther to the south. It is conceivable, however, that they might occur. They are not included in this monograph but a description of the former is given by Russell (1962) and of the latter by Russell & Rees (1960). More recently Repelin (1967) has described a new species *Stygiomedusa stauchi*.

A representative of the Ulmarinae, *Discomedusa lobata* Claus was recorded by me in the English Channel in 1936 and 1937 (Russell, 1937, 1938). This occurrence is, however, so rare that the medusa can hardly be regarded as a British species. It has been well described by Claus (1877).

Subfamily AURELIINAE

Ulmaridae with small marginal tentacles and lappet-like structures arising from exumbrella slightly above umbrella margin; with simple and branched radial canals with little or much anastomosis; with ring canal; with subgenital pits.

The subfamily contains the genus *Aurelia* Lamarck 1816. Earlier Péron & Lesueur (1809) had erected the genus *Aurellia*. The change of spelling introduced by Lamarck was used by all subsequent authors until *Aurellia* was resuscitated by Mayer (1910). Lamarck's spelling, however, continued to be so generally used that Péron & Lesueur's genus *Aurellia* was suppressed and replaced by the genus *Aurelia* Lamarck (see Rees, 1957, p. 26; Hemming, 1958, p. 117, no. 1157).

Genus **Aurelia** Lamarck

Aureliinae with umbrella margin divided by eight or sixteen marginal clefts; with four un-branched oral arms; with anastomoses between a few or all of the radial canal branches.

The only species of *Aurelia* occurring in British waters is *A. aurita* (L.). Many species of this medusa have been named, and this is an indication of the almost cosmopolitan distribution of *A. aurita*, of which most of the species named are probably geographical races with only small varietal differences. Kramp (1961) listed seven species, *A. aurita* (L.), *A. coerulea* von Lenden-feld, *A. colpota* Brandt, *A. labiata* Chamisso and Eysenhardt, *A. limbata* (Brandt), *A. maldivensis* Bigelow, and *A. solida* Browne. Later (Kramp, 1965), he regarded *coerulea, colpota* and *maldivensis* as varieties, and said that probably *labiata* and *solida* would prove also to be varieties (see Kramp, 1968). This would leave one good species *A. limbata* in addition to *A. aurita*.

The differences by which the species or varieties are recognized are: (1) the amount of folding of the margins of the lips of the oral arms, (2) the degree of branching of the radial canals and the extent of occurrence of anastomoses between branches, and (3) the presence of adradial incisions in the umbrella margin.

Even in each of these three characters there is a certain amount of intergrading; for instance in *A. aurita* on the North American coast of the Atlantic there is a tendency for more anastomoses between branches of the radial canals than in medusae from the eastern side of the Atlantic; adradial incisions in the umbrella margin may appear in living specimens but disappear on their preservation (Kramp, 1942).

The number of canal roots issuing from the margins of the gastric pouches has also been used as a diagnostic character, but this is a function of the size and rate of growth of the individual medusa (see e.g. Stiasny, 1922; Bigelow, 1938). We are thus left with the two species *A. aurita* (L.) and *A. limbata* (Brandt) in the North Atlantic and of these *A. limbata* is an Arctic species living in colder more northern waters than *A. aurita*. It is distinguishable from *A. aurita* by the profuse branch-ing of the perradial and interradial canals and the great degree of anastomosis between branches.

In the following synonymy list for *A. aurita* I have included only those used for Atlantic forms (see p. 171).

Aurelia aurita (Linnaeus, 1758) is the type species of genus.

Aurelia aurita (L.)

Plate XIV; Text-figs. 3, 74–86

Medusa aurita Linnaeus, 1746, *Fauna Suecica*, no. 1287.
   Linnaeus, 1747, *Westgöta Resa*, pl. 3, fig. 2.
   Linnaeus, 1758, *Syst. Nat.* ed. 10, **1**, 660.
   O. F. Müller, 1780, *Zool. Dan.* Fasc. **2**, pl. 76, 77; 1784, *ibid.* **2**, 109–13.
   Gaede, 1816, *Beitr. Anat. Phys. Med.* p. 12, pl. I.
   Baer, 1823, *Dt. Arch. Physiol.* **8**, 369, pl. IV.
   Eschscholtz, 1829, *Syst. Acaleph.* p. 61.
   Ehrenberg, 1837, *Abh. preuss. Akad. Wiss.* p. 186, pl. I–VII.
   Siebold, 1839, *Beitr. z. Naturgesch* p. 1, pl. I–III.
Urtica marina Borlase, 1758, *Nat. Hist. Cornwall*, p. 257, pl. XXV, figs. ix–x.
Medusa cruciata Baster, 1762, *Opuscula subseciva*, p. 123, pl. XIV, figs. 3, 4.
Aurellia flavidula Péron & Lesueur, 1809, *Annls Mus. Hist. nat. Paris*, **14**, 359.

Aurellia rosea Péron & Lesueur, 1809, *Annls Mus. Hist. nat. Paris,* **14**, 357.

Aurelia aurita, Lamarck, 1816, *Hist. anim. sans Vert.* **2**, 513.

Forbes, 1848, *Monogr. Brit. Med.* p. 75.

Haeckel, 1880, *System der Medusen,* p. 552 (long list of synonyms).

Vanhöffen, 1906, *Nord. Plankt.* **6**, XI, p. 60, figs. 27–31.

Kramp, 1937, *Danm. Fauna,* **43**, 184, figs. 77, 80–2.

Kramp, 1942, *Meddr Grønl.* **81**, nr. 1, p. 109, figs. 29, 30.

Kramp, 1961, Synopsis. *J. mar. biol. Ass. U.K.* **40**, 337.

Naumov, 1961, *Fauna of U.S.S.R.* no. 75, p. 73, figs. 50, 51.

Aurelia flavidula, L. Agassiz, 1862, *Contr. Nat. Hist. U.S.* **3**, pl. 6–9, 11 *a,* 11 *b* (Development. Pl. 10, figs. 18, 22, 31, 32, 36; pl. 10 *a,* figs. 4 *b,* 13, 15 *a,* 16–41; pl. 11 *c,* figs. 1–13): **4**, 12–17, 20–63, 75–86.

A. Agassiz, 1865, *N. Amer. Acaleph.* p. 42, figs. 65, 66.

Haeckel, 1880, *System der Medusen,* p. 555.

Fewkes, 1881, *Bull. Mus. Comp. Zool. Harv.* **8**, 172, pl. 7, figs. 2–4, 6.

Aurellia aurita, Mayer, 1910, *Medusae of the World,* **3**, 623, text-fig. 397, pl. 67, fig. 4; pl. 68.

SPECIFIC CHARACTERS

Umbrella with eight simple marginal lobes; oral arms as long as umbrella radius, with thick firm mesogloea and much-crenulated lips with many small tentacle-like processes along their margins; adradial canals unbranched; perradial and interradial canals with primary canal unbranched, but branches from their bases which branch successively towards umbrella margin have only few anastomoses; size usually up to 250–400 mm. in diameter.

DESCRIPTION OF ADULT

Umbrella flattened saucer-shaped, jelly fairly thick thinning evenly towards margin, surface smooth; margin entire except for eight small incisions situated at positions of marginal sense organs; very numerous small hollow marginal tentacles on exumbrella slightly above margin, one to three of which alternate with small lappet-like extensions of exumbrella; margin below tentacles in form of narrow velarium. Four perradial and four interradial rhopalia consisting of statocyst and sensory bulb, with exumbrellar ectodermal ocellus and subumbrellar endodermal ocellus, partially protected by small hood-like projection of exumbrella and by small lateral rhopalar lappets; exumbrellar sensory pit immediately above rhopalium. Four interradial folded gonads in form of horseshoe or almost complete circle, not projecting below subumbrellar surface. Embryos developing to planulae in brood pouches on mouth lips. Stomach in form of four circular interradial gastric pouches connected by grooves with mouth opening, with many gastric filaments arranged in four interradial horseshoe-shaped groups just centripetal to gonads; eight unbranched adradial canals, four branched perradial and four branched interradial canals, all connecting with marginal ring-canal, with few anastomoses between branches. Manubrium arising from four perradial regions of much thickened jelly to form four thickened oral arms each with median groove flanked by thinner folded lips with numerous small tentacle-like processes along their margins; length of oral arm about equal to umbrella radius. Thickened basal portions of manubrium fused to form four subgenital pits each with circular or oval subumbrellar orifice. Colour of radial canals, oral arms and gonads mauve, violet, reddish, pink or yellowish; umbrella colourless. Size usually up to 250 mm. in diameter, rarely 400 mm., one record of 1100 mm. (see p. 157).

ULMARIDAE

DISTRIBUTION

Aurelia aurita is almost cosmopolitan in its distribution and is to be found all round the coasts of the British Isles. It is in general an inshore species and may establish local races up estuaries and probably in harbours. It is very common in the Scottish sea lochs.

It occurs off all coasts of Europe as far north as Lofoten in Norway and round the coasts of Iceland, in the White Sea and Barents Sea (Naumov, 1961); it is found all along the Atlantic coast of North America, and in some years off West Greenland (Kramp, 1942, 1947). In Arctic waters its place is taken by *Aurelia limbata* (Brandt). It occurs in the Mediterranean, in the Black Sea and Bosphorus (Netchaeff & Neu, 1940), and Sea of Azov (Naumov, 1961), as well as in the Indian Ocean and Pacific. It appears in fact to be cosmopolitan from northern boreal to tropical waters.

It is unnecessary to list records of British occurrence as it is our commonest Scyphomedusa and may be found anywhere round the coasts of the British Isles. It should, however, be pointed out that it is sporadic in its appearance, being apparently absent altogether in some localities in some years (see p. 169). It may form local populations. For instance, in the neighbourhood of Saltash, on the Tamar estuary, Browne (1901) found large shoals which did not penetrate to Plymouth Sound. This population remained in existence for many years, the scyphistomas probably living on a local mussel bed (Percival, 1929). It has now disappeared and in recent years *Aurelia* has been scarce or absent off Plymouth. At St Andrews also the ephyrae have been very scarce recently although years ago they were very abundant (M'Intosh, 1926). The population in Loch Riddon which Browne (1905a) regarded as the headquarters of *Aurelia* in the Clyde sea area is however still in existence. There may possibly also be quite small local populations, such as that suggested for Portishead Dock, near Bristol (Purchon, 1937). *Aurelia* is an inshore species and is able to live in brackish water conditions in estuaries. Nowhere is this characteristic better shown than in the Baltic where it lives in a wide range of salinities. The species has been much studied in the Baltic (see e.g. Thill, 1937). Observations made in Finnish waters by Wikström (1921; 1925a, b; 1932–3) and Sjögren (1962) have indicated the lower limiting salinity for natural populations in the Baltic. Hela (1951) reviewed much published and local evidence which indicated that since 1940 the eastern border of the distribution of *Aurelia* had shifted about 50 km. eastward in ten years. Segerstråle (1951) produced an instructive chart showing this eastward extension along the South Finnish coast. Adult *Aurelia* were living in salinities as low as 3‰, but the limiting salinity for the scyphistoma appeared to be about 6‰. This may thus be the limit for successful reproduction of the species.

STRUCTURAL DETAILS

Umbrella margin and marginal tentacles

A characteristic feature of *Aurelia* is the form of the umbrella margin (Text-figs. 74, 75, 76). The margin between rhopalial clefts is entire, and thins out to form a narrow velar-like structure, the velarium. Issuing from the exumbrella, slightly above the velarium are many small hollow marginal tentacles. These tentacles are quite short and have numerous rings of nematocysts interrupted on the adaxial side by a longitudinal muscle. One to three tentacles issue from depressions in the umbrella margin whose projecting side walls form vertical laterally compressed lappet-like structures. The floor of the depression also projects beyond the roof to form a small lip-like protrusion.

142

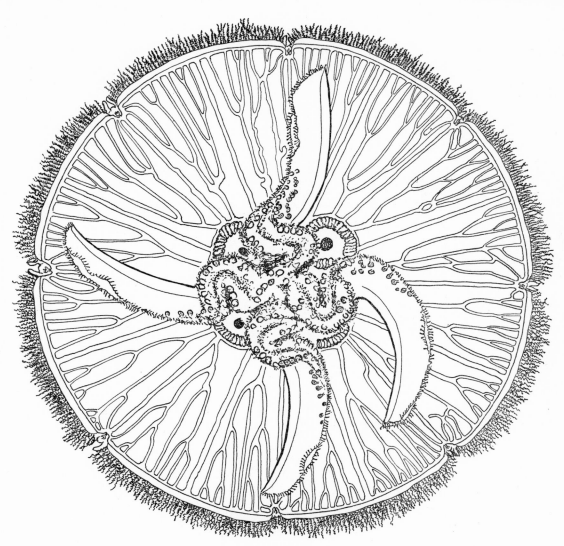

Text-fig. 74. *Aurelia aurita*. Subumbrellar view of adult specimen, to show general characters, Loch Melfort, June 1967. Female with brood pouches.

The exumbrellar surface along the margin and on the lappet-like projections has numerous slightly raised nematocyst clusters; the velarium also has nematocyst clusters along the distal half of its adaxial side.

Mesogloea

It has long been known that there are cells and fibres in the mesogloea (see e.g. M. Schultze, 1856; Kölliker, 1865). G. Chapman (1953) states that the living cells are approximately spherical and *ca.* $10\,\mu$ in diameter. They did not appear to be wandering cells or amoebocytes, and he could find no fine processes connecting the cells. He counted $5 \cdot 6 \times 10^6$ cells per c.c. near

143

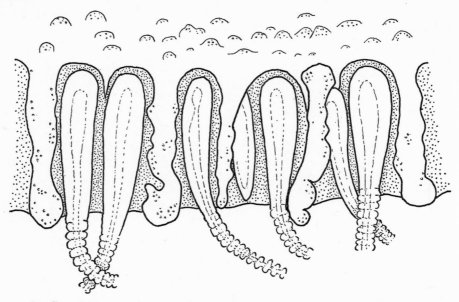

Text-fig. 75. *Aurelia aurita*. Margin of umbrella to show marginal tentacles inset in cavities between marginal lappets. Specimen *ca.* 30 mm. in diameter, Loch Riddon, 19 April 1967.

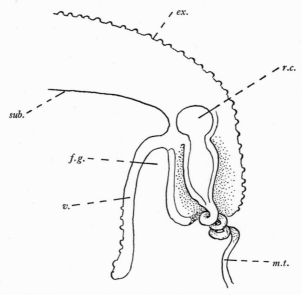

Text-fig. 76. *Aurelia aurita*. Diagrammatic section to show velarium and origin of marginal tentacle from ring-canal. *ex.* exumbrella; *f.g.* food groove; *m.t.* marginal tentacle; *r.c.* ring-canal; *sub.* subumbrella; *v.* velarium.

the centre of the umbrella, and calculated that on this basis there was about 0·3 % of living matter by volume in 1 c.c. of mesogloea.

Phase-contrast microscopy revealed slight traces of straight fibres, and Chapman thought that the network seen in preserved material and the spiral fibres recorded by Tretyakoff (1937) might be partly or entirely due to fixation and dehydration. He states that the jelly-like consistency could be partially restored by water after evaporation. The mesogloeal fibres resemble collagen (see p. 165).

Although Chapman did not think that the mesogloeal cells were amoebocytes, Wetochin (1930) reported cells moving from the mesogloea into the radial canal (see p. 161).

Marginal sense organs (Text-figs. 3, 77, 78)

Ehrenberg (1837) was the first to appreciate that what had previously been regarded as excretory organs were in fact sense organs. He gave a fairly accurate description and drawings, but omitted to mention or observe the subumbrellar endodermal ocellus on the rhopalium and the exumbrellar sensory pit. A good account of the early ideas of the functions of the marginal sense organs was given by the Hertwigs (1878, pp. 115–19). The first satisfactory descriptions of the sense organ were those of Eimer (1878) and Schäfer (1878) who omitted the endodermal ocellus, and these were supplemented by a very complete description given by Schewiakoff (1889), who described folds in the exumbrellar sensory pit subsequently confirmed by Maaden (1939).

Rhopalium. The rhopalium is short and bends slightly downwards at the junction of the statocyst with the bulb. It is protected on its exumbrellar side by a short hood formed by the umbrella margin.

D. M. Chapman has kindly supplied the following unpublished observations. The bulb is clothed with columnar sensory cells whose 9:2 flagellum is centred in a minute crater-like formation (Text-fig. 3B, p. 6). Strong flagellar currents in the exumbrellar sensory pit send dye particles centrifugally, and, except over the touch-plate (Text-fig. 3A), weak rhopalar currents are directed towards the subumbrellar sensory pits. The touch-plate in *Aurelia* and in *Cyanea* has static flagella. The ectodermal and endodermal ocelli contain round reddish to golden brown pigment granules, which are basic protein structures firmly bound to a carotenoid pigment and enclosed in a membrane. The endodermal ocellus is rarely cup-shaped as described by Schewiakoff.

The structure and composition of the crystals in the statocyst have been examined by a number of authors (e.g. Ehrenberg, 1837; Gosse, 1856; Eimer, 1878). According to Eimer the crystals are contained in amoeboid cells. (See also Spangenberg, 1968.)

Yamashita (1957a) examined the crystals chemically, with the electron microscope, and by X-ray and spectrum analysis.

The crystals, which are mostly six-sided, may also be four- or three-sided, or irregular in appearance. When fresh they appear brownish on the outside with transparent centres. They dissolve at once in HCl and distilled water, but more slowly in HCl and sea water. In distilled water alone the outer layer is dissolved in about 25 min., and after 1 hr only amorphous traces remain. In 10% formalin decomposition begins after a week and is complete in 2 weeks.

As regards the chemical composition, besides the dominating calcium, he found magnesium, silicon and strontium, and traces of copper and iron. Spangenberg & Beck (1968) also studied the statoliths by similar physical methods on specimens reared in artificial sea water. It had been found that ephyrae of *Aurelia* did not form statoliths in media lacking sulphate, but when the latter was present the crystals were calcium sulphate (gypsum = $CaSO_4.2H_2O$). They found no trace of calcium phosphate, the presence of which had been recorded by some previous authors.

Sensory pits. The exumbrellar sensory pit (Text-figs. 77, 78) was well described by Maaden (1939). It is a shallow depression with its deepest point immediately over the base of the rho-

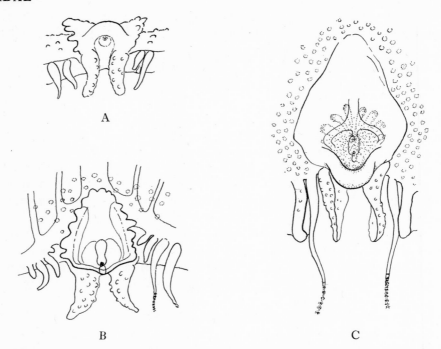

A

B   C

Text-fig. 77. *Aurelia aurita*. Marginal sense organ. A, B, specimen *ca.* 30 mm. in diameter, Loch Riddon, April 1967, frontal and exumbrellar views; C, adult specimen, exumbrellar view, Loch Melfort, June 1967.

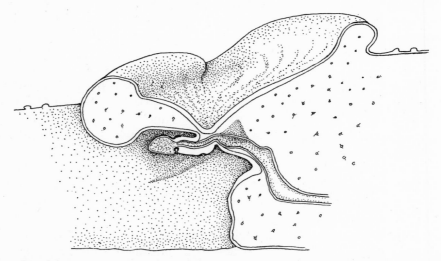

Text-fig. 78. *Aurelia aurita*. Solid diagrammatic section through marginal sense organ.

palium. It is often triangular in shape with the apex of the triangle pointing towards the umbrella centre. Its shape is, however, variable, and in some specimens there may be an outward embayment in the base of the triangle to produce a somewhat quadrangular shape. The walls of the depression rise gradually to its rim, which is slightly raised above the surface of the surrounding exumbrella; the edge nearest the umbrella margin is most raised and is variable in form.

146

Folds in the walls radiate from the deepest point of the pit. Varying around eight in number, they run outwards not quite to the rim, and may branch to form secondary folds. According to Maaden they can always be seen if the lighting is correct, although the walls of the pit may otherwise appear to be smooth.

The sensory pit is usually of a glass-like transparency, but it is often very finely granulated and has the appearance of ground glass.

At the base of the rhopalium there is an area of columnar flagellated ectoderm cells which also line the subumbrellar sensory pits. The rhopalar canal has at its base two short curved lateral branches which surround these subumbrellar sensory pits.

Ephyra. Friedemann (1902) found that the structure of the marginal sense organ in the ephyra was very similar to that of the adult; the crystals in the statocyst were greenish yellow. Horridge (1956*b*) gave the dimensions of the rhopalium in the ephyra as *ca.* 0·2 mm. high and 0·1 mm. broad at its base.

Manubrium

The oral arms are slightly shorter than the radius of the umbrella and have truncated ends (Text-figs. 74, 80).

The axis of each oral arm has fairly thick rigid mesogloea, along the subumbrellar side of which runs a median groove. This groove is flanked on either side by the wide thin lips of the oral arm. Along the margins of these lips the mesogloea is prolonged into very numerous short finger-like or tentacular processes about 2 mm. long (Text-fig. 84). The surfaces of these oral tentacles are richly endowed with nematocysts which are closely crowded together near the tentacle tips.

In mature females brood pouches are developed in the oral lips in which the embryos develop (see p. 150). In full-grown specimens the basal portion of each oral arm has an S-shaped curve (see p. 158).

Gastrovascular system

The gastrovascular system was first described by Ehrenberg (1837) who used the uptake of indigo to outline the radial canal system.

Stomach. The central stomach cavity is of very small capacity. On its upper side a cone-shaped mass of mesogloea projects downwards into the lower side which at the bases of the oral arms is closely pressed against almost the whole surface area of the cone. The four interradial gastric pouches (Text-fig. 79), in which the gonads develop, form the bulk of the gastric cavity. In each of these a serried row of gastric filaments lies just central to each horseshoe-shaped gonad.

The gastric pouches are connected to the central stomach cavity by trough-like areas between the apposed lower and upper surfaces of the stomach wall. These are called the gastro-genital grooves. Similar gastro-oral grooves run from the gastric pouches opposite each end of a gonad direct to the mouth opening. The gastro-genital and gastro-oral grooves have been called 'canals' by some authors. They are, however, not closed canals with cylindrical walls like radial canals, but furrows in the subumbrellar wall of the stomach whose upper open sides are occluded by the close apposition of the exumbrellar wall of the stomach.

The radial canal system consists of eight unbranched adradial canals running from each of the two opposite corners of the gastric pouches to the circular ring-canal at the umbrella margin; and of four perradial and four interradial canals each of which is increasingly branched towards the periphery but whose central canal is unbranched and runs direct to a rhopalium. The branches

10-2

arise by centripetal growth from the ring-canal, and there may be occasional anastomoses between branches, which may be fairly well marked or almost absent (see e.g. Browne, 1905*b*).

The root branches of the perradial canal systems connect directly into the mouth opening; those of the interradial canal system enter the gastric pouches midway between the two adradial canals.

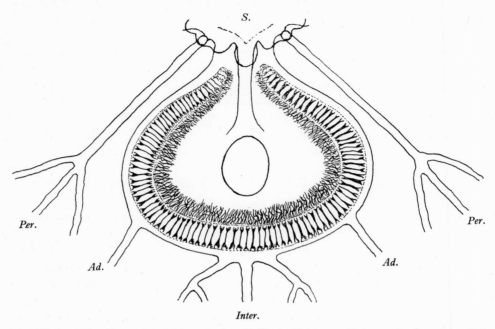

Text-fig. 79. *Aurelia aurita*. Male gonad and gastric filaments. *S.* stomach; *Per.* perradial canals; *Ad.* adradial; *Inter.* interradial.

As the medusa increases in size the gastric pouches enlarge and their peripheral extension eats into the bases of the interradial canals so that their basal trunks may disappear and the basal branches then issue direct from the gastric cavity. This characteristic is, however, very variable. For instance, Kramp (1942) recorded a specimen from Greenland waters 110 mm. in diameter in which in one quadrant there was still only the single basal interradial trunk issuing from a gastric pouch; while in a smaller specimen, 65 mm. in diameter, the basal branches were all completely cut off from the main trunk. In specimens 135 mm. and 145 mm. in diameter there were three to five interradial canal branches issuing from each gastric pouch between the adradial canals (see also Browne, 1905*b*; Bigelow, 1938).

The straight adradial canals may appear more brightly coloured than the other canals and gonads. The ring-canal, which runs on an uninterrupted course round the margin of the umbrella spans the rhopalar clefts by horseshoe-shaped loops.

Ehrenberg (1837) recorded the presence of excretory pores at the distal ends of the radial canals. The presence of such pores was mentioned by Widmark (1913), but there has never been a detailed description of them. It seems possible that Ehrenberg mistook food masses in the marginal food pouches for excreta (see p. 158).

Gonads

Claus (1883) described in detail the development of the gonad in *Discomedusa*. He said that the process was quite similar in *Aurelia* in which the gonads can already be recognized as thickened epithelial strips in medusae 12–15 mm. in diameter. During the first appearance of the genital strip cell proliferation is not continuous, but numerous small protrusions of cells grow centripetally. In specimens 18 to 20 mm. in diameter this discontinuity is evened out, and the margin of the ribbon-like strip, which is directed towards the centre, now appears smooth. Sections showed that the genital rudiments were endodermal cell proliferations growing inwards into the mesogloea. Later a longitudinal split appears in the cell mass, resulting in the separation of the sheet of genital cells from the parietal layer adjacent to the subumbrellar mesogloea. This cleavage space is the beginning of the genital sinus. In its final development the gonad becomes separated along its whole length from the subumbrella by the genital sinus except for a number of short connecting trabeculae.

Details of the structure of the gonad were also given by Widersten (1965). The genital epithelium consists of a single layer of tall columnar cells, among which the gonocyte-forming cells are distributed. The epithelium on the subumbrellar side of the genital sinus is composed of the usual flattened ciliated cells. The mesogloea of the gastro-genital membrane is thin and very fibrous. As the oocyte increases in size these fibres become compacted to form a dense surrounding envelope. The oocyte is supplied with nurse cells, as was already described by Claus.

It is characteristic of *Aurelia* that the oocytes arise along the whole length of the genital ribbon, whereas in *Pelagia* they first appear at the origin of the gonad and migrate outwards as they develop.

In its final form the gonad appears as a crescentic, horseshoe-shaped, or almost circular folded ribbon contained entirely in the gastric cavity (Text-fig. 79) and not protruding below the subumbrellar surface, as in *Pelagia*, *Chrysaora* and *Cyanea*.

The eggs have been figured by a number of authors, e.g. Wagner (1835, pl. III, figs. 4, 5), Siebold (1839), Claus (1877), Haeckel (1881*b*) and Widersten (1965). Their cytology was examined by Tsukaguchi (1914).

Hyde (1894) gives the diameter of the ripe egg as 120–150 μ and its colour as yellowish. Widersten (1965) states that the diameter does not exceed 160 μ and that its colour is pink. Thus there is variation in the colour of the egg from place to place and as it develops. For instance, Siebold (1839) gave the colour of the egg as pale violet, changing to dark yellow in the morula stage, and Lambert (1936*a*) gives the colour of the egg as mauve. It is of course the colour of the egg which gives the gonad its coloration, which may thus be violet, pinkish, reddish or yellow.

The egg is almost opaque and surrounded by a fine membrane (Hyde, 1894). Okada (1927), however, says that it is very transparent. Yolk bodies of different sizes are distributed throughout the cell plasma.

Siebold (1839) was the first to discover, in 1836, that *Aurelia* had separate sexes. He described the male gonad and figured the sperm. He described the colour of the male gonad as dirty yellow, brown-yellow, or rose. The male gonad develops similarly to that of the female and the spermatozoa are developed in follicles. Aders (1903) briefly described spermatogenesis (see p. 161).

Claus (1883) stated that in the Adriatic the genital products can already be ripe when specimens are 35–40 mm. in diameter, but that normally they are not mature until twice that size. Siebold found ripe spermatozoa in specimens 45 to 75 mm. in diameter; and on two occasions in specimens 25 mm. in diameter. M. E. Thiel (1959a) found that in Travemunde Bay females had brood sacs developed and that they matured when 60 to 65 mm. in diameter, while Hj. Thiel (1962) recorded specimens in the Öresund carrying planulae at a size of 40 mm. diameter. Siebold stated that the sexes could first be distinguished when the medusa was between 30 and 54 mm. in diameter.

Fertilization of the eggs takes place in the ovary or in the genital sinus where first cleavage takes place. The spermatozoa emerge into the genital sinus which functions as a seminal vesicle (Widersten, 1965). They are carried to the genital sinus of the ovary by inflowing feeding currents (Widmark, 1911). Southward (1955) saw two males spawn. Most of the spermatozoa passed through the gastro-oral arm grooves to the basal grooves of the oral arms. They then travelled to the distal end of the oral arms and were ejected in mucous masses. Goodey (1908, 1909) thought that the genital products passed to the grooves of the oral arms along what he described as gonadial grooves; but Widmark (1913) stated that they passed along the gastro-oral arm grooves (*eckcanalen*), and this was confirmed by Southward (1955) who observed spawning in two males and two females. Southward also found sperm masses in the marginal food pouches of a female which had been kept in a tank with a male.

Wetochin (1930) found that after maturation and fertilization the ciliary currents reversed from the stomach to the exterior and along the oral arms to the lip tentacles. He assumed that they ceased to feed and confirmed this by using blood cells on four females. He twice saw eggs passing through the gastro-genital canals and supported Goodey's ideas. He assumed that reversal of ciliary current was stimulated by hormonal secretion.

The segmenting ova after reaching the oral arms continue their development in special brood pouches. These were first described by Gaede (1816) and Ehrenberg (1837), and descriptions were also given by Siebold (1839), Agassiz (1862), and Claus (1883). Siebold remarked that the formation of the brood pouches proceeded while the egg was still in the ovary, and that the pouches were most strongly developed at the bases of the oral arms. Later Minchin (1889) redescribed the pouches.

The brood pouches (Text-fig. 80) first appear as small depressions in the walls of the oral arm lips with fairly thick mesogloea between the endoderm and ectoderm. Transverse folds form in the lips a short distance from the margin the centres of which deepen to form round or oblong pouches projecting above the surface of the lip with narrow entrances between the closely applied folds. As the pouches develop in depth the mesogloea becomes greatly attenuated and the openings of the pouches become narrow. The pouches may be tightly packed with embryos in all stages of development up to fully formed planulae. Minchin saw embryos with as few as four or eight segments. He stated that in addition to the embryos in the pouches many are also free in the oral arm grooves and lodged in folds of the margins of the arms. Presumably these are in transit to the pouches.

Hj. Thiel (1962) states that the colour of planulae en masse in the western Baltic is yellow or pale brown, and that Segerstråle had told him that on the Finnish coast they were reddish in colour.

From a study of the occurrence of brood pouches in medusae of different sizes M. E. Thiel (1959a) found that from a diameter of 65 mm. up to the largest size females could be found with brood sacs. On the assumption that males and females should be present in equal numbers he concluded from the occurrence of specimens of different sizes without brood pouches that an individual female might spawn twice. Between the two spawnings there was thus a rest period during which the brood pouches would disappear.

Nematocysts

There are two types of nematocysts in the adult medusa, atrichous haplonemes and microbasic heterotrichous euryteles. Weill (1934) gave measurements on preserved material as follows: atriches, 8 μ long; euryteles, 9–14 μ long.

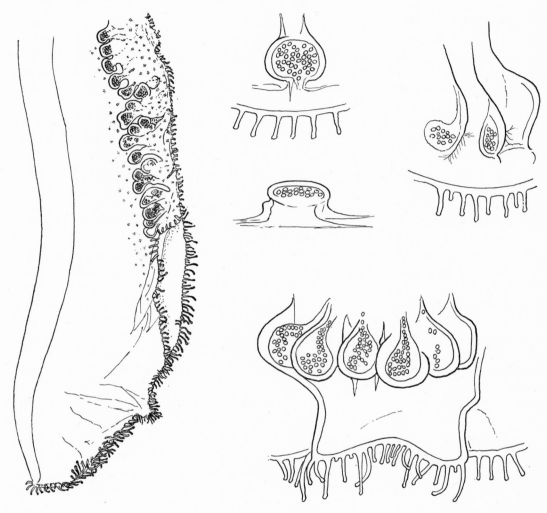

Text-fig. 80. *Aurelia aurita*. Left, 150 mm. in diameter, Loch Melfort, June 1967: oral arm showing distribution of brood pouches; right, various views of brood pouches.

Both types of nematocyst are present in the marginal tentacles and in the oral arm tentacles. In the latter there are also some slightly longer and more rounded atriches 9–11 μ long.

Weill also examined the nematocysts in one of the varieties of scyphistoma reared by Lambert at Leigh-on-Sea, Essex, and sent to him alive. In these scyphistoma there were atriches about 8 μ long and euryteles 10–12 μ long, the two types being abundant and distributed through all parts of the scyphistoma.

ULMARIDAE

In a young specimen of the medusa, in which there were about ten marginal tentacles and about five tentacles on each of the oral arms, he found the same types of nematocyst distributed throughout except that there were no atriches in the gastric filaments.

Weill drew attention to the absence of holotriches in *Aurelia* and its similarity in that respect to *Rhizostoma*.

DEVELOPMENT

Scyphistoma

Aurelia has been the object of many studies on development from egg to ephyra. A review of much of this has already been given in the general account of the scyphistoma on pages 15 to 22. It will be convenient, however, to list here those references which concern *Aurelia*: Siebold (1839) was probably the first to rear *Aurelia* from the planula to the scyphistoma with eight tentacles, and M. Sars succeeded in rearing it in 1841. Subsequent authors who studied the embryology and the development of the scyphistoma were Claus (1877, 1883, 1891), Goette (1887), Smith (1891), Hyde (1894), Hein (1900), Friedemann (1902), C. W. & G. T. Hargitt (1910), Percival (1923), Okada (1927), Ussing (1927), Hagmeier (1933), Lambert (1936*a*, *b*) and Hollowday (1951).

The formation of podocysts, which had been recorded by earlier authors, was described in detail by D. M. Chapman (1966, 1968). Cells originating from a narrow transitional zone of epithelium between the scyphopharynx-filament complex and the endoderm migrate singly or in small clumps to the aboral attachment region where they become encapsulated. Further unpublished work by D. M. Chapman has shown that this type of formation is probably an extreme in a spectrum which has at its other end the formation of the podocyst by the disk ectoderm only. Hérouard (1912) described the intermediate condition, in *Chrysaora*, in which both ectoderm and mesogloeal cells contribute to podocyst formation.

Verwey (1942) summarized much published information on the life history and strobilation and its relation to temperature and other factors. He considered that strobilation occurs between 4° and 8° C. Werner (see M. E. Thiel, 1959*a*, p. 101, footnote) stated that *Aurelia* starts to strobilate at Sylt in November to December when a critical low temperature limit of 7°–6° C. is passed. He could produce strobilation in midsummer by cooling to 5° C. Lytle (1961) kept scyphistomas in a cold room at 5° C. for six months with only occasional feeding; soon after bringing them back into the laboratory, at 18·5° C., they strobilated.

In this connection Naylor (1965) recorded the occurrence of *Aurelia* in a heated dock at Swansea only after the winter temperature at the intake had been reduced from 5° or more above that of the outside sea to only 1° or 2° above.

Custance (1964, 1966, 1967) and Kakinuma (1965) studied the effect of light and temperature on strobilation. Paspalev (1938) produced normal strobilation and ephyrae by treatment with potassium iodide solution (see p. 186), and Spangenberg (1967) used potassium iodide and thyroxin to induce strobilation. Gilchrist (1937), probably using *Aurelia*, studied methods of budding and regeneration, and Spangenberg (1965*a*, *b*) reared *Aurelia* at Little Rock, Arkansas, to study regeneration. Steinberg (1963) studied regeneration of the polyp from fragments of ectoderm.

M. E. Thiel (1959*b*, p. 997) listed the recorded substrata on which scyphistomas have been found which, apart from inanimate material, includes *Fucus*, *Zostera*, *Mytilus*, *Pecten*, *Ciona* and *Ascidia*. They have not been found deeper than 20 m. The most detailed and comprehensive observations made on the scyphistoma in the field are those of Hj. Thiel (1962) (see above, p. 17).

Cleavage of the fertilized egg is total and regular. Berrill (1949b) drew attention to the fact that gastrulation has been recorded as by ingression of cells and by invagination and that this might be correlated with the size of the egg. Hollowday (1951) gave the average length of the planula as 260 μ, but he found some 400 μ and one even 580 μ long. Widersten (1968) says that the planulae are yellowish brown and stick-like with dimensions 220 × 85–375 × 95–125 μ.

The scyphistoma may vary very much in size being usually between 2 and 7 mm. high, though Hagmeier (1933) records one 14 mm. in height. It may have as many as twenty-eight tentacles. The colour is also very variable from white to yellowish and pink, but this can also be affected by the food eaten.

The strobila is polydisk, though small ones may be monodisk, and between 20 and 30 ephyrae may be produced from a single scyphistoma.

In his quantitative study of plankton at Kiel, Lohmann (1908) found 40,000 planulae to one ephyra in a given volume of water.

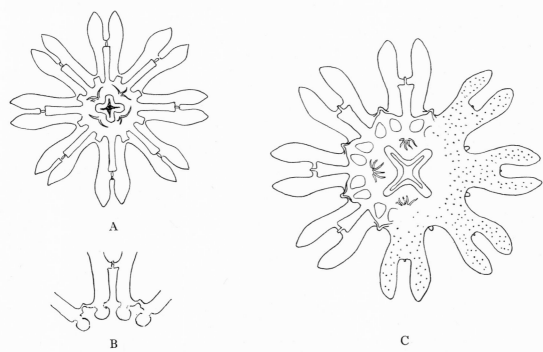

Text-fig. 81. *Aurelia aurita*. Ephyra. A, 3·6 mm. in diameter, Millport, 27 March 1967; B, 4 mm. in diameter, Plymouth, 25 February 1966, to show beginning of ring-canal; C, *ca.* 6·0 mm. in diameter, Plymouth, showing subumbrella on the left and exumbrella on the right.

Medusa

A detailed description of the ephyra (Text-fig. 81) and its development to the adult medusa was first given by Claus (1877) for *Aurelia* from the Adriatic. Except possibly in respect of rate of development and size at different stages this does not differ from the normal development in British waters.

When newly liberated from the scyphistoma the disk is 1·5 mm. in diameter. The central stomach area is almost circular in outline, and from it run the rhopalar canals, one along each marginal lobe. Between the bases of the rhopalar canals the stomach has very slight adradial bulges. There is only one gastric filament in each interradius. The coronal and radial muscles are already visible.

ULMARIDAE

The first signs of the marginal ring-canal appear as lateral bulges at the bases of the rhopalar canals (Text-fig. 81 B) and at the same time the adradial canals have increased slightly in length and have dilated ends. Just distal to the termination of the adradial canals the rudiments of the marginal lappet-like structures are appearing. At this stage there are three gastric filaments in each group and the ephyra is 2–2·5 mm. in diameter. At a diameter of 4·0 mm. the lateral bulges at the bases of the rhopalar canals and the dilations at the ends of the adradial canals have grown outwards so that they meet and join together; there is thus now a continuous ring-canal round the disk margin (Text-fig. 81 C). At this stage there are 6–8 gastric filaments in each group, and the first rudiments of the eight adradial marginal tentacles are appearing.

In the process of the formation of a marginal tentacle a convex ectodermal fold first develops, and at the end of an adradial canal an ectodermal pit-like depression appears on the exumbrellar side of the lappet which finally meets the end of the adradial canal.

At a diameter of 5 to 6 mm. canals grow centripetally from the ring-canal, one on each side of an adradial canal. These join the straight rhopalar canals, and this is the first appearance of the branching systems of the perradial and interradial canals. Rudiments of marginal tentacles now appear on either side of each developing adradial tentacle. The numbers of gastric filaments have increased, and the mouth lips are formed.

Claus said that at a diameter of 8 mm. there were five to six marginal tentacles on each side of a primary marginal tentacle making ten to twelve in a group. Two new centripetal canals have joined each primary branch. At a diameter of 10 mm. he regarded the medusa as no longer an ephyra; there were fourteen to sixteen marginal tentacles in each group.

At a diameter of 12 to 14 mm. the young medusae had twenty to twenty-four marginal tentacles in each group; the oral arms had papillae, or developing oral tentacles, near their bases; the rhopalar lappets were spread sideways.

At 16 mm. diameter centripetal canals of the third order were developing, and marginal tentacles had developed on the exumbrellar surfaces of the spreading bases of the rhopalar lappets.

In specimens 10 to 12 mm. in diameter the thickening of the endoderm to form the gonadial fold was already appearing.

There is variation in the size at which ephyrae are liberated according to the size of the scyphistoma, and rate of development is affected by temperature, quantity of food available, etc. The above data given by Claus must therefore only be regarded as showing the general order of development. Delap (1907), for instance, recorded that at Valencia a specimen 9 mm. in diameter had three marginal tentacles in each group, and one 12 mm. in diameter had five in each group. This is considerably less than found by Claus whose comparable figures were of the order of ten and twenty marginal tentacles in each group respectively.

These differences may be due to different ways of measuring the ephyra. It is not quite clear whether Delap's measurements are from tip to tip of marginal lappets. Claus mentions disk measurement, and his data are much more in keeping with those of Kramp (1942) on Danish medusae (Text-fig. 85) who measured the disk of the ephyra between lappets. Kramp gave the following numbers of marginal tentacles in each octant as development proceeds.

Diameter (mm.)	Number of tentacles per octant
3–7	1–13
8–10	15–25
12–14	ca. 31
22	ca. 40

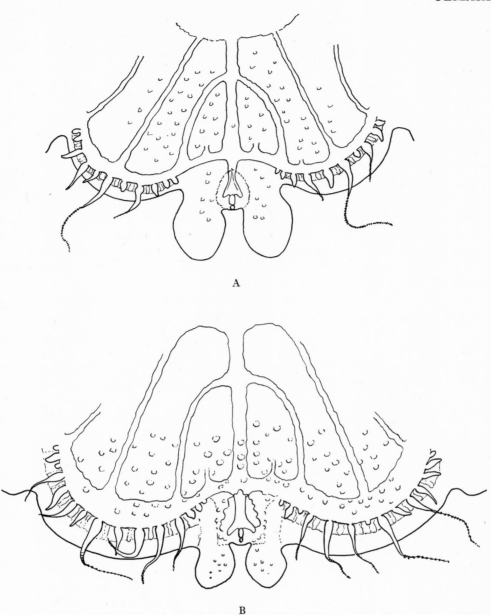

Text-fig. 82. *Aurelia aurita*. A, *ca.* 10 mm. Millport, 22 March 1967;
B, *ca.* 15 mm. Millport, 18 April 1967.

Kramp states that the first four centripetal canals begin to develop from each octant of the ring-canal when the diameter of the ephyra disk is 4 or 5 mm. He says that these join up with the primary pair of branches when the diameter is about 12 mm. or sometimes much later. The rate of development of succeeding radial canal branches is very variable and Kramp states that in large medusae there are very few blindly ending centripetal canals. There is, however, considerable variation in the amount of anastomosis between branches.

155

Drawings of specimens 10 mm. and 15 mm. in diameter from Millport are given in Text-fig. 82 The tentacles on the oral arms were already developing in a specimen 10 mm. in diameter (Text-fig. 83).

Rate of growth

There is little direct evidence of rate of growth in the sea. Information of sizes recorded by different workers has been summarized by Verwey (1942, p. 395).

D. C. M'Intosh (1910), in a comparison of his results in a study of the variations in *Aurelia* with those of Browne, pointed out that it was possible to conclude from Browne's measurements that the local race of *Aurelia* in the Tamar estuary grew from an average diameter of 18·8 mm. to a population with an average diameter of about 56 mm. in about five weeks. Later M. E. Thiel (1959a) analysed Browne's data in greater detail and added to them some observations of his own made on medusae from Travemunde Bay. He ranged the data in 5 mm. groups and his curves and table can be summarized as follows:

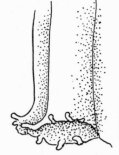

Text-fig. 83. *Aurelia aurita*. Oral arm of specimen 10 mm. in diameter, Millport, 22 March 1967

	Average diameter (mm.)	Extreme range of size (mm.)	Majority (mm.)
April	20	10–40	10–25
May/June	55	20–95	45–65
July	105	55–265	85–125

It is known that medusae can grow very quickly and this must be largely conditioned by food supply, but it is probable that *Aurelia* can become full grown within three months. In captivity

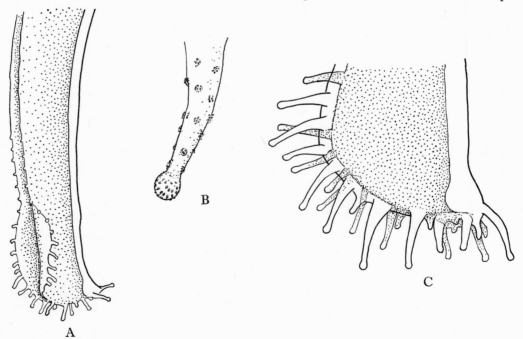

Text-fig. 84. *Aurelia aurita*. A, oral arm in specimen *ca.* 30 mm., Loch Riddon, 19 April 1967; B, oral arm tentacle; C, enlargement of tip of oral arm.

in an aquarium Delap (1907) had a specimen liberated on 24 March which had reached a diameter of 85 mm. by 4 July. Davidson & Huntsman (1926) found in Passamaquoddy Bay that the average size increased from 50 mm. at the beginning of July to 185 mm. by 20 August.

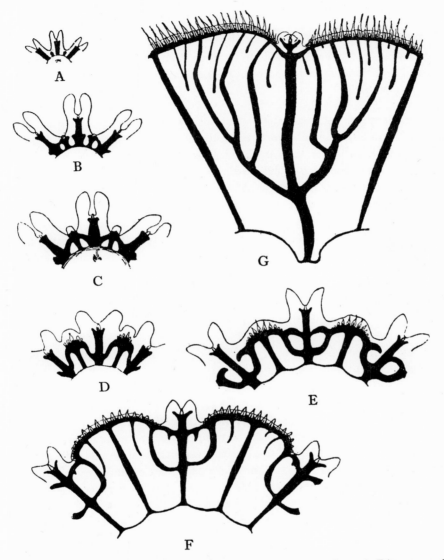

Text-fig. 85. *Aurelia aurita*. Stages of development (from Kramp, 1942, fig. 77). Diameters of specimens: A, 2·5 mm.; B, 4·0 mm.; C, 5·0 mm.; D, 6·0 mm.; E, 8·0 mm.; F, 12 mm.; G, 22 mm.

Reference has already been made (p. 150) to the opinion of M. E. Thiel that an individual may have two spawnings in its life, and there is evidence of occurrence of medusae in deep water in winter (p. 169). It is thus uncertain what the final growth of an individual or its length of life may be. Kramp (1939) records that Saemundsson observed a specimen 1100 mm. in diameter washed ashore in Iceland; this could have been a survivor from the previous year.

ULMARIDAE

Wetochin (1930) stated that at sexual maturity the mesogloea of the umbrella becomes markedly thicker in both sexes and especially the females. It is also noticeable that the oral lips carrying many brood pouches become thickened at the bases of the oral arms and thrown into short S-bends (Text-fig. 143; Pl. XIV).

ALIMENTARY SYSTEM

As *Aurelia* is a ciliary feeder the external system which collects the food and brings it to the mouth has here been included in the account of the alimentary system.

Adult

The adult *Aurelia* appears to feed usually, but not always (see p. 162), on plankton collected by ciliary currents. Orton (1922), while working on the oyster beds of the River Blackwater, Essex, kept an *Aurelia* 8 cm. in diameter 'as a pet'. He gave it oyster larvae to see what it would do with them and found that the larvae were quickly formed up into lines embedded in mucus on the surface of the exumbrella. He was thus the first to show the external transport of food, and the presence of the external 'food pouches' at the distal ends of the adradial canals. These had been figured by Ehrenberg (1837) but he thought that they were excretory masses (see p. 148). Gemmill (1920) had previously described ciliary feeding currents in the ephyra. Widmark (1911; 1913) was the first to describe in detail the internal ciliary currents in the stomach and gastrovascular canal system by placing the medusa in a vessel of sea water with Indian ink added to it. Ehrenberg (1837) had shown the gastrovascular system clearly outlined by using indigo. Wetochin (1926a, b; 1930) in an investigation of the radial canal system used fish and mammal blood, which had the advantage over ink and carmine in that it did not coagulate and was acceptable to the medusa as food.

At Orton's suggestion Southward (1949, 1955) made a further combined study of both external and internal ciliary current systems.

The cilia (which are really flagella 9–18 μ in length when preserved) on the exumbrellar and subumbrellar surfaces set up centrifugal currents carrying particles and food organisms trapped in mucus towards the umbrella margin. Here the cilia on the exumbrellar side of the marginal lappet-like structures continue the process, carrying material to their margins where cilia on the subumbrellar side transfer it round the bases of the marginal tentacles to a marginal groove running round the subumbrella between the marginal tentacles and the velarium. Material on the subumbrellar surface is carried outwards on the velarium whence it is transported into the marginal groove. Unwanted material is carried down the marginal tentacles from their bases to their tips when the tentacles are relaxed and there discarded (as observed also by Wetochin (1930) on isolated tentacles). When the tentacles are contracted and holding prey, the food organism is removed by the currents of the marginal lappet-like structures or of the velarium to the marginal groove.

At the position of the distal end of each of the eight adradial canals the marginal groove is slightly dilated to form a food pouch in which the mucus and its trapped food collect. These eight adradial food masses are picked up from the food pouches by the ends of the oral arm lips, which can also pick up food from the marginal tentacles (Wetochin, 1930).

The food so gathered is passed along the inner walls of the lips of the oral arms to the gastric pouches via the upper parts of the gastro-genital grooves (Goodey, 1908, 1909) and along the roof of the gastric pouches to the gastric filaments (Text-fig. 86). Here the food masses are broken

158

down and probably partly digested. The resulting material is then carried along the roofs of the gastro-circular grooves surrounding the gonads into the adradial canals.

The currents now run centrifugally along the roof and side walls of each of the eight adradial canals to the marginal ring-canal. Here the currents divide left and right and enter the terminal branches of the perradial and interradial canal systems, through which they return to the centre. The currents from the perradial canals flow directly into the grooves running along the roofs of the mouth arm cavity and so to the exterior. The currents from the interradial canal systems enter

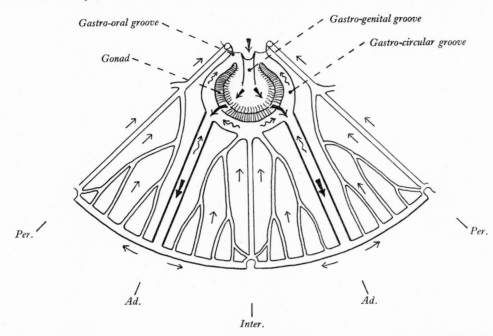

Text-fig. 86. *Aurelia aurita*. Diagram showing circulation in gastrovascular canal system (after Widmark & Southward). Simplified by omission of most branches of perradial and interradial canals. The wavy arrows indicate passage along bottoms of canals.

the gastric pouches along the floors of the gastro-circular grooves and thence via the gastro-oral grooves to the roofs of the oral arm grooves where they join the currents from the perradial canals.

While, as mentioned above, a centrifugal current flows along the roof and sides of the adradial canal, Southward found that there is a return current running centripetally along the floor of the canal. This may carry heavy particles or indigestible material back to the stomach to be excreted. In this connection he noticed that the adradial canals, in which there are thus two opposing currents, were higher in cross-section than the perradial and interradial canals in which he could only observe a current running in one direction. Wetochin (1926a) had also noticed a return flow in the adradial canals on the introduction of carmine or ink particles which stimulated the production of large quantities of mucus.

To summarize, the excretory currents flow along the floors of the adradial canals, the floors of the gastro-circular grooves, and via the gastro-oral grooves to the oral arm grooves.

Widmark (1913) measured the speed of circulation of fluid through the canals. Fluid is carried quickly to the marginal ring-canal along the adradial canals, but passes much more slowly back through the branches of the perradial and interradial canals to the centre. The speed in the perradial and interradial canals increases towards the centre as the canal branches decrease in number. Widmark recorded the following speeds:

Medusa diameter (cm.)	Gastric pouch to ring-canal (min.)	Ring-canal to gastric pouch (min.)	Total (min.)
12	4	15	19
14	2·5	18	20·5

Wetochin (1930) recorded that the speed of passage of blood cells along the oral arms was 2–6 mm. per min. and in places 10 mm. or more per min.

Earlier (1926a), Wetochin had experimented on *Aurelia* narcotized with CO_2 and injected them with fish or mammalian blood. He introduced dolphin's blood diluted with sea water (1:1) through the genital grooves into each gastric pouch of a specimen 12 cm. in diameter, and took photographs at successive intervals after the time of injection. After 12 min. blood had filled much of the ring canal and entered the peripheral branches; after 37 min. three or four sectors had all their branches filled with blood; and after 67 min. the whole canal system was filled.

Wetochin also estimated the total cross-sectional area of the canals at two radial distances from the umbrella centre by measuring the diameter of each of the canals along the circumference at each of the two distances. He found that the total cross-sectional area of the perradial and interradial branches decreased as the number of branches increased towards the umbrella margin. The sum total of the cross-sectional areas of the canal branches in any one sector was also less than that of the two adradial canals on either side of that sector.

He pointed out that in the narrow peripheral branches the cilia of the canal project almost to its axis; and thus that the work done by these cilia is more efficient than that done by the cilia in the larger root branches of the canal in which there is a dead area in the centre of the lumen. Indeed in the large canals he observed that the blood corpuscles were in turbulent motion.

As regards the effects of pulsation of the umbrella on current flow Wetochin stated that contraction might stop the current or cause temporary back flow, but that this did not affect the overall direction of flow.

Wetochin (1930) stated that in the examination of twenty medusae the blood cells passed in 74% of the observations into the gastric cavity lying to the left of the base of the oral arm when the medusa was lying with the subumbrellar surface upwards. He also observed that the direction of the current was reversed in the gastro-genital canals and on the oral arms at sexual maturity (see p. 150).

Ephyra

The ephyra is very voracious and can capture its prey by means of the stinging cells on the marginal lappets. Co-ordinated movements of the manubrium then transport the prey to the stomach. This is probably the ephyra's normal way of feeding as growth is rapid and must require large quantities of solid food.

There is, however, a ciliary mechanism capable of transporting small particles which was described by Gemmill (1921). On the exumbrellar surface of the disk, lobes, and lappets, and on the subumbrellar surface of the lobes the ciliary currents are directed centrifugally towards the periphery. On the subumbrellar surface of the disk they are directed centrifugally on a marginal zone about a quarter of the radius in width, but they flow centripetally over the rest of the disk surface. On the external surface of the manubrium they flow towards the mouth opening.

The internal currents flow centrifugally outwards over the roofs of the gastric cavity and of each of the radiating canals, but centripetally inwards on the floors of the gastric cavity and of each of the radiating canals. In the manubrium the currents are weak and ill-defined, but they are weakly exhalant in the trough of the oral arm grooves. Particles are passed from the bases of the gastric filaments to their tips. Percival (1923) found that the cilia lining the oral arm grooves could pass food in either direction. He said that the current passes along the grooves and sides in the same direction but he once observed them going in opposite directions.

Gemmill noted the capture of infusoria by this ciliary current system.

Scyphistoma

Percival (1923) demonstrated the passage of carmine particles by ciliary action up the sides of the polyp and over the oral disk to the mouth.

DIGESTION

The early experiments of Krukenberg on digestion in *Aurelia*, and also *Chrysaora* and *Cyanea*, are summarized by M. E. Thiel (1959*b*, p. 876). Krukenberg concluded that medusae have no powers of digestion, but that organisms found in their gastric cavities are autolysed by their own digestive juices and thus become available for assimilation by the medusa.

Lebour (1922) recorded the following observations on speed of digestion in small *Aurelia*. A small *Ammodytes* larva eaten at 09.30 hr. was not quite digested by 19.00 hr.; another eaten at 09.30 hr. had disappeared by 16.30 hr. A *Cottus* larva eaten at 10.00 hr. had disappeared next morning.

Wetochin (1930), using fishes' blood (usually *Myoxocephalus scorpius*), found that amoebocytes from the endoderm of the gastrovascular canals and from the neighbouring mesogloea penetrate into the lumen of the canal. There they take in as many as ten or twenty erythrocytes and increase visibly to five or six times their usual size. They then become smaller and migrate back into the endoderm and mesogloea where they can be seen to be larger in size than the amoebocytes already present, but they gradually decrease to normal size. Wetochin concluded that they had given up nourishment to the surrounding tissue, and that they themselves must be capable of some digestion. In healthy animals all erythrocytes are digested in 24 hr., but in less healthy animals they may still be seen after 48 hr.

Erythrocytes may also stick to the endodermal epithelium and are taken and ingested by pseudopodia-like processes. Wetochin stated that amoebocytes were capable of changing into flagellated cells. Aders (1903) described amoebocytes which carried nourishment to the male sexual elements.

Henschel (1935) examined the reactions of the oral arms to food. He found that they did not react positively to the carbohydrates, soluble starch, glycogen, saccharose and dextrose. Strong positive reactions were given to the proteins, white of egg, albumin and peptone, and the amino acids, glycine, asparagine, leucine, and tyrosine, and the breakdown product skatole. He suggested that the reason why, unlike other lower animals, the medusae did not react to carbohydrates might possibly be because short-lived medusae do not need to store glycogen.

ULMARIDAE

EFFECTS OF STARVATION

De Beer & Huxley (1924) studied the effects of starvation on *Aurelia*. Without food the medusa decreases in size, and the decrease is at first at the expense of the mesogloea. Thus, after 12 to 17 days, when the umbrella had become much smaller, the oral arms were still almost of their original size. After 23 days the marginal tentacles were reduced to about one-third of their original diameter, their mesogloea being by then no longer noticeable. After 38 days the oral arms were almost completely resorbed.

The medusa continues to pulsate until an advanced state of reduction in size.

Will (1927) found that starved scyphistomas and ephyrae were reduced in a few weeks to roundish or ellipsoidal planula-like forms, some free-swimming apparently by cilia and some gliding on the bottom.

Steiner (1935) kept ephyrae unfed until only the manubrium and gastric filaments remained. After eight days the gastric filaments were separated from the manubrium, but both organs still showed reactions to food. Henschel (1935) also recorded that oral arms removed from adults continued to show feeding reactions for some days.

Thill (1937) found that the volume of a medusa was reduced by starvation to 73 % of its original volume in 5 days and to 55 % in 10 days. Oxygen requirements ran parallel with volume size.

FOOD

Adult

As early as 1837 Ehrenberg (1837, pl. III, fig. 3) illustrated a medusa with a small fish in its stomach, and he mentions also finding *Gammarus* in them. The occurrence of food in the gastro-vascular system or ingestion of food by the medusa has since then been noted by many authors, but it is sufficient to refer only to the following authors: Delap (1907), Orton (1922), Lebour (1922, 1923) and Southward (1955). (But see also Fraser, 1969; and p. 263, below).

Orton demonstrated that *Aurelia* undoubtedly fed on small plankton organisms. He found the following in the masses collected in the marginal food pouches: diatoms, algal threads, eggs of invertebrates, young polychaetes, rotifers, cirripede larvae, copepods, gastropod larvae, and sand grains. On an occasion when the diatom *Nitzschia* was present as a natural culture in an oyster pit this diatom was taken as food. Although it was known that *Aurelia* also ate larger organisms Orton thought that plankton organisms probably formed its main diet.

Delap (1907) found that the young *Aurelia* preferred to feed on medusae; but if kept for a period without many available eagerly devoured *Clione* and *Limacina*. They also ate small cteno-phores, *Pleurobrachia* and *Bolina*. She said that above all they liked variety, and that two or three days with only one kind of food 'quite upset their digestion'.

Lebour (1922) demonstrated that small *Aurelia* in captivity would eat young fish. These were caught by the marginal tentacles or the oral arm lips themselves. She noted that, when young fish were not available, amphipods, crab zoeae, small copepods and medusae were eaten. Among the small fish she used as food were *Cottus*, *Blennius*, *Gobius*, *Ammodytes* and *Nerophis*. The *Aurelia* were very voracious, but Lebour notes that no fish were found in specimens over 30 mm. in diameter.

Southward (1955) found that a specimen 100 mm. in diameter cleared all the coarse plankton from 700 ml. of water in less than one hour. Thill (1937) fed *Aurelia* medusae on young of the fish *Lebistes reticulatus* 10 to 12 mm. in length and on enchytraeid worms. He found that the average daily requirement over a period of 39 days of one medusa was six fish or sixteen enchytraeid worms.

Aurelia can easily pick up small copepods, such as *Tigriopus*, on its external surfaces. Southward found that at 15° C. such food organisms could reach the marginal food pouches within 10 min., the oral arms in 12 min., and the gastric filaments in 50 min. Material picked up direct by the oral arms reached the gastric pouches more quickly, sometimes within 10 min.

Southward also examined the mucous masses in the marginal food pouches of freshly caught specimens. These contained plankton typical of the locality of capture. For instance, off Port Erin, Isle of Man, in early June the food was principally copepods; in the River Mersey, Liverpool, the food pouches contained large amounts of detritus, with diatoms, ciliates, flagellates and eggs of crustaceans; those from a pool in the Mersey estuary had *Noctiluca* and balanid cyprids.

He found that food particles above 5 mm. in diameter were rejected by the oral arms, as presumably the narrowness of the gastro-genital grooves limits the size of food entering the stomach.

The gastrovascular tissues are, however, no doubt capable of stretching, and earlier authors have recorded larger organisms in the stomach. Hüsing (1956) drew attention to the fact that in some years *Aurelia* fed on nereids, gammarids, *Corophium*, spiders and insects.

Ephyra

As already mentioned, Gemmill (1921) recorded ciliary feeding of the ephyra of *Aurelia*, and this was also seen by Southward (1955). But the ephyrae are so voracious that their normal food must consist of animals which they capture by stinging and then seizing with the manubrium. Anyone who has kept ephyrae in the laboratory will know how they will pack their stomachs with such animals as *Artemia* larvae.

Delap (1907) stated that newly liberated ephyrae in her aquaria immediately started to eat very young *Obelia* and other medusae such as *Phialidium*, and sometimes small copepods. She mentioned one ephyra which had 25 *Obelia* stowed away like rows of plates in the stomach cavity and along the canals. She remarked that they did not seem to thrive on anything except small medusae, but would eat copepods if nothing else was available.

Lebour (1922) found that they would eat young fish and crab zoeae. Southward (1955) saw *Balanus balanoides* larvae eaten, and recorded young medusae up to 10 mm. in diameter feeding on newly hatched plaice larvae in the Port Erin hatchery.

NEUROMUSCULAR SYSTEM

Classic experiments by Romanes (1876b, 1877b) on the locomotor system of *Aurelia* indicated that there must be nervous control, and Schäfer (1878) demonstrated the existence of a sub-umbrellar network of nerve fibres and their associated cells. Simultaneously, Eimer (1878) published the results of his well-known investigations on locomotion and the nervous system. It was Romanes and Eimer who first brought to notice the function of the marginal sense organs in the regulation of the contraction of the umbrella. Morse (1910) found that *Aurelia* regained pulsation some time after removal of all marginal sense organs.

Wetochin (1926b), in a study of the pacemaker function of the marginal sense organs, recorded that if all marginal sense organs but one are removed the medusa rotates three to four times a minute in the direction of the remaining organ. After about 30 hr., swimming movements gradually return to normal due to regeneration. Marginal tentacles regenerate first, and rotatory motion ceases before the rhopalia are regenerated, the excitatory contractile zone presumably regenerating first. In such specimens, if all regenerated sense organs and the original organ are removed pulsation remains normal.

Horstmann (1934a) observed that medusae over 50 mm. in diameter ceased pulsation after removal of all marginal sense organs, but that those smaller than 50 mm. still pulsated, the smaller the animal the less being the change from normal frequency.

Bullock (1943) studied facilitation in *Aurelia* at Woods Hole.

Horstmann (1934a, b) found that decreasing the light intensity lowered the rate of pulsation (confirmed by Horridge, 1959) and that increasing the light intensity raised the pulsation rate. He also did experiments on righting reactions with only two opposite marginal sense organs left. Horridge (1959) stated that *Aurelia* did not show compensatory movements and that it maintains its position in the sea by a kinesis.

More recently Horridge (1953, 1954a, b, 1956b, 1959) examined the nervous system in the ephyra and adult *Aurelia* by histological and experimental physiological methods.

In the adult medusa he found that the nerve fibres can be clearly seen in the living animal with phase-contrast microscopy and under oblique lighting with a dissecting microscope. Romanes had shown that a small bridge of tissue between two pieces of *Aurelia* will conduct a stimulus in either direction resulting in a contraction in each piece. Horridge showed for the first time that the excitation was carried by the nerve fibre, for there was no contraction when no nerve fibre could be seen in the bridge. He recorded one action potential from an individual fibre of a motor nerve at each spontaneous beat of the umbrella.

In the network concerned with propagation of the contraction wave, which Horridge called the giant fibre system, the nerve fibres were mostly between $6\,\mu$ and $12\,\mu$ thick, and in living animals the cell bodies appeared as breaks in the fibres; these were bipolar cell bodies $10-15\,\mu$ long. He recorded that the axons may twist around one another, cross, or run alongside one another for as long as $50\,\mu$.

In the ephyra, 3·0 to 5·0 mm. in diameter, the fibres of the giant fibre system were usually less than $1\,\mu$ thick. The giant fibre system overlies the radial and coronal muscles. The diffuse network spreads over the whole epithelium and consists of bipolar cells with short processes reaching the surface of the epithelium, which are probably sensory in function, of bipolar and multipolar cells forming a network over the whole exumbrellar and subumbrellar surfaces, and of bipolar cells with two long axons orientated along the lobes and round the mouth. The two nerve nets meet and interact at the marginal ganglion. There was an indication of double motor innervation.

Pantin & Dias (1952) examined the rhythmic activity and responses to stimulation of *Aurelia* at Rio de Janeiro and found that the behaviour resembled that of this species from north temperate regions.

Horridge (1959) found that *Aurelia* was much less sensitive to tryptamine than *Cyanea*.

Yamashita (1957b) studied the action potentials of the marginal sense organ in the adult *Aurelia*.

Marshall (1888), using gold preparations, confirmed the presence of striated muscle in *Aurelia* reported by earlier authors.

Scyphistoma

D. M. Chapman (1965) studied the microanatomy of the scyphistoma. In addition to the four longitudinal muscles running the length of the body there are (1) a lip sphincter (and weak circular fibres; unpublished observations) in the lining epithelium of the oral disk, (2) ectodermal radial muscle fibres in the oral disk, (3) longitudinal muscle fibres in the tentacles, and (4) a system of ectodermal myofibrils disposed longitudinally on the stolon, stalk and calyx. No endodermal myofibrils were revealed by the electron microscope. An idea of the disposition of nerve elements aligned parallel with the long axis of the muscle fibres was gained by using the electron microscope.

By mechanical and electrical stimulation, and by operative treatment, Chapman showed that the above four regions of the polyp, as well as the longitudinal muscle cords, appeared to be neurologically independent, yet during feeding the parts of the polyp are so situated that this activity is nevertheless co-ordinated mechanically.

CHEMICAL COMPOSITION

Water content

Krukenberg (1880) was the first to show that medusae from the Gulf of Trieste were composed of 95·34% water and 4·66% solids. Möbius (1880) recorded that in Kiel Bay *Aurelia* were 99·82% water, but later (1882) he repeated his experiments obtaining dried weights of 2·06% and 2·1%. These results are in keeping with those of Thill (1937) who found the water content of Baltic *Aurelia* to be 98% where the salinity was 7·3‰. Hyman (1938, 1943) found percentages between 95·9 and 96·6% for *Aurelia* from the Gulf of Maine where the salinity is 31·5 to 32·6‰. Lowndes (1942) weighed a living specimen 100 mm. in diameter at Plymouth and distilled it under xylol. He found 95·56% water, 3·87% residue, leaving 0·57% probably fats dissolved in the xylol. The 3·87% residue was made up of 3·2% salts leaving 0·67% dry protein. Lowndes concluded that about 4% of the medusa was protoplasm.

Bateman (1932), working with *Cyanea* and *Chrysaora* as well as *Aurelia*, using A. V. Hill's apparatus for vapour pressure measurements, concluded that the total water of the mesogloea is osmotically free.

Ionic composition

Macallum (1903, on *Cyanea* and *Aurelia*) and Koizumi & Hosoi (1936, on *Cyanea*) studied the ionic composition of medusae. Both found a high potassium content of 144–256% of that in sea water. Robertson (1949) thought that this might have been due to leakage from cells or inclusions of cells in the analyses. He removed a thin layer of mesogloea including the epithelium and then withdrew clear cell-free fluid from the cavity thus formed in the mesogloea. He said 'All the ions differed from those in the surrounding sea water, sulphate especially being particularly low, but the striking accumulation of potassium shown by previous investigators is absent. It did not exceed 108% of the value in sea water, and all the other ions except sulphate were much closer to equilibrium.'

He gave the percentage concentration in the dialysed fluid as follows: Na, 99%; K, 106%; Ca, 96%; Mg, 97%; Cl, 104%; SO_4, 47%. (See p. 84 above, Denton & Shaw, 1962.)

Composition of Mesogloea

Astbury (1940) by X-ray diffraction analysis stated that 'the structures now recognized as belonging to the collagen family include...jellyfish'.

G. Chapman (1953) found that the mesogloea contained alanine, glutamic acid, glycine, histidine, leucine, lysine, proline and hydroxyproline, serine and valine: a composition resembling that of collagen.

ULMARIDAE

Macallum (1903) remarked on the healing powers of jellyfish in that in damaged specimens the mesogloea acted as a barrier to diffusion and allowed new epithelium to develop over its surface. Metchnikoff (1892, p. 70) recorded the aggregation of amoebocytes in the mesogloea around foreign bodies and their ingestion of carmine particles.

METABOLISM, ETC.

Raymont et al. (1967) investigated the succinic dehydrogenase activity in *Aurelia* and found it to be low compared with that of other animals studied. Wiersbitzky & Scheibe (1966) showed the presence of substances with immunological behaviour similar to that of human blood group substances B, H and P_1.

EFFECTS OF TEMPERATURE

Mayer (1914) suggested that *A. aurita* was originally a boreal species which had extended its distribution into warmer waters to which it has become adapted although living there within 9° C. of its lethal temperature. In the tropics a change of 2° or 3° above their normal temperature leads to slowing of the rate of pulsation. He found that medusae from Halifax, Nova Scotia, ceased to pulsate at an upper limit of 29·4° C. and still moved at $-1·4°$ C. with ice floating in the water; the animals were most active at about 18° to 23° C. At Tortugas the medusae were most active at 29° C., and would cease to pulsate at about 37° C., but could still pulsate at 8° C.

Thill (1937) studied the effects of temperature on rate of pulsation and oxygen requirements. He found that the frequency of pulsation did not alter much between 9° and 19° C.; there was still movement at 1° C. and 29·5° C., but all movement ceased at $-0·5°$ C. and over 30° C. He said that the optimum temperature for Baltic *Aurelia* thus lay between 9° and 19° C. He found the oxygen consumption to be 34% at 4° C., 100% at 16° C. and 145% at 24° C.

EFFECTS OF SALINITY

Thill (1937) studied the effects of salinity change on *Aurelia* from the Danish Wiek in the Baltic where the salinity was 7·3‰. Lowering the salinity decreased the rate of pulsation and the medusa assumed the shape of an inverted umbrella. At a salinity of 5‰ the medusa lay on the bottom and only pulsated weakly five times a minute compared with the twenty or more strong pulsations in the normal medium.

On increasing the salinity the rate of pulsation increased but the medusa could not swim down to the bottom of the vessel. The body of the umbrella thickened and 1–2 cm. of the margin was upturned. Thill attributed the changes in body form to swelling or shrinkage of the musculature.

He did similar experiments on planulae, and concluded that in brackish water the medusa can withstand gradual change of salinity from 6‰ to 17‰, and that the planulae could develop normally into scyphistomas from 5‰ to 16‰.

He regarded the Baltic medusa as an ecological modification of the North Sea form.

Further information on the limiting salinities in the Baltic is given on page 142.

EFFECTS OF PRESSURE CHANGES

Rice (1964) found that the ephyrae of *Aurelia* swam upwards on increase of pressure and downwards when the pressure was reduced. Digby (1967) also found that this medusa was sensitive to pressure changes.

SOME GENERAL OBSERVATIONS

Swarming

At times *Aurelia* occurs in immense numbers and a few recorded examples will suffice. Möbius (1880) recorded them as so abundant in Kiel Bay that boats could only be brought through them with difficulty and oars pushed down between them remained standing upright. Browne (1890) recorded passing through shoals between Plymouth and Cork. Johnstone (1908, p. 98) said that

'One may sail for miles through a swarm of *Aurelia* so densely packed together that the sea has a uniform reddish colour.' Bigelow (1926) mentions that it is characteristic that they may occur in lanes or windrows often miles in length as where two currents meet.

Effects on commercial fish

While many of the swarms in inshore areas are brought about by wind and tide action there is much evidence that swarms in the open sea are sufficiently large to interfere with commercial fishing. Hela (1951) for instance records clogging of nets in the Baltic, and fishermen's reports of an inverse relationship between the occurrence of *Aurelia* and that of herring and sprat. These difficulties seem often to arise in winter months when it appears that survivors collect near the bottom. I am indebted to J. Mauchline for the following information obtained by him while working in Aberdeen. He says: 'on the fishing grounds around the Faeroes, shoals of *Aurelia* appear in late summer and become so numerous that the fishermen are forced to move to other grounds'. Information obtained from fishermen indicates that in late autumn the *Aurelia* sink down from the surface so that they get caught in the trawl. 'These jellyfish can be present in immense quantities and some skippers state that almost immediately the trawl goes down it becomes filled; when hauled to the surface the cod-end is usually burst open due to the weight of jellies. Not only is the cod-end filled but a good part of the bag as well, and the square is often also burst open.' The fishermen themselves think that the fish move away from the shoals of *Aurelia*, but poor catches may well be due to the clogging of the trawl.

Stinging

All the evidence is that the stinging powers of *Aurelia* are not noticeable. Ehrenberg (1837) could get no sensation from touching the medusa with his tongue. Weill (1934, p. 243), however, cites a curious use of *Aurelia*: 'at the thermal establishment of Sandifjord, in Norway, neuralgia and rheumatic pains are treated by the application of the medusa *Aurelia aurita* (L.). It is taken by the convex face of the umbrella which is inoffensive and the inferior face is touched on to the spots on which it is desired to produce a revulsion. This unexpected treatment should produce excellent results in certain cases.'

Fluorescence

Ries & Ries (1924) noted that in the otherwise glass-clear *Aurelia* there is frequently a light blue-violet shimmer. They found that if the jellyfish was illuminated with sunlight which had passed through an ultraviolet filter the fluorescence disappeared. They suggested that the coloration in some medusae was a protection against ultraviolet light.

ULMARIDAE

Influence on productivity

Davidson & Huntsman (1926) noted the heavy mortality of *Aurelia* after spawning in late summer in Passamaquoddy Bay. They suggested that the disintegration products might influence the autumnal diatom increase. Pieces of *Aurelia* were added to sea water and the resulting numbers of diatoms were noted as follows:

Aurelia pieces (gm.)	o	1·25	2·5	5
Nitszchia numbers	1	91	116	155

Commensals and parasites

Young fish are occasionally found in association with *Aurelia*, first noted by Lawless (1877) and Romanes (1877c). Hjort & Dahl (1900, p. 118) mention one- and two-year-old whiting (*Gadus merlangus*) swarming beneath jellyfish, and the occurrence of large numbers of *Aurelia* caught with many whiting. It seems doubtful, however, whether there is association of young fish with *Aurelia* to anywhere near the extent that there is with *Cyanea*.

The association of hyperiid amphipods with *Aurelia* is frequently recorded. Romanes (1877a) thought that diminution in size of *Aurelia* towards the end of August might be due to *Hyperia galba* 'which appeared to devour with avidity all the coloured parts of the hosts', but later he altered his opinion on observing a similar decrease in size when *Hyperia* were much fewer. He noted (1876) that the *Hyperia* lodged chiefly in the ovaries and canals. Lambert (1936a) noted that after the liberation of the planulae the subumbrella becomes thickly covered with a sticky glair; in such medusae it was common to find the gastric pouches filled with *Hyperia* feeding on the tissues. Tattersall (1907) says that *Aurelia* is a common host for *H. galba* on the east coast of Ireland.

Bowman *et al.* (1963) record the occurrence of *H. galba* with *Aurelia aurita* in Narragansett Bay.

Moribund condition

Walcott (1898) quotes Louis Agassiz (1862, 4, 62) who says that after the spawning period *Aurelia* is often to be seen floating in the sea. Its disk has become thin and almost leathery and is more elastic, though at the same time more brittle than before. The tentacles are for the most part gone, as well as the marginal sense organs, and decomposition of the margin extends to include the ring-canal and marginal anastomoses. The lips of the oral arms break away and with them parts of the oral arms themselves, especially towards their extremities which become blunt.

Walcott himself found that stranded medusae were firm and hard, and that they would not break or tear when tossed on to the wharf. The shrunken oral arms were tough and were hard to pull apart.

In this condition they might fossilize, and Walcott made plaster casts of *Aurelia* and reproduced photographs of them. Nathorst (1881) had earlier made sand and plaster impressions and casts for his study of fossil medusae.

Effects of weather

Claus (1877) stated that *Aurelia* goes deep when it begins to rain.

SEASONAL OCCURRENCE

Around British shores the ephyrae of *Aurelia* may be expected in the plankton any time from January to early April, after which specimens are either young or adults. The adults may occur from the end of April to the end of August and they generally start to disappear, or have already disappeared, in September. They are nearly always not to be seen in October.

The first to give accurate data on the occurrence of planulae, ephyrae, and adult medusae in relation to temperature was Lohmann (1908); his data were given in diagrammatic form by Thill (1937, p. 92, fig. 18).

Most observations on the seasonal occurrence of the species in European waters have been reviewed in detail by Verwey (1942) and his conclusions have been summarized in tabular form by M. E. Thiel (1959 *b*, p. 1015). In general ephyrae tend to make their first appearance somewhat later in the more northern or colder waters.

Although along the shores *Aurelia* usually disappear in September it seems probable that the medusae may linger on in deeper water as late as December. There are many records of their occurrence in fishermen's nets in the winter. For instance, Evans & Ashworth (1909) give a reference to records of jellyfish in fishermen's trawls as late as 18 January 1893; Hela (1951) says that *Aurelia* is taken in quantities in fishing nets in the Baltic in deeper water in November and December; as early as 1837 Ehrenberg had mentioned such fishermen's remarks, and Möbius (1880) records them frozen in the ice in Kiel Bay in December. J. Mauchline tells me that in Loch Etive there were many small specimens present in November 1968 which lived on right through the winter.

Aurelia is also very sporadic in its appearance. For instance, Browne (1896) said that none was seen at Valencia during April and May 1895, but that they appeared in mid-June. Again, Browne (1895 *b*) recorded that specimens 50 to 125 mm. in diameter suddenly appeared in Port Erin Bay on 2 June 1893, though he had seen none in May.

This sudden appearance, so often remarked upon, may be due to wind effects. Hj. Thiel (1962) found that, while *Aurelia* have normally disappeared by the end of August in the western Baltic, in 1958 numbers appeared towards the end of November. He attributed this to strong south-west winds. Maaden (1942 *a*) in his detailed observations on the occurrence of medusae on Dutch beaches found that *Aurelia* were most frequent in northerly winds, but *Cyanea* in easterly winds. There is no doubt that in some areas medusae which are in offshore deeper water may be brought inshore in an undercurrent by offshore winds.

Not only are *Aurelia* sporadic in their appearance during any one year, but they also show considerable year-to-year variations in abundance. Vallentin (1900) recorded that he could find no *Aurelia* at Falmouth in 1898 and only one battered specimen in 1899, the last year of abundance being 1895. At Plymouth in 1965 and 1966 it was not possible to obtain specimens for research purposes, and enquiries showed that they were also scarce at Torquay, and even at Millport and St Andrews, though there were plenty at Whitstable.

M. E. Thiel (1962 *a*) quotes an observation on the abundance of ephyrae in Kiel Bay when his son Hjalmar Thiel caught over 2,500 in half an hour.

ABNORMALITIES

The abnormalities shown by *Aurelia* have been investigated or recorded by many workers from the earliest times. A basic list of references is the following:

Adult: Ehrenberg (1837); Romanes (1876 *a*, 1877 *a*); Bateson (1892); Unthank (1894); Sorby (1894); Browne (1894, 1895 *b*, 1901); Ballowitz (1899); C. W. Hargitt (1905); D. C. M'Intosh (1910, 1911); Stiasny (1930); Lambert (1936 *a*).

Ephyra: Browne (1895 *b*, 1901); C. W. Hargitt (1905); Low (1921); Hj. Thiel (1963 *a*, *b*).

ULMARIDAE

A historical summary and detailed survey of abnormalities was given by M. E. Thiel (1959*b*), while earlier summaries were given by Bateson (1894) and A. Agassiz & Woodworth (1896).

The most common form of abnormality is meristic variation above or below the normal characters of four oral arms, four gastric pouches, eight marginal sense organs, and sixteen radial canals. Meristic variation is much more common in the peripheral characters than in the oral arms and gastric pouches.

The extreme range recorded for variation in number of marginal sense organs is 4 to 18; that for oral arms and gastric pouches is usually 2 to 7, but Browne recorded one adult with 8 and one with 9 gonads out of 5,000 specimens. It is usual for the numbers of oral arms to correspond with that of the gastric pouches; and when there is variation in numbers of these organs there is usually parallel variation in number of peripheral organs.

The percentage occurrence of meristic variations in natural populations is fairly high; and comparisons have been made between their occurrence in ephyrae and early stages of the medusa and the adult to see if they have selective value. There appears to be no evidence that within the above ranges of variation these abnormalities prevent normal successful growth. There is, in fact, no reason why increased number of organs should not be beneficial provided no other accompanying characters lead to decreased viability. Decrease in numbers of organs which play a part in feeding might, however, be harmful and there is possibly slight evidence that this may be so.

Observations on large numbers of specimens have been made by Browne, Bateson, D. C. M'Intosh, C. W. Hargitt, and Hj. Thiel.

Their results may be summarized as follows:

Percentage variation in number of marginal sense organs

	Locality	Number examined		Percentage
Browne	Tamar estuary (1893)	Ephyra	359	22·6
	Tamar estuary (1894)	Ephyra	1,156	20·9
		Adult	383	22·8
	Plymouth (1898)	Ephyra and small	2,000	20·2
		Adult	1,000	22·9
Hargitt	Woods Hole	Ephyra⎱ Adult⎰	2,500	24·9
M'Intosh	Clyde	Under 20 mm.	1,000	16·0
		Adult	281	15·3

Percentage variation in number of mouth lips and gonads

	Locality	Number examined		Percentage
Browne	Plymouth	Ephyra⎱ Adult⎰	3,000	2·4
Bateson	Northumberland	Adult	1,763	1·47
Hargitt	Woods Hole	Ephyra⎱ Adult⎰	2,500	2·76
M'Intosh	Clyde	Under 20 mm.	1,000	1·8

Low (1921) studied the characters of individual ephyrae as they were liberated from 27 scyphistomas. In all he had 278 ephyrae of which 25·18% were abnormal. Of these 70 abnormal specimens 38 had less than eight marginal lobes and 32 had more than eight. Thus he had a much higher proportion of ephyrae with fewer than eight lobes than other investigators had found among specimens collected from the sea. Browne, for instance, out of 3,000 specimens found 4·94% with fewer than eight marginal sense organs, and 16·16% with more than eight; and Hargitt's corresponding figures were 1·96% and 22·97% respectively. It is possible, therefore, that among specimens with a low number of meristic characters there is a higher natural mortality in the earlier stages than among those with higher numbers of such characters.

Another possibility is shown by the above summarized results. It is noticeable that the percentage variation shows differences from place to place, being lowest in Bateson's results and highest in Hargitt's. The sequence probably runs parallel with environmental conditions which would be most constant off Northumberland and least so up estuaries. In this connection Hargitt recorded a slightly higher abnormality in the polluted 'eel pond' than elsewhere, though he did not regard his results as conclusive.

Meristic variation may be symmetrical, or it may involve only half of a medusa or only one or more segments.

In addition to meristic variation many other abnormalities have been recorded. These include twinning of the marginal lobes and of the rhopalia themselves, absence of marginal sense organs, incomplete number of gastric filament groups, and so on. Duncker (1894) described a specimen with an inverted umbrella; and Ballowitz (1899) found one with a balloon-shaped body.

More recently Hj. Thiel (1963 a, b) has described unusual abnormalities in the ephyra which he called part-ephyra (Teil-ephyra) and spiral ephyra.

The part-ephyra was an individual which might have only one to seven, or even eight, marginal lobes. In one-lobed specimens there was no manubrium; in those with three to seven lobes a manubrium was present but situated on the open margin of the umbrella. Thiel found that such abnormal specimens could re-form themselves into a normal shape. The spiral ephyrae might have as many as forty-nine to seventy-six marginal lobes. These ephyrae were produced by spiral strobilation of the scyphistoma. Such spiral ephyrae, if they had only a few more lobes than normal, could re-form themselves into normal ephyrae with a higher than normal number of lobes.

Thiel found that part-ephyrae and spiral ephyrae could re-form themselves into individuals having four to twenty marginal lobes. He thought that spiral ephyrae having more than twenty lobes were unable to regulate their transformation into the normal ephyra form. In this connection it is to be noted that the highest recorded number of lobes by those studying variation is eighteen.

Vanucci (1959) described what she called a spiral ephyra, but Thiel considered that this consisted of two ephyrae which had become spiral by a regulation process.

HISTORICAL

Aurelia aurita was first named *Medusa aurita* by Linnaeus (1746) and repeated in the tenth edition of his *Systema* in 1758. It was later described and figured under the same name by O. F. Müller (1780) and Gaede (1816), and in considerable detail by Ehrenberg (1837). It was described as *Medusa* sp. by Borlase (1758), and under the name *Medusa cruciata* by Baster (1762).

Péron & Lesueur (1809) created their genus *Aurellia* and placed in it, among many other species, their species *Aurellia rosea*, which they said was the same as the *Medusa aurita* of O. F. Müller. Their genus *Aurellia* has been suppressed (see p. 139) and *Aurellia rosea* has become a synonym of *Aurelia aurita* (L.) (see Rees, 1957; Hemming, 1958).

I have not included in my list of synonyms all the names introduced by Péron & Lesueur and other authors. They have been listed by L. Agassiz (1862, **3**, 159) and by Haeckel (1880, p. 552).

ORDER

RHIZOSTOMEAE

Medusae in the order Rhizostomeae are characterized by the absence of marginal tentacles and by the structure of the manubrium which has its lips branched to form eight oral arms and on which there are numerous mouth openings.

The systematic division of the order has been a subject of much discussion in the past, but the major division at present adopted is that of Stiasny (1921 *b*, 1923), which was modified as regards its subdivisions by Uchida (1926). This is based primarily on the arrangement of the gastro-vascular system. There are two sub-orders, Kolpophorae and Dactyliophorae, distinguished as follows.

Kolpophorae (ὁ κολπός = sinus) having a large primary wheel-like sinus consisting of a network of anastomosing canals with many direct connections with the stomach.

Dactyliophorae (ὁ δακτύλιος = finger ring) having a small primary ephyral sinus consisting of a peripheral network of anastomosing canals sometimes separated from the stomach by a secondary main ring-canal and only connected with the stomach by radial canals.

The only British rhizostomid is included in the sub-order Dactyliophorae. This sub-order is further subdivided into the Inscapulatae and Scapulatae, that is those medusae without 'scapulettes' or 'epaulettes' on the manubrium and those with.

The Scapulatae has two families, the Rhizostomatidae and the Stomolophidae, distinguished mainly by the structure of the manubrium.

The British representative is a member of the former family and belongs to the genus *Rhizostoma* Cuvier.

FAMILY RHIZOSTOMATIDAE

Scapulate rhizostomid medusae with manubrium with single terminal club on each oral arm.

Genus **Rhizostoma** Cuvier

Rhizostomatidae with small epaulettes on base of manubrium; each oral arm with single club-shaped terminal appendage; usually with clearly defined main ring-canal from which arise intracircular coarse-meshed arcade networks of canals.

Three species of *Rhizostoma* are maintained by some authors, *R. luteum* (Quoy & Gaimard), *R. pulmo* (Macri), and *R. octopus* (L.).

R. pulmo occurs in the Mediterranean and adjacent seas, *R. octopus* in north-west Europe, and *R. luteum* off the coasts of Portugal, the Straits of Gibraltar, and the west coast of Africa.

R. luteum is regarded by most authors as a valid species. The terminal appendages or clubs have very long thin stalks and there is an egg-shaped or bean-shaped protuberance from the subumbrellar wall in each subgenital pit.

In both *R. octopus* and *R. pulmo* the terminal clubs have no thin basal stalk and the subgenital papillae are in the form of thickened valves on the marginal edges of the subgenital pits.

In fact the only valid character by which *R. octopus* and *R. pulmo* might be separated at present is the number of marginal velar lappets. In *R. octopus* these average ten in number in an octant, while in *R. pulmo* they are always stated to be eight per octant. The number of these velar lappets is subject to variation and has been examined in large numbers of specimens of *R. octopus* by M. E. Thiel (1965). No such examination has been made on *R. pulmo* in the Mediterranean. Prof. E. Ghirardelli kindly examined a few specimens in the Adriatic; these all had eight velar lappets per octant.

Mayer (1910) and Kramp (1961) regard *R. octopus* as a variety of *R. pulmo*, but it is a curious fact that as far as is known the distribution of these two forms is discontinuous, *R. luteum* coming between them. Until detailed observations have been made on *R. pulmo* it seems wiser to retain *R. octopus* as a separate species. Should later research, however, show that they must be regarded as the same species then the name *pulmo* has priority.

Because the two species are so nearly related I have included in the following account all information on the general biology of and experimental research on *Rhizostoma*, stating where the observations were made.

Rhizostoma octopus (L.)

Plate XV; Text-figs. 87–100

'*Urtica marina...octopedalis*' Borlase, 1758, *Nat. Hist. Cornwall*, p. 258, pl. XXV, figs. 15–17.
Medusa octopus Linnaeus, 1788, *Syst. Nat.* ed. 13, p. 3157.
Cassiopea borlasea Péron & Lesueur, 1809, *Annls Mus. Hist. nat. Paris*, **14**, 357.
Rhizostoma cuvieri Péron & Lesueur, 1809, *Annls Mus. Hist. nat. Paris*, **14**, 362.
   Eschscholtz, 1829, *Syst. Acaleph.* p. 45 (in part).
   Gosse, 1856, 'Tenby', p. 37, pl. I.
Cassiopea rhizostomoidea anglica Tilesius, 1831, *Nov. Act. acad. Leopold*, **15** (2), 273, pl. LXXI.
Rhizostoma pulmo, Forbes, 1848, *Monogr. Brit. Med.* p. 77.
Rhizostoma cuvierii, L. Agassiz, 1862, *Contr. Nat. Hist. U.S.* **3**, 150.
Pilema octopus, Haeckel, 1880, *System, der Medusen*, p. 593.
Rhizostoma octopus, Vanhöffen, 1906, *Nord. Plankt.* **6**, XI, 63, fig. 33.
   Stiasny, 1921*b*, *Capita Zool.* **1** (2), 160, pl. II, fig. 14; pl. IV, fig. 33.
   Stiasny, 1928, *Zool. Meded.* **11**, 177, figs. 1–10.
   M. E. Thiel, 1965, *Abh. Verh. naturw. Ver. Hamburg*, N.F. **9** (1964), 37, pl. I–II.
Rhizostoma pulmo var. *octopus*, Mayer, 1910, *Medusae of the World*, **3**, 703.
   Kramp, 1937, *Danm. Fauna*, **43**, 192, figs. 83–5.
   Kramp, 1961, Synopsis, *J. mar. biol. Ass. U.K.* **40**, 378.

SPECIFIC CHARACTERS

Usually an average of ten velar marginal lappets in each octant. Terminal club without thin basal stalk.

DESCRIPTION OF ADULT

Umbrella dome-shaped; mesogloea very solid, thick in central region, thinner beyond periphery of stomach; exumbrellar surface covered with small nematocyst warts; usually about eighty hemispherical marginal velar lappets, and sixteen smaller pointed rhopalar lappets situated one on either side of each marginal sense organ. No marginal tentacles. Four perradial and four interradial rhopalia consisting of statocyst and sensory bulb without ocellus, protected by small hood-like projection of exumbrellar margin and by lateral rhopalar lappets; exumbrellar sensory pit immediately above rhopalium. Four interradial irregularly lobed and much-folded gonads, not projecting below subumbrellar surface. Stomach four-sided, with many gastric filaments arranged in rows centripetal to gonads; sixteen radial canals, four perradial, four interradial and eight adradial, running to umbrella margin; peripheral halves of radial canals connected by fine-meshed network of anastomosing canals situated centrifugally to well-defined, or sometimes ill-defined, ring-canal; blindly ending coarse-meshed arcade network of anastomosing canals arising centripetally from ring-canal in each octant. Manubrium massive, consisting of short pillar-like basal stem followed by sixteen three-winged epaulettes with frilled mouth openings, then dividing into four pairs of oral arms, eight in all, each oral arm consisting of three-winged portion with frilled mouth openings followed by smooth three-winged elongated terminal appendage or club without mouth openings; central mouth opening rudimentary or absent; four canals running from perradial corners of floor of stomach through basal stem of manubrium and forking into each of eight oral arms; oral arm canals with branches to mouth openings along margins of each wing of each epaulette and of each wing of each oral arm, and into its terminal club. Thickened basal portion of manubrium extended over subumbrella to enclose four subgenital pits, each with horizontal slit-like centrifugal subumbrellar orifice partially closed by knob-like protrusion of mesogloea from subumbrellar surface of stomach floor. Colour of bell milky or opalescent blue or green; possibly sexual dimorphism, marginal lappets and gonads bright blue or violet in male, reddish brown in female; mouth fringes similarly bluish violet or pinkish brown, canals in terminal clubs blue or violet. Size up to 900 mm. in diameter.

DISTRIBUTION

Rhizostoma octopus is a southern warmer-water species and in general its distribution is confined to the southern and western shores of the British Isles where it may in some years be extremely abundant. It has been commonly recorded from Essex, through the English Channel and Irish Sea to Belfast, the Solway Firth and Clyde sea area, and no doubt occurs farther up the west coast of Scotland. It appears to occur only rather rarely off the north-east coast of Scotland, although it was more than usually abundant in 1960 (Fraser, 1961). It is rare off south-east Scotland and was recorded in 1913 off North Berwick and in 1902 off the Isle of May (Evans, 1916).

It is indigenous in the southern North Sea where very young medusae have been recorded at times in the German Bight (Stiasny, 1928; Künne, 1952; Kühl, 1964, 1966; M. E. Thiel, 1966a) and off the Belgian and Dutch coasts (see Verwey, 1942, p. 441). M. E. Thiel (1966a) from observations at Cuxhaven thinks that the species may have only rather recently become endemic there.

Its distribution extends normally to the coast of Jutland, but it may be carried farther north in unusual conditions; Kramp (1934) has collated data on its extension into the Skagerak and Kattegat in 1933, and cites references of earlier observations. It has occasionally been recorded as far north as Bergen and Oslofjord (see Rustad, 1952, for records in Norwegian waters), and Naumov (1961) includes the Lofoten Islands in its area of distribution.

STRUCTURAL DETAILS OF ADULT

Umbrella (Text-fig. 87)

The thick central dome-shaped portion of the umbrella and the thinner marginal area have a very solid tough mesogloea; its consistency has been likened to that of cartilage, or more graphically by Gosse (1856) as 'resembling the texture of the skin of a boiled calf's head when cold'. The mesogloea contains abundant cells and fibres.

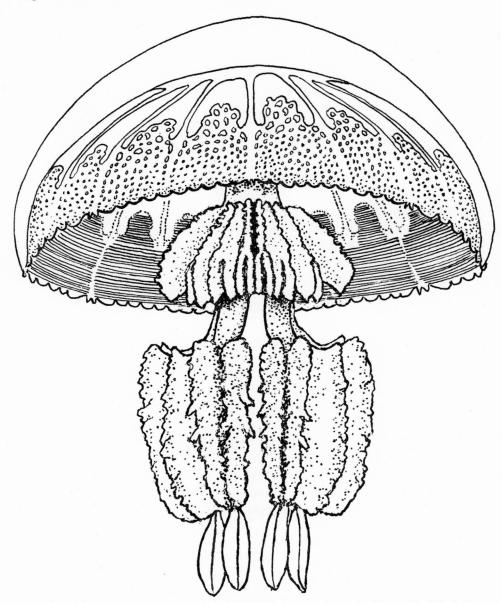

Text-fig. 87. *Rhizostoma octopus*. Slightly diagrammatic drawing of whole animal to show the more important general characters (only two pairs of oral arms are shown).

175

Metchnikoff (1892, p. 70) recorded the aggregation of amoebocytes in the mesogloea around foreign bodies.

The exumbrellar surface is covered with small closely packed nematocyst warts. In preserved specimens when exposed to the air these warts give the umbrella surface a matt appearance.

There are no marginal tentacles, but the umbrella margin is divided into a number of semi-circular marginal lappets. These lappets are known as velar lappets to distinguish them from the sixteen rhopalar lappets which are the original lappets of the ephyra, and are small, pointed and narrow by comparison with the broad, rounded velar lappets.

M. E. Thiel (1965) has made a detailed study of the numbers of the velar* lappets in medusae of different sizes from Cuxhaven.

It is known from the observations of Claus and Stiasny that the formation of the velar lappets is fairly rapid during growth. The number of velar lappets may vary from octant to octant. For instance I have seen small specimens from the Solway Firth which showed the following ranges of numbers of velar lappets per octant:

Diameter (mm.)	Lappets per octant
16	4–6
23	5–6
29	7–9
30	6–7
35	7–9

Thiel found that after the medusa has grown to a diameter of about 50–60 mm. the average number of velar lappets per octant is fairly constant between eight and ten.

He only examined specimens up to 210 mm. in diameter, but within that size limit there was no indication that further increase in number of lappets might be expected. The largest number of velar lappets in any one octant was fourteen. His observations can be summarized in the following table.

Frequency of occurrence of velar lappet numbers per octant in size ranges 80 to 209 mm

Number of velar lappets	5	6	7	8	9	10	11	12	13	14
Frequency	1	2	24	170	257	314	82	17	3	1

Thus as a diagnostic character one might conclude that the average total number of marginal velar lappets in an adult medusa is seventy-two to eighty, or including the sixteen rhopalar lappets that the total number of both kinds of marginal lappets is eighty-eight to ninety-six on average, with a maximum limit of the order of 112.

Observations on specimens larger than those examined by Thiel have been made by C. Edwards (personal communication) in the Clyde sea area. He confirmed the results that Thiel obtained for the small specimens. In the larger specimens there was a tendency to higher numbers, but he pointed out that it was not easy to decide when to count a lappet as fully formed as some always appeared to be in process of dividing into two. His observations were as follows:

Diameter (mm.)	Lappets per octant
330	11–12
350	12–16
450	80 in all
450	*ca.* 100 in all

* In his paper Thiel speaks only of 'marginal' lappets, which might include both velar and rhopalar lappets. He has told me that he only counted the velar lappets, and did not include the rhopalar lappets which of course do not vary in number.

Thus Edwards' results and those of Thiel confirm the range given by Vanhöffen (1906) for specimens up to 600 mm. in diameter, namely ten to twelve velar lappets in each octant which with the rhopalar lappets make 96–112 in all.

Muscle

The subumbrellar coronal muscle (Text-fig. 88) is very well developed, being thrown into a number of large folds which extend to the umbrella margin at the bases of the velar lappets. The muscle is divided into eight separate fields between marginal sense organ radii, and extends centripetally over the region of the arcades as short folds.

There are no radial muscles.

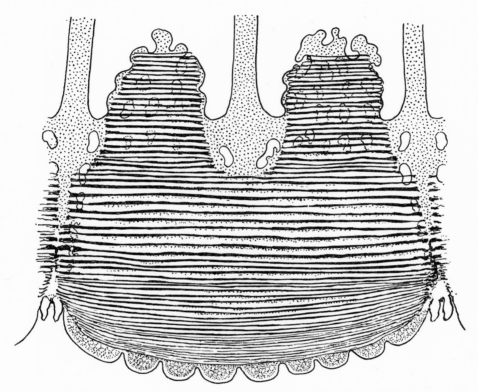

Text-fig. 88. *Rhizostoma octopus*. Coronal muscle field in specimen
ca. 250 mm. in diameter, Millport, 1966.

Marginal sense organs

The marginal sense organs (Text-fig. 89) were described by Eimer (1878) and Hesse (1895) in specimens from the Mediterranean.

The rhopalium consists of a basal stalk and statocyst. It has no ocelli. In section it is somewhat flattened dorsoventrally throughout its length. The basal portion is about twice as wide as it is high, while in the middle portion the upper surface is fairly flat and the lower surface is rounded. The basal portion is knee-shaped so that the rhopalium is directed obliquely downwards at an angle of about 10° with the horizontal when the medusa is swimming in the upright position.

RHIZOSTOMEAE

The dimensions given by Hesse were: overall length, *ca.* 850 μ; length of statocyst, dorsal 340 μ, ventral 440 μ; height of statocyst at base 290 μ, and 340 μ farther forward; greatest width of stalk 410 μ and of statocyst 475 μ.

The exumbrella margin has a cushion-like thickening over the rhopalium in which is a fairly deep exumbrellar sensory pit, situated immediately over the base of the rhopalium. The umbrella margin distal to the sensory pit forms the hood of the rhopalium, which is protected laterally by

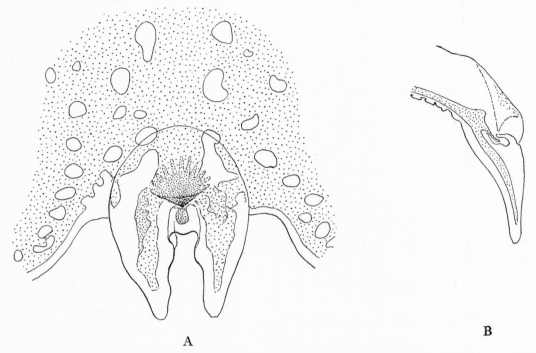

A

B

Text-fig. 89. *Rhizostoma octopus*. Marginal sense organ. A, exumbrellar view; B, diagrammatic radial section to show position of exumbrellar sensory pit.

the edges of the flanking rhopalar lappets. The exumbrellar sensory pit is broadest and deepest on the side nearest to the umbrella margin, and gradually shallows and narrows centripetally to form an elongated cone. It is covered with neuroepithelium which is folded so as to produce furrows in the mesogloea. Some of these furrows run outwards like rays from the deepest part of the pit; others form the boundaries of the pit. As first recorded by Huxley (1849) the pit is ciliated, the direction of ciliary currents being down the sides of the furrows and then centripetal in the bottoms of the furrows.

Gastrovascular system

Stomach. The stomach is four-sided, the four corners being perradial; its sides are slightly concave; its roof projects downwards slightly in the centre and its floor is divided into four inter-radial triangular areas formed by the thickened ingrowth of the mesogloea between the original four oral lips during the course of their development. Each of the four triangular areas thus formed has along its outer margin a slightly convex row of gastric filaments. Between each of these four areas is a perradial slit-like depression which leads into the basal pillars of the manubrium. The triangular mesogloeal thickenings may be completely fused to one another at their apices, so that the original central opening of the stomach is completely occluded; or there may still be some vestiges of the original aperture left, as shown in Text-figure 98 A.

Radial canal system. The radial canal system has been studied in detail by Stiasny (1929) on specimens from the North Sea. Sixteen straight radial canals run from the stomach to the

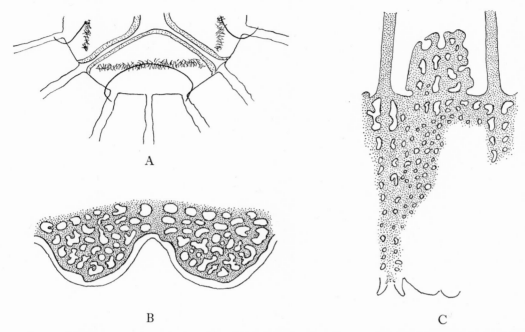

Text-fig. 90. *Rhizostoma octopus.* A, specimen 18 mm. in diameter, to show beginning of formation of subgenital cavity and thickening of interradial areas to form perradial grooves. There are about 100 gastric cirri in each group (Solway Firth, coll. T. G. Skinner); B, 140 mm. diameter, Millport, to show ring-canal in velar lappet; C, general view of canal system.

umbrella margin (Text-fig. 87). Of these four are perradial and four interradial running to their respective marginal sense organs, and eight are adradial. At the umbrella margin the radial canals meet the original primary ring-canal of the ephyral stage which can sometimes be traced as a continuous thin canal round the umbrella margin (Text-fig. 90 B), but which in large specimens may no longer appear continuous.

Each radial canal is joined to its neighbour at a point slightly more than half the length of its course from the umbrella margin to the stomach by a broad comparatively straight canal curving parallel with the umbrella margin. These sixteen canals together constitute a secondary ring-canal, which is usually regarded in the adult medusa as the main ring-canal. Between this ring-canal and the umbrella margin lies the extra-circular fine-meshed network of slender anastomosing canals which connect with the radial canals and with both the marginal and the main ring-canals.

In each of the areas between the main ring-canal and the stomach there is an intra-circular arcade consisting of a coarse-meshed network of anastomosing canals which connect only with the central portions of the ring-canal and not with the radial canals on either side (Text-fig. 90 C).

RHIZOSTOMEAE

There may be much variation in the arrangement of the canal systems within one or more sectors and these were the subject of a special study by Stiasny (1929). They are described later under 'Abnormalities' on p. 197.

Slonimski (1926) described a method of showing up the canal system in *Rhizostoma* from the Mediterranean by the presence of gas after immersion in a weak solution of hydrogen peroxide.

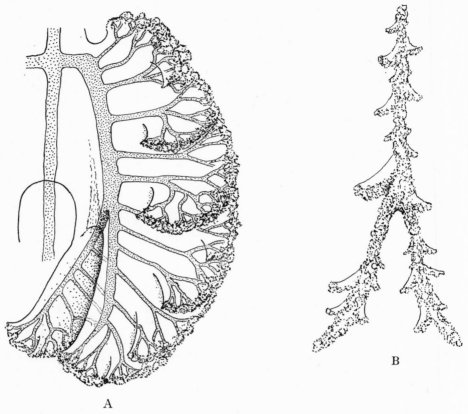

Text-fig. 91. *Rhizostoma octopus*. Epaulette. A, side view of one wing;
B, edge view showing supernumerary lateral folds.

Manubrium

The massive and complicated manubrium is a characteristic feature of *Rhizostoma*.

Its base is very thick and solid and cross-shape in section, the four vertical grooves corresponding to the interradial portions of the original simple four-lipped mouth of the young medusa. This basal portion is quite short and immediately succeeded by the sixteen scapulettes or epaulettes which are grouped in eight pairs.

Each epaulette consists of an upper leaf-like structure which divides at its lower end into two similar structures. Each epaulette can thus be called three-winged. Along the outer edges of each of the three wings run the fringes of folded mouth lips. Each wing itself has additional side folds (Text-fig. 91).

In this area the manubrium is still in the form of a continuous solid pillar and is therefore equivalent to the oral tube, but just below the epaulettes the oral arms begin. There are eight oral arms, arranged in four pairs. Each pair has arisen from a bifurcation of one of the four original oral lips of the young medusa.

Each oral arm consists of an upper portion with folded mouth lips, and a lower portion without mouth openings forming a terminal appendage. The upper portion is triradiate in section except just near its origin where there are mouth openings only along the adaxial edge. In its main portion it thus has three wings, one of which is adaxial the other two being abaxial. As in the epaulette there are frilled mouth openings along the edge of each wing, which also has supplementary folds.

The lower portion, hereafter called the terminal club, which has no mouth openings, is similarly triradiate in section.

As pointed out by M. E. Thiel (1965) it is very difficult to measure the proportions of the different parts of the manubrium. It is difficult to decide on the actual boundaries between the different parts and the length of any one part is dependent on its state of preservation and degree of contraction. The terminal club also is often damaged or broken off. As stated on page 197 there is in addition much abnormality in growth of different parts of the manubrium which may sometimes be very short and broad (Stiasny, 1928).

In general, probably the total length of the manubrium is equal to or greater than the diameter of the umbrella in the European species.

Thiel made measurements of the lengths of the oral arms excluding the terminal club, and of the terminal clubs, of medusae of different sizes from Cuxhaven. The oral arms increase in length as the medusa grows. By the time the medusa has grown to about 80–90 mm. in diameter, measured between opposite marginal sense organs, the length of the oral arm has reached a fairly constant average proportion of 40 % of the diameter of the umbrella in specimens up to 200 mm. in size. The terminal club, however, appears to go on increasing a little in proportionate length, although its curve also flattens out when the medusa is 80–90 mm. in diameter. While the length of the terminal club is in the neighbourhood of 15 % of the umbrella diameter at a size of 80–100 mm., it is about 17 % at a diameter of 140–150 mm.

Stiasny (1928) reviewed the early descriptions of the canal system of the manubrium given by Eysenhardt, Brandt, and Milne Edwards. He regarded these as incorrect and substituted his own interpretation in his figure 8. Specimens that I have examined do not agree with Stiasny's diagram and resemble rather the drawing given by Brandt (1870). I have shown the canal system diagrammatically in Text-figure 92, omitting those canals which run also into the subsidiary folds.

The four canals leading from the perradial slit-like depressions in the floor of the stomach pass into a cavity just above the basal stem of the manubrium (Text-fig. 92 A). Four perradial canals issue from this cavity; on a level with the epaulettes these canals bifurcate, one branch running into each of the eight oral arms. Side branches run to each of the sixteen epaulettes. Each of these side branches runs axially through the length of an epaulette sending off lateral branches to the mouth openings ranged along the margins of each of the three wings of the epaulette (Text-fig. 92 B).

In each of the oral arms (Text-fig. 92 B) the main canal at first runs down axially sending branches to the single adaxial series of mouth openings; but when it reaches the point at which the oral arm becomes triradiate it sends off two main branches to each wing. These branches run parallel with the margins of the wings, receiving small branches from the mouth openings. The main axial canal continues on to the end of the terminal club.

A somewhat similar though less regular arrangement of canals is present in the three-winged terminal clubs, but there are of course no small mouth branches as the clubs have no mouth

181

openings. Stiasny noted that some canals in the terminal clubs apparently opened to the exterior as injected fluid flowed out at these points (see p. 192 on excretion).

The structure of the mouth openings along the wings of the epaulettes and of oral arms is the same throughout. The first correct drawing of the detail of a portion of a lip was made by Huxley (1849, pl. 38, fig. 28) on a medusa which he called *Rhizostoma mosaica*.

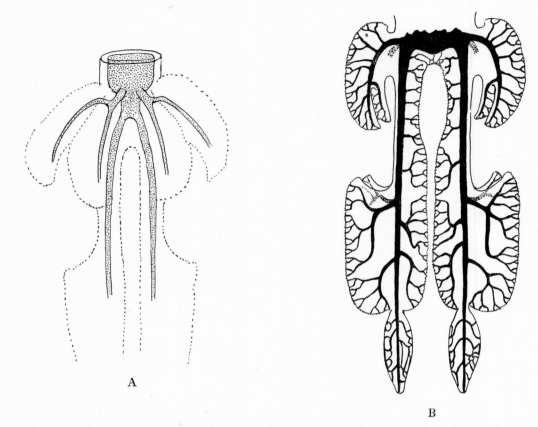

A

B

Text-fig. 92. *Rhizostoma octopus*. Manubrium. A, diagram to show branching of canal roots from central stomach cavity in one perradius; B, oral-arm canal system; one only of each pair of epaulettes shown, and only two wings of terminal club.

The fringe along the margin of a wing of an epaulette or an oral arm consists of a groove flanked on either side by a lip (Text-fig. 93). Branches from the canal system open at irregularly spaced intervals into this groove. Each lip has along its margin many minute tentacular processes, each up to 0·5 mm. in length. As in *Aurelia*, these consist of a core of mesogloea with an ectodermal covering. But unlike *Aurelia*, in which the processes are comparatively large and tapering, in *Rhizostoma* the processes are very small and numerous, have more parallel sides and terminate in a more distinct knob. Hamann (1881) called them *digitellen*.

As the medusa grows the lips become increasingly folded, pleated and convoluted until their description by early authors such as Gosse (1856) as resembling the head of a cauliflower is very apt.

182

Subgenital pits

As in other medusae, at the base of the manubrium the subumbrellar mesogloea is thickened interradially to form the subgenital pits (see p. 189). This thickening extends to the margin of the stomach leaving a long horizontal opening into the subgenital cavity itself. The mesogloea of the floor of the stomach which forms the roof of the subgenital pit is thickened into a knob-like shape.

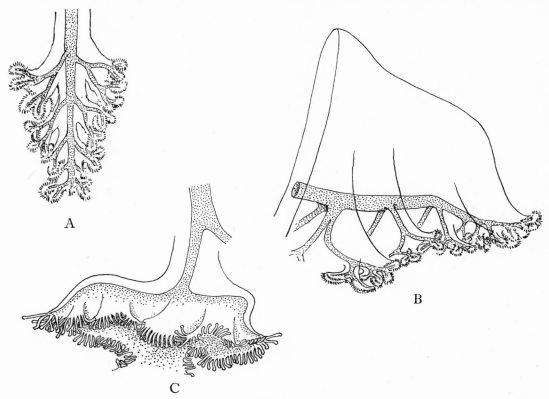

Text-fig. 93. *Rhizostoma octopus*. A and B, details of lateral folds of epaulette; C, enlarged view of oral lip at the end of a canal branch; tentacle 0·25 mm. long.

Gonads

The gonads are somewhat similar in form to those of *Aurelia* but are much more irregular in outline (Text-fig. 94). They are composed of a number of lobes directed towards the umbrella margin along whose edges the folded genital tissue runs. As in *Aurelia* they do not project down into the subumbrellar cavity. They have been studied by Claus (1877, Adriatic), Paspalev (1938, Black Sea) and Widersten (1965, Sweden). The mesogloea of the gonad, like that of the umbrella, contains cells. The genital epithelium consists of a single layer of columnar cells interspersed with gonocyte-forming cells. The genital sinus is not so convoluted as in *Cyanea* and *Aurelia*.

In old animals they have a pattern of transverse grooves along their length (Text-fig. 94B).

There appears to be sexual dimorphism in colour, the male gonads being blue and the female reddish brown (see p. 185).

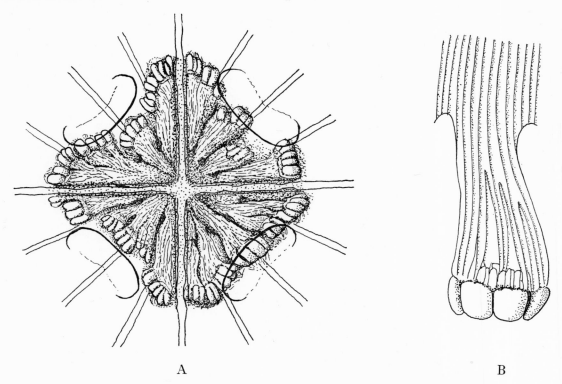

A
B

Text-fig. 94. *Rhizostoma octopus*. Gonad. A, subumbrellar view; manubrium removed; B, detail to show ridges with troughs between them on basal portion.

Nematocysts

The nematocysts were first described by Iwanzoff (1896) and later more accurately by Weill (1934) in specimens from Wimereux.

There are two kinds, atriches and microbasic euryteles.

The atriches are 7–8 μ long; and the euryteles measure between $8 \times 5 \mu$ and $11 \times 6 \mu$.

Both kinds occur on the marginal lappets and in the tentacular processes along the margins of the mouth lips.

COLORATION

Although the mass of the umbrella is unpigmented it has a translucent appearance which has been variously described as pale greenish blue, bluish, greenish, pinkish brown, milk-white to somewhat bluish, generally bluish, opalescent, and rose-red shimmering. The mesogloea has a very great number of cells in it and I wonder whether they may not be the cause of the general opalescent appearance of the mesogloea, which in a medusa like *Cyanea*, which has no cells in the mesogloea, is perfectly transparent.

The marginal lappets are generally deeply pigmented and edged with an intense violet blue colour; the rhopalar lappets may however be less strongly coloured, or even colourless.

The mouth frills of the epaulettes and oral arms may have a bluish, violet, yellowish or reddish tinge.

184

In the terminal clubs the canals may be bright blue, yellowish brown, pink, deep purple-brown, or colourless.

M. E. Thiel (1965) made a detailed study of the colour of the terminal clubs of more than 2,200 specimens from Cuxhaven. He found the colour to be blue, yellow-brownish to brown-red, or brown-violet; a number were colourless. Those which were blue were not much more numerous than those which were brown. The numbers of blue were about equal to the numbers which were not blue including the colourless ones, i.e. 422:427. Thiel suggested that there might be sexual differentiation.

He found that the male gonads were blue and the female brown or red. He had to leave the question open as to whether the colours of the terminal clubs agreed with those of the gonads. He published a plate reproducing coloured photographs of the two sexes.

In the above respect it is interesting to record that Eysenhardt (1821, p. 388) noted that in specimens from Cette the mouth fringes were often violet, but that in old egg-bearing individuals the mouth parts are always brown and 'everything gleams dirty reddish'.

Paspalev (1938), however, could find no sexual colour differentiation in medusae in the Black Sea.

DEVELOPMENT

The scyphistoma has not yet been found in nature, though a determined attempt to find it was made by M. E. Thiel (1966a).

The development from the egg was studied by Paspalev (1938) in *Rhizostoma* from the Bulgarian coast of the Black Sea. He stated that the ripe egg had a fine-grained structure with a clearly formed egg membrane. He gave the size of the egg as *ca*. 0·1 mm., and said that it was yellowish in colour. Widersten (1965) working with Swedish specimens gave the diameter of the egg as up to 140–180 μ and the colour as bluish white. M. E. Thiel (1966a) found at Cuxhaven that the eggs were brown in colour.

Paspalev thought that fertilization occurs either in the gastric cavity or in the surrounding water, but Widersten (1965) found sperms in the female genital sinus, indicating that fertilization might occur either when the egg is still in the mesogloea of the ovary, or in the genital sinus.

According to Paspalev segmentation is total and equal and gastrulation is by invagination. The planula, at first oval, becomes pear-shaped *ca*. 0·5 mm. long. The mouth is formed at the anterior end while the planula is still free-swimming, and at the posterior end the ectoderm is rich in gland cells.

After settlement four tentacles develop first and, with favourable conditions and a good food supply, growth is fairly rapid, the scyphistoma reaching a size of 12 mm. with 32 tentacles. The basal attachment is covered with a yellow perisarc.

The scyphistoma budded and stolonated normally. It produced podocysts not sealed by chitin on their upper sides which developed in one or two days into scyphistomas. In unfavourable conditions more permanent podocysts were formed completely enclosed in chitin.

Small free-swimming hollow buds, 0·2–0·3 mm. in diameter, were also produced, and these developed as far as the 16-tentacle stage with the foot region still rounded, but then attached themselves to the substratum.

Paspalev's scyphistomas, although he succeeded in keeping them for a long time, did not produce ephyrae. He tried combining variations of temperature, salinity, pH, and periods of starvation and of rich feeding, but could produce no strobilation. On the assumption that perhaps

RHIZOSTOMEAE

in nature strobilation was stimulated by the autumn run-off of water from the land and the breakdown of algae rich in iodine he tried treating his scyphistomas with potassium iodide solution. He kept scyphistomas of *Aurelia* for ten minutes in a 2% KI/sea water solution. The scyphistomas were then placed in clean sea water for two hours, after which they were subjected to a similar potassium iodide treatment for another ten minutes. The *Aurelia* scyphistomas strobilated and produced normal ephyrae.

He also tried adding 16–30 drops of 2% potassium iodide solution to his cultures, but by this method only 30–40% strobilated.

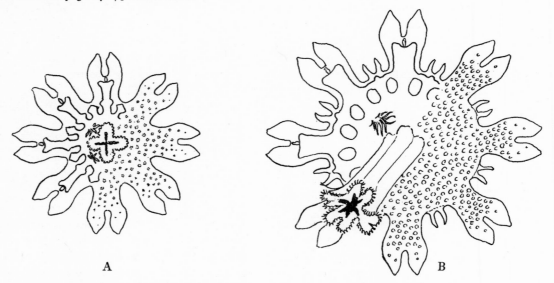

A   B

Text-fig. 95. *Rhizostoma octopus*. Ephyra. A, 3·5 mm. in diameter; B, *ca.* 5·0 mm. in diameter. Subumbrellar surface on left and exumbrellar surface on right to show nematocyst warts. Specimens from Solway Firth collected by T. G. Skinner on 11 July 1966 and 16 July 1965 respectively.

Paspalev then did similar experiments with the scyphistomas of *Rhizostoma*, but although the scyphistomas strobilated they did not produce normal ephyrae. Strobilation was polydisk, 12–18 segments being formed but, instead of normal ephyrae, disk-like free-swimming scyphistomas were produced. These had no ephyral lobes, no marginal sense organs, and no gastric filaments. At first there were no septal divisions in the stomach, but when the free-swimming scyphistoma had developed sixteen tentacles there were signs of endodermal septa. The scyphistoma then elongated, developed a foot, and sank to the bottom.

The newly liberated ephyra of *Rhizostoma* thus remains undescribed. I have, however, seen very young ephyrae from the plankton kindly sent to me by T. G. Skinner who collected them in the Solway Firth. In the smallest specimen, 2·7 mm. in diameter, the lateral outgrowths from the radial canals had not yet joined to form the primary ring-canal. Small tentacles were, however, already appearing along the distal margins of the oral lips (Text-fig. 96A). In another specimen, 3·5 mm. in diameter (Text-fig. 95A), the primary ring-canal was still not formed; there were about five gastric filaments in each group. The oral lip tentacles were forming, and on the exumbrella there were already distinctive prominent nematocyst warts. In Text-figs. 96B and C, drawings are given of oral lips of specimens 4·0 and 4·5 mm. in diameter respectively; it can be seen that each lip is already dividing at its extremity to form the eight oral arms. In a specimen 5·0 mm. in diameter (Text-fig. 95 B) the primary ring-canal was fully formed, the manubrium was quite long and the lips dividing, and there were 11–12 gastric cirri in each group. The first pair of velar lappets in each radius was already formed and the exumbrella was covered with nematocyst warts.

The ephyra and young stages of *Rhizostoma* are thus easily separable from those of other British Scyphomedusae.

Stiasny (1928) described a specimen 2·5–3·0 mm. in diameter from Helgoland (Text-fig. 97A). This was flat and ephyra-like and the primary ring canal was already formed. The perradial and

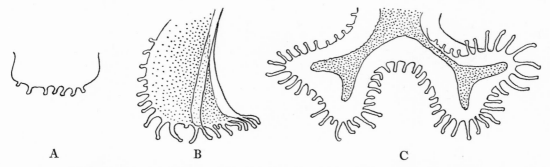

Text-fig. 96. *Rhizostoma octopus*. Oral lips of ephyra. A, 2·7 mm. diameter; B, 4·0 mm. diameter, side view. C, 4·5 mm. diameter. Collected from Solway Firth by T. G. Skinner.

interradial canals were only slightly longer than the adradial canals and they were about twice as wide. Two marginal velar lappets had formed in each octant between the rhopalar lappets but they were only separated by shallow clefts. The rhopalar lappets were still longer and more pointed than the velar lappets. The exumbrella had relatively few nematocyst warts. There were three to four gastric filaments in each group. The margins of the mouth lips were already fringed with tentacular processes. The lips were still simple and undivided.

An essentially similar description had already been given by Claus (1884) of a specimen of *Rhizostoma* 3·5 mm. in diameter from the Adriatic. In his drawing, however, the basal portions of all the radial canals were shown of the same width.

Stiasny described a specimen 3–4 mm. in diameter which had about ten gastric filaments in each group (Text-fig. 97 B). He also gave a drawing showing a side view of this specimen (Text-fig. 97 C) with the exumbrella and marginal lappets closely covered with nematocyst clusters. Already at this stage the medusa was well rounded and no longer flat and ephyra-like. Its oral arms were 1·5–2·0 mm. long and had indications of the beginning of dichotomous branching at their ends. The spaces between radial canals and the primary ring-canal were now more elongated and this made the stomach appear relatively smaller. All the radial canals were nearly of the same width. The ring canal was more bent preparatory to the formation of the first anastomoses.

In a specimen 7 mm. in diameter (Text-fig. 97D) the marginal velar lappets were clearly separated. The epaulettes were forming and the oral arms were deeply dichotomously divided, but still with open troughs and a gaping mouth opening into the stomach. There were numerous gastric filaments arranged in bow-like groups and the subgenital pits were already formed. The first marginal ring of the network of anastomosing canals had already developed in a number of sectors. The canals running along the inner sides of this network would eventually form the secondary or main ring-canal. The coronal muscle was broad and well developed.

Stiasny found specimens 9 and 10 mm. in diameter which agreed in all respects with descriptions given by Claus of specimens of the same size from the Adriatic.

He also described one 18 mm. in diameter from Cuxhaven (Text-fig. 97 E). It was somewhat rounded and the exumbrella was covered with countless small nematocyst warts, which were present also on the marginal lappets. There were four rounded velar lappets in each octant; the

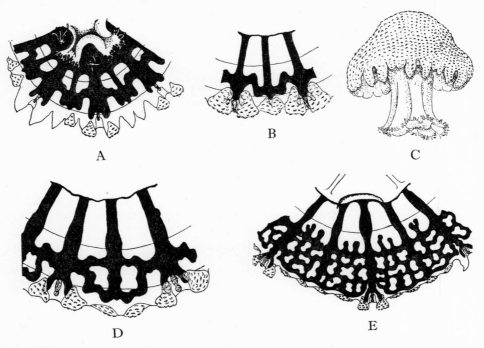

Text-fig. 97. *Rhizostoma octopus*. Development of canal system. A, ephyra 2·5–3·0 mm. in diameter, Helgoland; B and C, young specimen 3·0–4·0 mm. in diameter and 2·5 mm. high; D, 7 mm. in diameter; E, 18 mm. in diameter (from Stiasny, 1928, figs. 1–4; 1927, fig. 5).

rhopalar lappets were narrower and more pointed. The oral arms were deeply forked, but still with open troughs and without terminal clubs. The epaulettes were already well formed. There were numerous gastric filaments. The openings of the subgenital pits were wide. The gastro-vascular canal system was well developed. The extra-circular network of anastomosing canals was fairly broad and consisted of three concentric rows of irregularly shaped meshes. In each sector there were one or two short centripetal outgrowths from the main ring-canal, some of which were forking; these were the beginnings of the systems of arcade networks.

Drawings of ephyrae 4·5 mm. and 5·6 mm. in diameter have also now been given by Kühl (1967), together with reproductions of photographs of living post-ephyral and young specimens.

Detailed descriptions of young *Rhizostoma* of different sizes from the Adriatic were also given by Claus (1877; 1883) who described very fully the general development of the medusa and of the manubrium.

In order to understand the development of the different parts of the manubrium and its canal system it is easier if we begin with the growth and development of the stomach cavity. The central stomach in the very small medusa is at first circular in outline with the radial canals issuing from it at equal intervals like the spokes of a wheel. The margin of the stomach gradually becomes eight-sided, with the perradial and interradial canals issuing one from each side of the stomach and the adradial canals issuing from its eight corners. The perradial and interradial canals thus become slightly longer than the adradial canals.

At the same time the four triangular interradial sectors of the lower walls or floor of the stomach push towards the roof of the stomach cavity leaving between them four perradial canal-like grooves on the stomach floor (Text-fig. 90A). This development is accompanied by the gradual thickening of the subumbrellar mesogloea at the base of the manubrium and its coalescence to form the crescent-shaped subumbrellar walls of the subgenital pits, which are left with horizontal slit-like openings along their sides directed towards the umbrella margin.

The perradial grooves in the floor of the stomach lead into the troughs of the oral arms through the central mouth opening. At a later stage, as the mesogloea of the basal stem of the manubrium thickens and coalesces, the passages leading from the stomach to the oral arms become separate canals embedded in the mesogloea, and only vestiges of the original mouth opening may remain.

While these developments are taking place the differentiation of the parts of the manubrium is proceeding. Already when the medusa is only about 7 mm. in diameter the epaulettes begin to appear. These start as sixteen gelatinous processes or papilla-like swellings which later grow into cylindrical tubes whose endoderm-lined lumens arise from the troughs of the oral arms. These outgrowths appear in pairs, four perradial and four interradial in position. But they are in fact not true perradial and interradial pairs in origin. They are really all perradial in origin. Four pairs grow first; the rudiments of the second four pairs then develop, each pair being between the outgrowths of one of the original pairs with their lumens connecting with the basal canals of that pair. Each of the epaulettes of a first-formed pair thus becomes pushed sideways and comes to lie nearer its neighbour in the adjacent pair, so that the two appear to compose an interradial pair. As the epaulettes increase in size with the growth of the medusa they become closely packed together to form a compact ring round the base of the manubrium.

Even before the first appearance of the epaulettes the ends of each of the four oral lips have begun to divide to form the eight oral arms. The canals to the oral arms are thus formed above the branches to the epaulettes (Text-fig. 98). By the time the epaulettes are developing each bifurcation of the oral lips begins to divide again to form the sixteen divisions. At this stage the end of each lip starts to turn outwards and upwards, dividing to form the two abaxial wings of the triradiate oral arm.

In medusae 16 mm. in diameter these triradiate oral arms are already distinctly recognizable, but there are as yet no indications of the terminal clubs. According to M. E. Thiel (1965) specimens less than 20 mm. in diameter have no terminal clubs, but they begin to appear on some of the oral arms in medusae between 20 and 29 mm. in diameter. They are generally present in medusae over 30 mm. in diameter. My own observations on specimens sent me by T. G. Skinner from the Solway Firth agree with this, the first signs of the terminal clubs appearing on some oral arms in medusae 22–25 mm. in diameter; specimens 25 and 26 mm. in diameter had clubs less than 1 mm. in length. When the medusa was 28–30 mm. in diameter the clubs might be 2 mm. long; and at diameters of 31 and 35 mm. the clubs were 3 mm. long. Average figures for lengths of terminal clubs given by Thiel were: 20 mm. diameter, 0·6 mm.; 30 mm. diameter, 2·6 mm.; 40 mm. diameter, 3·6 mm.

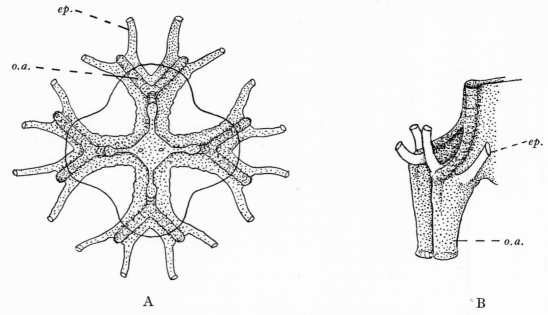

Text-fig. 98. *Rhizostoma octopus*. A, cut end of base of manubrium of small specimen, *ca.* 30 mm. in diameter seen from exumbrellar side to show origins of oral arm canals and canals to epaulettes. The mouth is practically closed except for the openings of two small canals; B, side view. *ep.* epaulette branch; *o.a.* oral arm branch. (Coll. T. G. Skinner, Solway Firth, July 1966.)

RATE OF GROWTH

The only detailed observations on the rate of growth are those of M. E. Thiel (1966 *a*), made on medusae from the Elbe estuary near Cuxhaven. His results showed that a size greater than 300 mm. in diameter could be reached in two and a half months; they can be summarized as follows.

Date	Diameter in millimeters		
	Smallest individual	Largest individual	Average
25 June 1964	3	38	18
13 July 1964	5	98	28
23 July 1964	18	103	49
6 August 1964	14	173	65
21 August 1964	22	228	94
3 September 1964	60	320	225

Thiel assumed that it was questionable whether very large specimens up to 600 mm. diameter would occur in the Elbe estuary area, as with the lowering of temperature in the autumn they would not grow so quickly and would sink to the bottom and die.

In British waters very large specimens are to be found; for instance Vallentin (1907) said that such large specimens, none less than 12 in. in diameter, were brought to the surface in St Ives Bay and Falmouth harbour by the starting of a steamer's propeller, and recorded one 36 in. across the umbrella. This is about 900 mm.

It seems to be likely that the very large specimens are survivors of the previous year's production of ephyrae (see p. 197).

ALIMENTARY SYSTEM

While Eysenhardt (1821), Brandt (1870), and others had shown the gastrovascular system of canals by injecting coloured matter, milk, or air, from the earliest times onwards there was much speculation as to how *Rhizostoma* obtains its nourishment.

Eysenhardt (1821) only found slimy matter in the stomach. He was correct in concluding that the mouth openings were too small to allow other than slimy matter in the water, infusoria, etc., to reach the stomach. He concluded that, unlike other medusae which eat from time to time larger organisms, *Rhizostoma* could get sufficient nourishment because the taking in of water with its contained matter and assimilation of food was continuous.

He thought that movement of fluid through the canal system was effected by the activity of the medusa and that the canals acted like capillaries. He also imagined that the gastrovascular network was for the working up and transportation of nutrient matter, the network supplying nourishment to the marginal part of the umbrella, and the stomach to the upper part. But he was surprised that the marginal canals did not communicate with the exterior, and explained this on the assumption that there was so little solid matter to be excreted.

Matters became more confused later when others recorded the findings of digested fish remains in the medusa, and the idea was also introduced that the oral arms had suckers. Hamann (1881) found small animals half-digested between the oral lips and assumed that digestion could take place there. Winterstein (see M. E. Thiel, 1964, p. 261) estimated that the oxygen requirements of the medusa were such that it could not possibly obtain sufficient nourishment from solid plankton food, and introduced Pütter's theory that dissolved substances in the water could be used (see also Krüger, 1968).

The full history of these discussions, to which I have made only brief references, has been well summarized by M. E. Thiel (1964), to whom we owe the first satisfactory account of feeding and excretion in *Rhizostoma*.

Thiel kept medusae for a week in the Helgoland aquarium, to the water of which finely minced mussel flesh was added. The medusae appeared the whole time to be in good health, and he concluded that these small particles could be taken in without difficulty. He thought that detritus might also be used as food, since Bodanski & Rose (1922) had shown that the gastric filaments of *Stomolophus meleagris* contained enzymes for the digestion of plant substances.

Thiel observed that small animals placed in the clefts between the mouth lips were first enclosed in mucus and then slowly carried along the canals to the stomach. He placed *Rhizostoma* in water to which a thick supply of plankton had been added and found that after a short time the canals in the oral arms contained many plankton organisms. After a couple of hours many of these had arrived in the stomach. Later he found much detritus in the stomachs of freshly caught *Rhizostoma* from Cuxhaven. He could not get them to take fish or ctenophores, but as his specimens were only 80–120 mm. in diameter he thought it possible, though unlikely owing to their structure, that very large specimens might be able to eat small fish.

Thiel injected carmine particles into the stomach through the apex of the umbrella, or through an oral arm canal. By turning the medusa quickly on its side with the umbrella margin turned back he was able to watch the passage of the particles through a binocular microscope. The particles passed along the adradial canals to the umbrella margin and into the main ring-canal and anastomosing networks. They returned to the stomach along the perradial and interradial canals, just as in *Aurelia*. The passage through the canals was rapid and it was difficult to establish the steps by which it proceeded; injected coloured gelatine filled the whole canal system before it cooled.

191

Thiel saw, and reproduced photographs showing, the method of excretion. When the stomach was filled with injected fluid, after a time the oral arm canals became filled. Threads of slime were then extruded chiefly from the upper ends of the oral arm mouth fringes, but also from the ends of the terminal clubs. He saw similar brown slime threads from specimens which had been in the aquaria containing minced mussel flesh. These were chiefly extruded at the upper ends of the oral arm mouth fringes.

Thiel noted that the presence of excretory products with partly digested remains in the oral arm mouth fringe would explain the suggestion by Hamann that digestion might take place there. His conclusion was that *Rhizostoma* feeds on plankton, which contains at times copepods, isopods, zoeae, nauplii, polychaete larvae and *Sagitta*, but that it is not improbable that detritus and dissolved substances carried in with the water current could also be used as food.

It is evident that *Rhizostoma* must be an efficient plankton feeder. Unlike a medusa such as *Cyanea* whose tentacles extend for many feet through the water while it slowly pulsates, *Rhizostoma* swims actively through the water. Its massive manubrium with its hundreds of small mouth openings presents an area to the water covered with literally thousands upon thousands of nematocyst-carrying tentacular processes which fringe the mouth lips. Swimming steadily through a plankton-rich layer in the sea it must capture enormous numbers of small prey which strike against the manubrium.

EFFECTS OF STARVATION

Vernon (1895) kept *Rhizostoma* in captivity at Naples for over five weeks. Presumably the medusae could get little or no food, and he found that after this period one had decreased to 11·6% of its original weight and another to 7%.

In this connection it might be noted that *Rhizostoma* appears to be rather a hardy medusa and easy to keep in captivity. C. W. Hargitt (1904) especially remarks on this. He found that he could keep specimens at Naples quite healthy for four to six weeks. For this reason he considered it a good subject for experiments, but found that small specimens 40–70 mm. in diameter were easier to keep healthy and recovered from damage more quickly.

FOOD

As already stated *Rhizostoma* is a plankton feeder and is therefore likely to ingest any organism small enough to pass through a mouth opening.

PHYSIOLOGY

Neuromuscular system

Hesse (1895), who described the marginal sense organ of *Rhizostoma* from Naples, showed that the nerve fibres running from the sense organs were congregated into tracts running below the perradial and interradial canals. Each tract was wider than the canal, covering a width of 1·7–2·3 mm., and had about a hundred fibres. The nerve fibres spread out diffusely from these tracts over the coronal muscle.

Bethe (1903, 1908) considered that the nerve fibres spreading over the coronal muscle formed a true nerve net, and that the fibres were continuous. He noted that the coronal muscle was not continuous, but that it was divided into sixteen fields separated by muscle-free areas over which the nerve fibres ran. By suitable preparations he showed that on stimulation of the marginal sense organ the stimulus is conducted by the nerve net, and not by the muscle.

Von Uexküll (1901) thought that the stimulus to the sense organs for producing pulsation was purely mechanical; Fränkel (1925) thought, however, that the sense organs came into action from physiological stimuli, such as that of light. Bozler (1926a) concluded that automatic rhythmic pulsation is apparently controlled by the ganglion cells lying at the bases of the rhopalia. He studied the compensatory movements by which the medusa can bring itself back to an upright position when on its side, and concluded that these are regulated by pressure on the sensory cells of the rhopalium.

Rhizostoma has no subumbrellar radial muscles and Bozler thought that the high rounded apex of the umbrella rendered these unnecessary, since owing to the shape of the medusa the uniform contraction of the marginal portion by the coronal muscle alone can effect a decrease of the subumbrellar cavity.

Bozler (1926b) found that rhythmic pulsation started again after removal of all marginal sense organs, but at a slower rate of 20 pulsations per min. compared with the normal 80 per min.

Hykes (1928) examined the effects of the ion content of sea water on movement; it is not clear from his account whether *Rhizostoma* could luminesce.

Horridge (1955), working at Naples, found evidence of the shortening of the refractory period, in some muscle fibres following frequent stimulation of the bell of *Rhizostoma*.

Skramlik (1945), as did Vernon (1895), noted a decrease in the pulsation rate after capture of the medusae and concluded that experiments should be done during the first two hours after the specimens are brought into the laboratory.

Bethe(1908, 1909) studied the effects of electrolytes on pulsation (discussed by Mayer, 1910, p. 701); and Schaefer (1921) observed the effects of adrenalin, atropine, nicotine, strychnine and curare.

Vernon (1895) noted, as did Romanes for other species, that the rate of pulsation of *Rhizostoma* bears an inverse relation to the size of the animal.

Effects of temperature

Vernon (1895), working at Naples, found that *Rhizostoma* was paralysed and killed at temperatures much above 24° C. He found that a specimen 93 gm. in weight had a pulsation rate of 43·5 per min. at 10·6° C. and 73·5 per min. at 23·5° C.; and that one 150 gm. in weight pulsated 43 times per min. at 9·7° C., and 79 per min. at 27·5° C.

Uexküll (1901) found that raising the temperature from 13° to 22° C. resulted in a doubling of the pulsation rate from one to two per sec.

Schaefer (1921), using *Rhizostoma* from the North Sea, found that they were heat paralysed at 28°–32° C., and cold paralysed between 4° and 6° C. Skramlik (1945), working at Naples, found similar effects: at 5° C. pulsation was rare, and at most arhythmic; at 8° C. upwards pulsation was rhythmic, and between 8° and 26° C. the pulsation rate increased with temperature; at 30° C. pulsation became periodic, and it ceased at 35° C. but could be restored on cooling; at 48° C. pulsation could not be restored.

Benazzi (1933), at Naples, recorded the rate of pulsation as 88 per min. at 22° C. When the temperature was raised during four hours to 28° C. the rate reached 98 per min., and the next day at 30·5° C. it was 100 per min. At 34° C. pulsation stopped completely.

J. & M. Ries (1924) made temperature measurements inside the mesogloea of the disk and found that the temperature was the same as that of the surrounding water. If, however, the thermometer was placed under the pigmented margin so that it was shielded from the sunlight by the pigment the temperature dropped by 1–1·5° C.

RHIZOSTOMEAE

Effects of salinity

Benazzi (1933) studied the effects of dilution of sea water on *Rhizostoma* at Naples. He found that if the change was gradual the medusa could tolerate dilution to about 30 parts sea water and 70 parts fresh water. The rate of pulsation decreased with decreasing salinity. For instance, one small specimen in 40% sea water decreased its pulsation rate from 150 per min. to 20 per min., gradually increasing again to 100 per min.; in 35% sea water the rate was 89–90 per min., and in 30%, 70–75 per min. Another very small specimen in 25% sea water ceased pulsation, but on being put in 50% sea water it regained regular pulsation at 130–160 per min. A larger specimen ceased pulsation in 40% sea water, and a medium-sized one in 30%. Another in 40% sea water showed a pulsation rate of 85 per min. reducing to 50, and in 30% to 35 per min. but spasmodic.

CHEMISTRY

The pigment of *Rhizostoma*, which was called 'cyanein' by Krukenberg (1882), and 'zoo-cyanin' by Colosanti (1888) and von Zeynek (1912), has been more recently studied by Fox and Pantin (1944) and Christomanos (1954). It is probably a conjugated protein with a prosthetic group belonging to, or related to, the carotenoids.

Haurowitz (1920), at Trieste, studied the gonad fats and remarked on the presence of trimethyl-amine hydrochloride as a constituent of the phospholipids of the gonad, and on the complete absence of stearic and oleic acids. To his knowledge this occurred in no other animal fat.

Aarem *et al.* (1964) found that *Rhizostoma* could not synthesize sterol from (1-^{14}C)-acetate, but could build up higher fatty acids from this precursor. Palmitic acid preponderated. Lipids comprised only 1% of the dry weight.

Christomanos (1954) found in the mesogloea, glutamic acid, leucine, *iso*leucine, valine, aspartic acid, arginine, perhaps lysine, proline, methionine, alanine, glycine, serine, an amino-acid accompanying threonine, and at least two unidentified amino-acids of which one might have been glucosamine.

Danmas & Ceccaldi (1965) studied amino-acids in *Rhizostoma pulmo* chromatographically. Tyrosine was present in great concentration, also taurine in the combined state, and sarcosine.

Von Zeynek (1912) commented on the pungent radish-like smell given off by the medusa and water in which it had been.

Krüger (1968) measured oxygen consumption in relation to body weight, as did Vernon (1895).

GENERAL OBSERVATIONS

Swimming

Eysenhardt (1821) pointed out that when quiescent *Rhizostoma*, at Cette, were vertical in position, but that with a following current they were horizontal: he never saw one swimming completely against the current, but quite well across the current.

Uexküll (1901) and Fränkel (1925) drew attention to the rudder effect that the massive manubrium has on the swimming movements of *Rhizostoma*. This results in the medusa swimming in straight lines or in curved trajectories. Fränkel watched *Rhizostoma* for quarter of an hour from a boat and noticed that it swam at a depth of about 1 m. in an exactly horizontal position. It often swam along close under the surface so that its margin disturbed the water surface. Bauer (1927) noted that when sinking passively it always remains upright.

Behaviour

The following observations throw some light on the vertical movements of *Rhizostoma* in the sea.

Eysenhardt (1821) at Cette noted that in fine weather they were to be seen in quantities at the water surface, but that in rain and storm they sought deeper water.

Bozler (1926*b*) said that they were positively geotactic in sunlight, although previously (Bozler, 1926*a*) he had stated that they were negatively geotactic.

Bauer (1927) repeatedly observed that swarms of *Rhizostoma*, not previously evident, appeared on the surface in overcast skies. In this connection Vallentin (1907), commenting on the infrequent observation of the medusa, noted that in St Ives Bay and in Falmouth harbour he had more than once seen large specimens brought to the surface by the sudden starting of a steamer's propeller, and thought that it was not often seen because of the depth at which it usually swam.

Verwey (1966) used a large collecting net with an opening 4 × 6 m. at three different depths (0–4, 10–14, and 22–26 m.) in water 28 m. deep at den Helder during the days of 1 and 2 July and 3 and 4 August 1964. He found that *Rhizostoma* became more abundant in the water column during ebb and flood tides than during slack-water periods of high and low tides, when the medusae were presumably on the bottom. He attributed this to a need to avoid the actual neighbourhood of the bottom or of sand brought into suspension during the high current velocities with accompanying strong turbulence. Although there was a rise of the medusae towards the surface at dusk, Verwey had indications that current velocity changes dominated in their effect over changes of light intensity.

Rhizostoma appears also to be very sensitive to vibration. Nagabhushanam (1959) said that in the Irish Sea they appeared to be sensitive to the noise of an approaching boat and sank into deeper water.

J. S. Colman has told me that sometimes when crossing from Liverpool to Douglas, Isle of Man, one may go for 20 or 30 miles through water with a *Rhizostoma* every few feet ranging in size from that of a tennis ball to 2 or 3 feet across. In calm weather many of them could be seen to be at the surface 50 yards away, and bigger ones sometimes up to 100 yards off; but when the ship reached them they had sunk several feet below the surface still in the upright position. Presumably on feeling the vibrations of the ship's engine they ceased pulsating and sank passively.

Maaden (1942*a*) remarked that on the Dutch coast medusae in any one swarm were often approximately of the same size. It gives the impression that they are of about the same age or come from the same place of origin. He could find no connection between the strandings of *Rhizostoma* on the beach and the direction of the wind.

Stinging

Eysenhardt (1821) referred to the stinging sensation produced if *Rhizostoma* were touched and its severity if touched by the lips.

Von Zeynek (1912) comments on sneezing and catarrh produced when dissecting the medusae, which can also be caused by the dried slime particles getting into the air off clothing. He found that the slime and mouth fringes caused intense burning on the tongue some seconds after it was touched.

Krumbach (1925) says that the Atlantic *Rhizostoma* stings only weakly, but that people with sensitive mucous membranes could not remain in rooms in which these medusae were dissected.

RHIZOSTOMEAE

Muir Evans (1943), however, refers to nuisance caused by this medusa in the Pas de Calais area. Stings may give rise to urticaria, or to constitutional symptoms such as shortness of breath and a feeling of anxiety, which usually disappear in 24 to 36 hours. He records in full an instance of severe stinging at Dunkirk in the First World War.

Economic

Netchaeff and Neu (1940) said that in the Black Sea *Rhizostoma* clogged the fishing nets.

Borlase (1758) said that medusae were eaten by men: it is not clear whether he was referring to Cornishmen or was making a general statement that might refer to other parts of the world.

Parasites and commensals

Monticelli (1897) recorded the occurrence of cysts on the umbrella of *Rhizostoma* at Naples, but was unable to determine what the organism might be. He called it *Pemmatodiscus socialis*.

The amphipod *Hyperia galba* is very commonly found in association with *Rhizostoma*, as it is with other Scyphomedusae. Hollowday (1947) gives a short review of earlier observations. He found in specimens stranded on the Cheshire coast that the amphipods might be very tightly packed together in the subgenital cavities. The smaller medusae harboured very young *Hyperia*, indicating that they formed the association at a fairly early age.

A more detailed study was made by Nagabhushanam (1959) in the Irish Sea. From one medusa 430 mm. in diameter he took 22 amphipods varying in length from 2 to 18 mm. The smaller *Hyperia* were mostly found inside the canals of the host medusa. These were still alive and might have entered after capture. The larger amphipods were invariably found on the exterior of the medusa including the exumbrella, marginal lappets, sense organs and the subumbrella. A few were on the terminal clubs. He could find no set pattern in their distribution, but they could of course have changed position after capture.

Nagabhushanam also made observations on the association of young whiting, *Gadus merlangus*, with *Rhizostoma*. With the same medusa referred to above he also had 59 whiting, and others might have escaped. Fourteen of these were found to be in a semi-digested state, and he thought that they might have got into the stomach accidentally through rupture of the gastro-genital membrane. He noted that young whiting associated with *Rhizostoma* had been feeding on *Hyperia*, and that this amphipod was practically absent in the stomachs of whiting which were not associated with the medusa.

The following table summarizes his results (Nagabhushanam, 1959, table III).

Date	*Rhizostoma* diameter (mm.)	Whiting Number	Size range (mm.)	*Hyperia* Number	Size range (mm.)
25 May 1957	380	27	21–40	18	2–6
29 May 1957	430	46	25–46	22	2–18
17 June 1957	410	11	30–51	16	3–14
28 May 1958	340	1	21	3	4–10
29 May 1958	420	2	21, 25	16	4–15
2 June 1958	340	1	31	19	2–12
Totals	6	88	21–51	94	2–18

He also took three young cod, *Gadus morhua*, with a *Rhizostoma* in May, one of which had ingested *Hyperia*; and two *Gadus luscus*, 33 and 45 mm. long, with a *Rhizostoma ca.* 600 mm. in diameter on 14 November 1958.

SEASONAL OCCURRENCE

In general, in coastal British waters *Rhizostoma* appears in late summer and autumn. This is the time probably when they are most noticeable as they are cast ashore in numbers by the autumn storms. At times they may also be recorded through the winter. Some published records are: Valencia, south-west Ireland, August, September, October (Browne, 1896; M. & C. Delap, 1907); Whitstable, summer (Newell, 1954); coast of Pembroke, January, June, July, December (Bassindale & Barrett, 1957); Cheshire coast, November (Hollowday, 1947).

But specimens are recorded in western waters at other times of year. Browne (1895*a*) recorded large specimens in April in the Irish Sea off Port Erin, and mentions one stranded on the beach 23 in. (*ca.* 580 mm.) in diameter. Vallentin (1900) said that they were to be seen from May to October in Falmouth harbour. Nagabhushanam (1959) recorded specimens in the Irish Sea up to 430 mm. in diameter in May. I am indebted to C. Burdon Jones for information of their occurrence in the Anglesey area in February and May 1966. J. P. Hillis informs me that a hundred or more specimens were caught per half-hour trawl in January 1966 in the Irish Sea.

Many specimens were stranded at Minehead in February and March 1968 and a large specimen was caught off Plymouth in the last week of February 1968.

It thus appears that in some areas *Rhizostoma* may occur throughout the year. Verwey (1942) stated that such early dates as April were never recorded for the North Sea. He thought that the ephyra formation off the Dutch coast probably extended from May to September, the larger specimens appearing from August to November. M. E. Thiel (1966*a*) in his careful study of the growth of the medusa in the Elbe estuary, concluded that strobilation of the scyphistoma occurs from mid-June to the end of July.

There are no comparable published records for the occurrence of very young stages in British waters, but I am grateful to T. G. Skinner for a collection of *Rhizostoma* taken in July 1966 in the Solway Firth. These were all quite small specimens 16–35 mm. in diameter, and this, supported by the above evidence from the North Sea, indicates that strobilation in western British waters may also occur in June and July.

Since all the evidence is that strobilation is restricted to the summer months, the assumption must be that those specimens recorded from January to June are survivors from the previous year. Many are, of course, cast ashore in autumn and winter storms, but those living in deeper offshore waters would survive.

Information is needed on the occurrence of very small specimens round our coasts. The regions in which the scyphistoma lives are not known. It is interesting that while there must evidently be scyphistomas in the Solway Firth, C. Edwards tells me that he has never seen very small specimens of *Rhizostoma* at Millport.

Like other scyphomedusae *Rhizostoma* is spasmodic in occurrence from year to year, e.g. Vallentin (1907) writes 'for some unexplained cause this species has been very scarce'; during 1906 he had not seen one and had only heard of two caught by local trawlers when trawling at nightfall.

ABNORMALITIES

There have not been many records of meristic variation in *Rhizostoma*. Brandt (1870) recorded a specimen with seventeen radial canals. C. W. Hargitt (1905) in an examination of fifty specimens at Naples found that 15% showed variation in one or more organs. The most frequent variations were in the numbers of marginal sense organs, in which there were as few as five in two specimens and as many as twelve in one specimen. There was also variation in the numbers of gonads, gastric lobes and oral arms, but all three always agreed in number.

Owing, however, to its very complicated structure the medusa shows many features which may be considered rather as abnormalities than as variations.

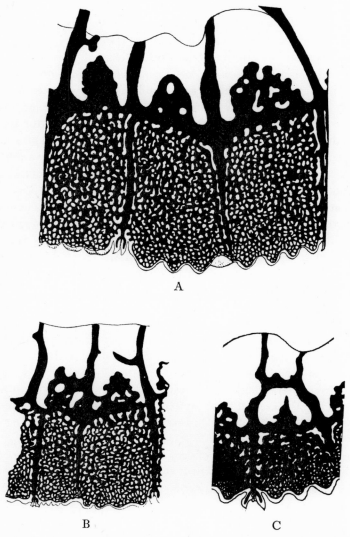

Text-fig. 99. *Rhizostoma octopus.* Some abnormalities in gastrovascular canal system (from Stiasny, 1929, pl. I, 2; IV, 8; VII, 14). (For A, B, & C see text below.)

Stiasny (1929) made a detailed examination of such abnormalities in the gastrovascular canal system (Text-fig. 99) and found the following irregularities:

1. Canals from the intracircular arcade network may connect with the perradial and interradial canals (Text-fig. 99 B).
2. There may be similar connections between the arcade network and the adradial canals.

3. Longer or shorter outgrowths from any radial canal may sometimes reach the arcade network.

4. Occasionally, in place of an arcade network, there may be large club-shaped or sack-shaped centripetal outgrowths from the main ring-canal with very few, or single, meshes, or even none at all (Text-fig. 99 A).

5. The arcade network may connect direct with the stomach by a canal which grows centripetally from it.

6. The arcade network arrangements may be quite different in each sector.

7. Radial canals may often run with an irregular course, be more or less bent, or have lateral protuberances of various shapes.

8. Radial canals may fork or even have three roots. Very rarely there may be a cross-anastomosis between neighbouring radial canals on the centripetal side of the main ring-canal running parallel with the margin of the stomach (Text-fig. 99 C).

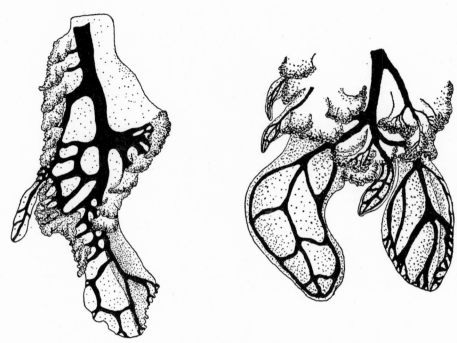

Text-fig. 100. *Rhizostoma octopus*. Abnormal oral arms and supernumerary terminal clubs
(from Stiasny, 1928, figs. 2, 5).

9. There may occasionally be a large sinus in the course of a radial canal or of the main ring-canal.

10. A sinus-like widening of an adradial canal can assume considerable dimensions.

11. The main ring-canal is usually well defined, but it may occasionally be ill-defined in one sector, or even not apparent at all.

12. The number of meshes in the arcade networks is rather variable. Usually each network has three or four more or less regular rows. Longer arcades may have four or five mesh rows and shorter ones one or two.

Stiasny (1928) also described a number of abnormalities in the structure of the manubrium in specimens from the Dutch coast. The whole manubrium including the terminal clubs was often very irregularly formed. A number of incipient appendages, like small terminal clubs, were often present on the mouth fringes of the abaxial wings of the oral arms. Sometimes there were accessory clubs close to the main terminal club. Twinning of the terminal club was also seen (Text-fig. 100).

The end club might sometimes be excessively large and broad, forming a direct extension of the mouth-fringe part of the oral arm. It might even have indications of mouth openings along its edges.

The upper part of the manubrium might also vary considerably in length.

RHIZOSTOMEAE

REGENERATION

C. W. Hargitt (1904) did some experiments on regeneration in *Rhizostoma* at Naples. He found that they could regenerate their rhopalia; on two occasions regeneration resulted in the formation of twin rhopalia, and once an additional rhopalium was developed on the side of the excision notch in the margin of the umbrella.

HISTORICAL

The north European *Rhizostoma octopus* was first called *Medusa octopus* by Linnaeus (1788). Earlier Borlase (1758) had given a description and illustrations of what was obviously this species in his *Natural History of Cornwall*. He called it 'Urtica marina ex trunco octopedalis limbo imbricatim undante'.

Macri (1778) described the Mediterranean *Rhizostoma* under the name *Medusa pulmo*.

Subsequently the only detailed descriptions of *Rhizostoma* were on the Mediterranean medusa by Eysenhardt (1821) and Brandt (1870) as *R. cuvieri*. The first really detailed descriptions of the structure of the European *Rhizostoma* were those of Stiasny (1923, 1928, 1931) and M. E. Thiel (1965). As stated on p. 173 some authors have regarded *R. octopus* as only a variety of *R. pulmo*, but until there are more up-to-date observations on the structure of the Mediterranean medusa it is better to keep them distinct.

NOTE (added in proof)

See also Browne (1900) for *Rhizostoma* swimming against the tide and for association with young *Trachurus*.

INCERTAE SEDIS

The genus *Tetraplatia* was erected by Busch (1851) to include an aberrant form of cnidarian whose exact systematic position is still in doubt. Its determination will probably depend upon the possibility of obtaining successive early stages of development by rearing in the laboratory.

The problem has been a matter for discussion by a number of authors. Busch, who obtained the first specimen off the Mediterranean coast of Spain, did not discuss its affinities. Subsequent investigators have placed it in the following categories.

Hydromedusan, half medusa and half polyp, Claus (1878).
Anthomedusan, Dantan (1925).
Either larval narcomedusan, or intermediate between hydroid polyp and craspedote medusa (Haeckel, 1879).
Trachyline medusa, Viguier (1890).
Narcomedusan, Leuckart (1866, aeginid); Weill (1934).
Narcomedusan ancestry with trachymedusan sense organ, Hand (1955).
Scyphomedusan, Krohn (1865, young scyphozoan); Delage & Herouard (1901); Krumbach (1925, larval coronate medusa); Komai (1939, close to Cubomedusae and Stauromedusae); Ralph (1960, coronate medusa).
Between hydromedusa and ctenophore, Fewkes (1883).

Carlgren (1909) created a new order for *Tetraplatia*, the Pteromedusae, which was accepted by Hand (1955) and included by Kramp (1961) in his 'Synopsis of the Medusae of the World'.

The various opinions have been based on different structural characters, as follows:

1. The marginal edge has been regarded by some as equivalent to the hydrozoan velum, and by others as the ephyra-like marginal lappets of a scyphozoan.
2. The marginal sense organs have been regarded as either trachyline or scyphozoan in structure.
3. The gonads have been described as either of ectodermal or of endodermal origin.
4. Gastric filaments have been considered to be absent by all except Ralph (1960) who considers that they contain the gonads.

On the whole I regard the characters to be more scyphozoan than hydrozoan and include it in this monograph as of uncertain position.

Genus **Tetraplatia** Busch

Scyphomedusae with body divided into upper (aboral) and lower (oral) portions separated by circumferential groove in which is umbrella marginal tissue divided into eight pairs of marginal lappets. Aboral and oral portions of body may or may not be connected by four flying buttresses alternating with marginal lappets. Eight marginal sense organs, situated one in each cleft between paired marginal lappets. No marginal tentacles. Four gonads each with paired aboral and oral lobes.

The genus *Tetraplatia* Busch contains two species, *T. volitans* Busch and *T. chuni* Carlgren. Of these *T. volitans* occurs over deep water west of the British Isles.

T. chuni has been recorded only from the South Atlantic and differs from *T. volitans* in being without flying buttresses, and in the proportions of the body, the aboral and oral portions of which are approximately equal in length.

INCERTAE SEDIS

Tetraplatia volitans Busch
Text-figs. 101, 102

Tetraplatia volitans Busch, 1851, *Beob. Anat. Entwick. Wirbell. Seetiere*, p. 1.
  Krohn, 1865, *Arch. Naturgesch.* **31**, 337, pl. XV.
  Viguier, 1890, *Archs Zool. exp. gén. Paris*, sér. 2, **8** 101, pl. VII–IX.
  Carlgren, 1909, *Wiss. Ergebn. dt. Tiefsee-Exped. 'Valdivia'*, **19**, 78, pl. X, figs. 1–9; pl. XIII, figs. 6–15.
  Dantan, 1925, *Annls Inst. océanogr., Monaco*, n.s. **2**, 429, text-figs. I–XII, pl. I–II.
  Komai, 1939, *Jap. J. Zool.* **8**, 231, text-figs. 1–11, pl. XXXI.
  Hand, 1955, *Pacif. Sci.* **9**, 332, figs. 1–8.
  Ralph, 1960, *Proc. R. Soc.* B, **152**, 263, figs. 1–4, 5 *a*, *b*, 6 *a*, *c*, 8 (pl. 19).
Tetrapteron (Tetraplatia) volitans, Claus, 1878, *Arch. mikrosk. Anat. EntwMech.* **15**, 349, pl. XXII.
Tetraptera volitans, Fewkes, 1883, *Am. Nat.* **17**, 426.

SPECIFIC CHARACTERS

Oral portion of body longer than aboral portion; with four flying buttresses; width of marginal lappet zones usually equal to intervening zones.

DESCRIPTION OF ADULT

Aberrant body form consisting of upper aboral pyramidal portion separated from lower oral pyramidal portion by slight circumferential groove; oral portion twice, or up to five times, length of aboral portion; oral and aboral portions connected by four flying buttresses; four continuous nematocyst tracts running from mouth to apex over flying buttresses, four shorter nematocyst tracts, one in each interspace between adjacent flying buttresses. Eight marginal lobes arising in pairs from four common bases, one in each space between flying buttresses, each dividing into two ephyra-like marginal lappets each further subdivided terminally into upper and lower finger-like processes; margin continued as narrow undivided band of tissue running between adjacent lobes under flying buttresses. Eight marginal sense organs, one in or near base of cleft between each ephyra-like marginal lappet; sense organ consisting of statocyst, without ocellus. No marginal tentacles. Gastrovascular system consisting of continuous cavity in aboral and oral portions of body, usually with continuous connecting canals, one through each flying buttress; simple terminal mouth opening without lips. Four gonads each dividing into four elongated lobes, two aboral and two oral. Size range 4–9 mm. in length. Colour whitish or bluish white.

DISTRIBUTION

Tetraplatia volitans is a pelagic oceanic medusa. It has been caught in large numbers over deep water off the western approaches to the English Channel (Ralph, 1959), and one specimen has been recorded north-west of Scotland (Rees & White, 1957). Owing to its peculiar shape it has the appearance of some wormlike animal rather than a medusa and it may be overlooked. It is evident that it is probably quite common over deep water off the western coasts of the British Isles.

Its distribution is worldwide, *Tetraplatia* having now been recorded in all three oceans. In the South Atlantic it has been recorded as far south as 52° S.; in the North Atlantic the northernmost record at present is that north-west of Scotland at 58° 38' N.; 8° 21' W.

An analysis of its distribution in relation to environmental conditions by Rees & White (1957) indicates that the species is eurythermal and euryhaline.

STRUCTURAL DETAILS OF ADULT

Umbrella

The aberrant shape of *Tetraplatia* precludes the use of normal terms in its definition. The general shape is that of an octahedron whose lower pyramid is longer than its upper pyramid. Where the bases of the two pyramids adjoin there is a circumferential groove in which lie the marginal structures. Thus the upper pyramid may certainly be regarded as umbrella. The homology of the lower portion is, however, less certain. It might be regarded as a completely protruding manubrium whose basal portion was derived from the subumbrella, but there is no indication of any demarcation between subumbrella and manubrium tissue. Both the upper umbrella portion and the lower manubrium portion appear to be essentially similar in histological structure. In order to avoid being too definite as regards homologies with a normal medusa I use the terms aboral and oral for the two portions of the body lying above and below the circumferential groove. The oral portion is almost always markedly longer than the aboral portion, ranging between 2:1 and 5:1.

The narrow circumferential groove which runs round the body where the aboral and oral portions meet might be regarded as a coronal groove. But the siting of the marginal lobes and their connecting marginal folds within the groove itself makes any homology here also uncertain.

The aboral and oral portions are connected by the four remarkable flying buttresses which arch over the circumferential groove in the four regions between adjacent marginal lobe bases. In addition to the four long nematocyst tracts running from mouth to apex over the flying buttresses, and the four shorter tracts between them, there are scattered nematocysts over the whole body surface except on the marginal lobes.

Between adjacent flying buttresses there are two pairs of marginal lappets, making sixteen in all. Each pair of lappets rises from a single marginal lobe, and the pair of marginal lobes arises from a common fleshy base. Each marginal lappet resembles a marginal lappet of an ephyra, except that its distal margin is further subdivided into a number of finger-like processes. There are usually four such processes on each lappet, two above and two below.

The fleshy bases of the marginal lobes are joined in the circumferential grooves under the flying buttresses by a narrow flap-like band of tissue which might be regarded as umbrella margin.

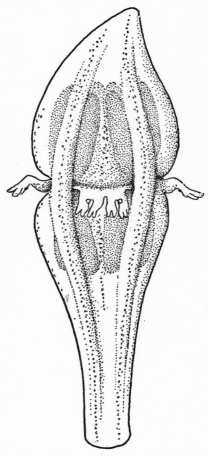

Text-fig. 101. *Tetraplatia volitans.* Slightly diagrammatic drawing of specimen 8 mm. high to show the main features.

Marginal sense organs

There are eight marginal sense organs, one in each of the eight clefts between paired marginal lappets. They are situated on the lower, or subumbrellar, side at the base of a cleft, or just lateral to its base. Each sense organ consists of an endodermal core having a short basal stalk and expanding distally to form a statocyst. The whole is surrounded by a thin layer of mesogloea covered externally by ectoderm. It thus might be regarded as having the form of a short simple rhopalium, over which the junction of the bases of the two lappets bordering the cleft forms a slight protective hood.

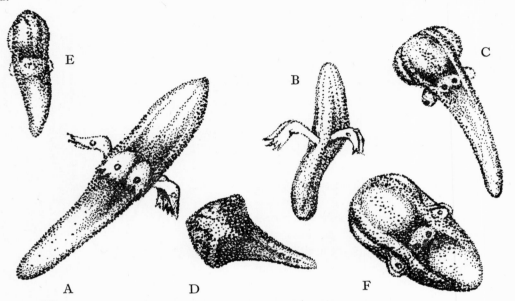

Text-fig. 102. *Tetraplatia volitans*. Drawings made from photographs of living specimens (from Viguier, 1890, pl. VII, fig. 1). A, B, swimming; C, D, creeping; E, F, at rest.

Gastrovascular system

The whole body of the animal from apex of the aboral portion to the mouth opening is hollow, and all this cavity must be regarded as stomach. Gland cells are distributed throughout the endodermal lining epithelium, and they are especially numerous in the lower third, namely at the oral end.

There is nothing in the nature of a marginal gastrovascular system. The aboral and oral portions of the stomach cavity are usually connected by a canal running through each of the four flying buttresses. In some specimens these canals are blocked by an endodermal plug and a transverse mesogloeal system may be present. It is thought, however, that this may be due to incomplete junction of the canals during the development of the flying buttresses.

Gonads

Each of the four gonads is composed of four elongated processes joined at their bases in the endoderm of the stomach wall just above the level of the circumferential groove. Two processes project upwards into the aboral portion of the stomach cavity, and two, usually longer, project downwards into the oral portion of the cavity. A duct leads from the region of junction of the bases of the four processes to a pore opening to the exterior under a flying buttress just below the marginal band. Hand (1955) observed groups of sperm cells in these pores in males, and in one female he found a large ripe egg half extruded through a pore.

In mature specimens the swollen gonads may fill much of the cavity of the stomach so that it appears much reduced in sections.

DEVELOPMENT

Ralph (1959) recorded a small specimen 1·0 mm. in length. The aboral and oral portions of the body were approximately equal in length. The flying buttresses were just appearing as four equidistant swellings on the aboral portion of the body situated just above the circumferential groove and four complementary swellings on the oral portion of the body just below the groove. Three of these pairs were in contact, and in the fourth pair there was a linking strand of tissue. The marginal lappets were of equal widths to those of the intervening spaces. Patches of incipient gonadial tissue could be seen just near the base of each of the developing flying buttresses of the aboral portion of the body.

The medusa thus already has the characteristic form at a very small size, and its further development consists of increase in size with change in the proportional lengths of the aboral and oral portions of the body, as well as presumably of the growth of the finger-like processes of the marginal lappets.

GENERAL OBSERVATIONS

Swimming

Viguier (1890) made observations on living *Tetraplatia volitans* which he collected in the Bay of Algiers.

When at rest in a container the marginal lappets are folded inwards and the animal moves by ciliary action over the bottom of the container. It can also move in a more active manner by contractions of the body. From time to time the body assumes its regular form and the marginal lappets slowly expand. The medusa then swims by brisk flapping of the marginal lappets at 80 to 120 beats per min. An animal of average size traverses about 40 mm. in one minute with 120 beats. The medusa swims in a straight line, and the direction is changed by bending the aboral portion of the body. Marginal lappets usually beat simultaneously.

Krohn (1865) found that excised marginal lappets still continued to beat, indicating that they contained the excitatory elements.

Food

Dantan (1925) recorded the occurrence in the stomach of the remains of an annelid larva, parts of a crustacean, and a small chaetognath. Hand (1955) also recorded a chaetognath, copepods, and a young euphausid.

The medusa is thus evidently carnivorous, feeding on small plankton animals.

Vertical distribution

Tetraplatia volitans is apparently a pelagic species living over deep water. Observations on its vertical distribution are given by Hand (1955) and Beyer (1955), and this is further discussed by

Rees & White (1957) in the light of collections with closing nets. The medusa has been taken at all levels down to between 1,000 and 1,500 m. Hand found evidence of vertical migration at night. The medusa has been taken in quite shallow inshore waters (Dantan, 1925; Komai, 1939).

SEASONAL OCCURRENCE

Tetraplatia volitans is unlikely to be seasonal in occurrence, the species having been recorded at all times of the year except in January and October (Rees & White, 1957).

HISTORICAL

The history of *Tetraplatia* has been outlined on p. 201 and it is unnecessary to discuss it further.

BIBLIOGRAPHY

AAREM, H. E. van, VONK, H. J. & ZANDEE, D. I., 1964. Lipid metabolism in *Rhizostoma*. *Archs inst. Physiol.* **72**, 606–13.

ADERS, Walter M., 1903. Beiträge zur Kenntnis der Spermatogenese bei den Cölenteraten. *Z. wiss. Zool.* **74**, 12–38, text-figs. 1–8, pl. V–VI.

AGASSIZ, Alexander, 1865. Illustrated Catalogue of the Museum of Comparative Zoology at Harvard College. No. II. *North American Acalephae.* pp. 1–234, 360 text-figs.

AGASSIZ, Alexander & MAYER, Alfred Goldsborough, 1899. Acalephs from the Fiji Islands. *Bull. Mus. comp. Zool. Harv.* **32**, no. 9, pp. 157–89, pl. 1–17.

AGASSIZ, Alexander & WOODWORTH, W. McM., 1896. Studies from the Newport Marine Laboratory communicated by Alexander Agassiz. No. XL. Some variations in the genus *Eucope. Bull. Mus. comp. Zool. Harv.* **30**, 121–50, pl. I–IX.

AGASSIZ, Louis, 1860–2. *Contributions to the Natural History of the United States of America*, 1–4.

ALEXANDER, R. McN., 1964. Visco-elastic properties of the mesogloea of jellyfish. *J. exp. Biol.* **41**, 363–9, figs. 1–5.

ALLMAN, G. J., 1874. On the structure and systematic position of *Stephanoscyphus mirabilis*, the type of a new order of Hydrozoa. *Trans. Linn. Soc. Lond.* (Zool.), **1**, 61–6, pl. 14.

ALVARADO, Salustio, 1923. Contribución al Conocimiento histológico de las Medusas. 1. Los epitelios y la musculatura. *Trab. Mus. nac. Cienc. nat., Madr.* (Zool.), núm. 47, pp. 1–96, pl. I–XII.

ASTBURY, W. T., 1940. The molecular structure of the fibres of the collagen group. *J. int. Soc. Leath. Trades Chem.* **24**, 69–92, figs. 1–12.

ASTBURY, W. T. & BELL, Florence O., 1939. X-ray data on the structure of natural fibres and other bodies of high molecular weight. *Tabul. biol.* **17**, 90–112.

AURIVILLIUS, C. W. S., 1896. Das Plankton des Baltischen Meeres. *K. svenska VetenskAkad. Handl.* **21**, Afd. IV, no. 8, pp. 1–82, pl. I–II.

BAAN, S. M. van der, 1967a. *Pelagia noctiluca* (Forskål) collected off the Dutch coast. *Netherlands J. Sea Res.* **3**, 601–4, pl. I.

—— 1967b. Invasie van de Parelkwal, *Pelagia noctiluca. De Levende Natuur*, **70**, 66–9, fig. 1.

BAER, K. E., 1823. Ueber *Medusa aurita. Dt. Arch. Physiol.* **8**, 369–91, pl. IV.

BAILY, W. H., 1865. Notes on Marine Invertebrata, collected on Portmarnock Strand. *Proc. nat. Hist. Soc. Dublin*, **4**, 251–8.

BALLOWITZ, E., 1899. Über Hypomerie und Hypermerie bei *Aurelia aurita* Lam. *Arch. EntwMech. Org.* **8**, 239–52, pl. V.

BARBUT, J., 1783. *The Genera Vermium exemplified by various specimens of the animals contained in the Orders of the Intestina et Mollusca (Testacea, Lithophyta, and Zoophyta Animalia) Linnaei, etc., Genus 16. Medusa.* Tab. IX, pp. 76–83. London.

BARNES, J. H., 1960. Observations on jellyfish stingings in North Queensland. *Med. J. Aust.* **2**, no. 26, 24 December, pp. 993–9.

—— 1967. Extraction of cnidarian venom from living tentacle. In *Animal Toxins* (ed. Findlay E. Russell & Paul R. Saunders), pp. 115–29, figs. 1–7.

BARNES, W. J. P., 1964. Nervous, muscular and chemical transmission in coelenterates. *Biol. J. Univ. St Andrews*, **4**, 12–24.

BARNES, W. J. P. & HORRIDGE, G. A. 1965. A neuropharmacologically active substance from jellyfish ganglia. *J. exp. Biol.* **42**, 257–67, figs. 1–4.

BASSINDALE, R., & BARRETT, J. H., 1957. The Dale Fort marine fauna. *Proc. Bristol Nat. Soc.* **29**, 227–328, figs. 1–3.

BASTER, J., 1762. Opuscula subseciva, observationes miscellaneas de Animalculis et Plantis quibusdam marinis, eorumque ovariis et seminibus continentia. **1**. Harlem.

—— 1765. Natuurkundige Uitspanningen, behelzende eene Beschrijving, van meer dan vier hondert planten en insekten, keurig naar het leven Afgebeeld. **2**, 1–167, pl. I–XIII. Utrecht.

BATEMAN, J. B., 1932. The osmotic properties of medusae. *J. exp. Biol.* **9**, 124–7.

BATESON, William, 1894. *Materials for the study of variation treated with especial regard to discontinuity in the origin of species.* 598 pp., 209 figs. London: Macmillan. (*Aurelia*, pp. 426–9, fig. 128.)

BIBLIOGRAPHY

BAUER, V., 1927. Die Schwimmbewegungen der Quallen und ihre reflektorische Regulierung. *Z. vergl. Physiol.* **5**, 37–69, figs. 1–3.

BEER, G. R. de, & HUXLEY, J. S., 1924. Studies in dedifferentiation. 5. Dedifferentiation and reduction in *Aurelia. Q. Jl. microsc. Sci.* **68**, 471–9, text-figs. 1–10, pl. 22.

BENAZZI, Mario, 1933. L'influenza dell'acqua di mare diluita sui movimenti ritmici delle Meduse. *Boll. Zool.* anno **4**, n. 6, pp. 211–19.

BENEDEN, P. J. van, 1860. La strobilation des Scyphistomes. *Bull. Acad. r. Belg. Cl. Sci.* 2ᵉ sér. 7, no. 7, pp. 1–11.

BENNETT, Isobel, 1967. *The Fringe of the Sea.* 261 pp., 179 pl. London & Adelaide.

BERRILL, N. J., 1949a. Form and growth in the development of a Scyphomedusa. *Biol. Bull. mar. biol. Lab., Woods Hole,* **96**, 283–92, figs. 1–4.

—— 1949b. Developmental analysis of Scyphomedusae. *Biol. Rev.* **24**, 393–410, figs. 1–7.

BETHE, Albrecht, 1903. *Allgemeine Anatomie und Physiologie des Nervensystems.* 488 pp., 95 text-figs., 2 pl. Leipzig.

—— 1908. Die Bedeutung der Elektrolyten für die rhythmischen Bewegungen der Medusen (I. Theil). *Pfluegers Arch. ges. Physiol.* **124**, 541–77.

—— 1909. Die Bedeutung der Elektrolyten für die rhythmischen Bewegungen der Medusen (II. Theil). *Pfluegers Arch. ges. Physiol.* **127**, 219–73.

—— 1937. Experimentelle Erzeugung von Störungen der Erregungsleitung und von Alternans- und Period bildungen bei Medusen in Vergleich zu ähnlichen Erscheinungen am Wirbeltierherzen. *Zeit. vergl. Physiol.* **24**, 613.

BEYER, Fredrik, 1955. A new record of *Tetraplatia* (Pteromedusae). *Nytt Mag. Zool.* **3**, 106–12, figs. 1–2.

BIGELOW, Henry B., 1909. Reports on the Scientific Results of the Expedition to the Eastern Tropical Pacific, in charge of Alexander Agassiz, by the U.S. Fish Commission Steamer 'Albatross', from October, 1904, to March, 1905, Lieut. Commander L. M. Garrett, U.S.N. commanding. XVI. The Medusae. *Mem. Mus. comp. Zool. Harv.* **37**, 1–243, pl. 1–48.

—— 1913. Medusae and Siphonophorae collected by the U.S. Fisheries Steamer 'Albatross' in the North-western Pacific, 1906. *Proc. U.S. natn. Mus.* **44**, no. 1946, pp. 1–119, pl. 1–6.

—— 1926. Plankton of the offshore waters of the Gulf of Maine. *Bull. Bur. Fish., Wash.* **40**, 1924, pt. II, doc. no. 968, pp. 1–509, figs. 1–134.

—— 1928. Scyphomedusae from the Arcturus Oceanographic expedition. *Zoologica, N.Y.* **8**, no. 10, pp. 495–524, figs. 180–4.

—— 1938. Plankton of the Bermuda Oceanographic Expeditions. VIII. Medusae taken during the years 1929 and 1930. *Zoologica, N.Y.* **23**, pt. 2, no. 5, pp. 99–189, figs. 1–23.

BIGELOW, Robert Payne, 1910. A comparison of the sense-organs in medusae of the family Pelagidae. *J. exp. Zool.* **9**, 751–85, figs. 1–38.

BLAINVILLE, H. M. de, 1834. *Manuel d'Actinologie ou de Zoophytologie.* 1–2, 694 pp., 103 pl. Paris.

BODANSKY, Meyer & ROSE, William C., 1922. Comparative studies of digestion. I. The digestive enzymes of Coelenterates. *Am. J. Physiol.* **62**, 473–81.

BOGUCKI, Mieczyslaw, 1933. Sur le cycle évolutif de l'*Aurelia aurita* L. dans les eaux polonaises de la Baltique. *Fragm. faun.* **2**, nr. 12, pp. 117–19.

BORLASE, William, 1758. *The Natural History of Cornwall.* 326 pp., 29 pl.

BOUILLON, J., CASTIAUX, P. & VANDERMEERSSCHE, G., 1957. Quelques aspects histologiques de *Limnocnida tanganyicae.* Observations de coupes ultrafines au microscope électronique. *Expl Cell Res.* **13**, no. 3, pp. 529–44, figs. 1–18.

BOUILLON, J. & VANDERMEERSSCHE, G., 1956–7. Structure et nature de la mesoglée des Hydro- et Scypho-méduses. *Annls Soc. r. zool. Belg.* **87**, 9–25, figs. 1–11.

BOURNE, Gilbert C., 1890. Report of a trawling cruise in H.M.S. 'Research' off the south-west coast of Ireland. *J. mar. biol. Ass. U.K.* **1**, n.s., 306–27.

BOWMAN, Thomas E., MEYERS, Caldwell C. & HICKS, Steacy D., 1963. Notes on associations between hyperiid amphipods and medusae in Chesapeake and Narragansett Bays and the Niantic River. *Chesapeake Sci.* **4**, 141–6, figs. 1–2.

BOZLER, Emil, 1926a. Sinnes- und Nervenphysiologische Untersuchungen an Scyphomedusen. *Z. vergl. Physiol.* **4**, 37–80, figs. 1–8.

—— 1926b. Weitere Untersuchungen zur Sinnes- und Nervenphysiologie der Medusen: Erregungsleitung, Funktion der Randkörper, Nahrungsaufnahme. *Z. vergl. Physiol.* **4**, 797–817, figs. 1–5.

—— 1927. Untersuchungen über das Nervensystem der Coelenteraten. I. Teil: Kontinuität oder Kontakt zwischen den Nervenzellen. *Z. Zellforsch. mikrosk. Anat.* **5**, 244–62, figs. 1–13.

BRANDT, Alexander, 1870. Ueber *Rhizostoma cuvieri* Lmk. Ein Beitrag zur Morphologie der vielmündigen Medusen. *Zap. imp. Akad. Nauk*, ser. 7, **16**, no. 6, pp. 1–29. 1 pl.

BRANDT, J. F., 1838. Ausführliche Beschreibung der von C. H. Mertens auf seiner Weltumsegelung beobachteten Schirmquallen. *Zap. imp. Akad. Nauk*, ser. 6, **2**, 237–411, pl. I–XXXI.

BROCH, Hjalmar, 1913. Scyphomedusae from the 'Michael Sars' North Atlantic Deep-Sea Expedition 1910. *Rep. 'Michael Sars' N. Atl. Deep Sea Exped.* **3**, pt. 1, pp. 1–20, pl. I, text-figs. 1–12. (Reprinted 1933.)

BROWNE, Edward T., 1894. 'Aurelia aurita.' *Nature, Lond.*, **50**, 524.

—— 1895*a*. Report on the medusae of the L.M.B.C. district. *Proc. Trans. Lpool biol. Soc.* **9**, 243–86.

—— 1895*b*. On the variation of the tentaculocysts of *Aurelia aurita*. *Q. Jl microsc. Sci.* **37**, pt. 3, n.s., pp. 245–51, pl. 25.

—— 1896. The medusae of Valencia harbour, County Kerry. *Ir. Nat.* July, pp. 179–81.

—— 1900. The fauna and flora of Valencia harbour on the west coast of Ireland. I. The pelagic fauna (1895–98). II. The medusae (1895–98). *Proc. R. Ir. Acad.* 3rd ser. **5**, no. 5, pp. 667–736, pl. XX–XXI.

—— 1901. Variation in *Aurelia aurita*. *Biometrika*, **1**, 90–108, 5 figs.

—— 1905*a*. A report on the medusae found in the Firth of Clyde (1901–1902). *Proc. R. Soc. Edinb.* **25**, pt. IX, pp. 738–78.

—— 1905*b*. Scyphomedusae. *Fauna Geogr. Maldive & Laccadive Archipelagos*, **2**, suppl. 1, pp. 958–71, pl. XCIV.

—— 1910. Coelentera. V. Medusae. *Rep. scient. Results natn. Antarct. Exped. 1901–1904, Nat. Hist.* **5** (Zool. Bot.), pp. 1–62, pl. I–VII.

—— 1916. Medusae from the Indian Ocean. *Trans. Linn. Soc. Lond.* 2nd ser. Zool., **17**, pt. 2, pp. 169–210, pl. 39. (Percy Sladen Trust Exped. H.M.S. 'Sealark', 1905; **6**, no. IV.)

BRUCE, J. R., COLMAN, J. S. & JONES, N. S. (Ed.) 1963. Marine fauna of the Isle of Man and its surrounding seas. *L.M.B.C. Mem. typ. Br. mar. Pl. Anim.* Mem. no. 36, 307 pp.

BULLOCK, Theodore H., 1943. Neuromuscular facilitation in Scyphomedusae. *J. cell. comp. Physiol.*, **22**, 251–72, figs. 1–6.

BULLOCK, T. H. & HORRIDGE, G. A., 1965. *Structure and function in the nervous systems of invertebrates*, **1**, chap. 8, Coelenterata and Ctenophora, pp. 459–534.

BUSCH, Wilhelm, 1851. *Beobachtungen über Anatomie und Entwickelung einiger wirbellosen Seethiere.* pp. 1–143, Taf. I–XVII. Berlin.

CARGO, David G., & SCHULTZ, Leonard P., 1966. Notes on the biology of the sea nettle, *Chrysaora quinquecirrha*, in Chesapeake Bay. *Chesapeake Sci.* **7**, 95–100, figs. 1–2.

CARLGREN, Oscar, 1909. Die Tetraplatien. *Wiss. Ergebn. dt. Tiefsee-Exped. 'Valdivia'*, **19**, Heft 3, pp. 78–121, text-figs. 1–3, pl. X–XIII.

CHAMISSO, A. & EYSENHARDT, C. G., 1821. De animalibus quibusdam e classe Vermium Linneana, in circumnavigatione terrae, auspicante Comite N. Romanzoff duce Ottone de Kotzebue, annis 1815–1818 per acta, observatis...*Nova Acta Acad. Caesar. Leop. Carol.* **10** (2).

CHAPMAN, David M., 1965. Co-ordination in a Scyphistoma. *Am. Zoologist*, **5**, 455–64, figs. 1–3.

—— 1966. Evolution of the Scyphistoma. In 'The Cnidaria and their evolution'. *Symp. zool. Soc. Lond.* no. 16, pp. 51–75, figs. 1–5.

—— 1968. Structure, histochemistry and formation of the podocyst and cuticle of *Aurelia aurita* (L.). *J. mar. biol. Ass. U.K.* **48**, 187–208, text-figs. 1–3, pl. I–V.

CHAPMAN, Garth, 1953. Studies of the mesogloea of coelenterates. I. Histology and chemical properties. *Q. Jl. microsc. Sci.* **94**, 155–76, figs. 1–6.

—— 1958. The hydrostatic skeleton in the invertebrates. *Biol. Rev.* **33**, 338–71, figs. 1–5.

—— 1959. The mesogloea of *Pelagia noctiluca*. *Q. Jl microsc. Sci.* **100**, 599–610, figs. 1–7.

—— 1966. The structure and functions of the mesogloea. In 'The Cnidaria and their evolution'. *Symp. zool. Soc. Lond.* no. 16, pp. 147–68, figs. 1–4.

CHRISTOMANOS, A., 1954. A violet pigment from the Mediterranean medusa *Rhizostoma pulmo*. *Nature, Lond.* **173**, 875–6, figs. 1–2.

CLAUS, C., 1877. Studien über Polypen und Quallen der Adria. I. Acalephen (Discomedusen). *Denkschr. Akad. Wiss., Wien*, **38**, 64 pp., pl. I–XI.

—— 1878. Ueber *Tetrapteron* (*Tetraplatia*) *volitans*. *Arch. mikrosk. Anat. EntwMech.* (Anat.), **15**, 349–58, pl. XXII, figs. 1–9.

—— 1883. *Untersuchungen über die Organisation und Entwicklung der Medusen.* Prague & Leipzig, pp. 1–96, 9 text-figs. pl. I–XX.

—— 1884. Die Ephyren von *Cotylorhiza* und *Rhizostoma*. *Arb. zool. Inst. Univ. Wien*, **5**, 169–78, pl. I–II.

BIBLIOGRAPHY

CLAUS, C., 1891. Ueber die Entwicklung des Scyphostoma von *Cotylorhiza*, *Aurelia* und *Chrysaora*, sowie über die systematische Stellung der Scyphomedusen. *Arb. zool. Inst. Univ. Wien*, **9**, 85–128, pl. IV–VI.

CLELAND, Sir J. B. & SOUTHCOTT, R. V., 1965. Injuries to man from marine invertebrates in the Australian region. *National Health & Medical Research Council. Special Rep. Ser.* no. 2, 282 pp., text-figs. 1–19, 10 pl.

COLE, F. J., 1952. *Pelagia* in Manx waters. *Nature, Lond.* **170**, 587.

COLOSANTI, J., 1888. Das blaue Pigment der Hydromedusen. *Unters. Naturl. Mensch. Tiere*, **13**, 471.

CONKLIN, Edwin G., 1908. The habits and early development of *Linerges mercurius*. *Publs Carnegie Instn*, **103**, 153–70, pl. 1–8.

COWLES, R. P., 1930. A biological study of the offshore waters of Chesapeake Bay. *Bull. Bur. Fish., Wash.* **46**, 277–381, figs. 1–16.

CUSTANCE, D. R. N., 1964. Light as an inhibitor of strobilation in *Aurelia aurita*. *Nature, Lond.* **204**, 1219–20.

—— 1966. The effect of a sudden rise in temperature on strobilae of *Aurelia aurita*. *Experientia*, **22**, 588–9, 1 fig.

—— 1967. Studies on strobilation in the Scyphozoa. *J. biol. Educ.* **1**, 79–81.

DAHL, Erik, 1959a. The amphipod, *Hyperia galba*, an ectoparasite of the jellyfish, *Cyanea capillata*. *Nature, Lond.* **183**, 1749.

—— 1959b. The hyperiid amphipod, *Hyperia galba*, a true ectoparasite on jellyfish. *Univ. Bergen Årb.* nr 9, pp. 1–8, figs. 1–5. (Publs biol. Stn Espegrend, no. 28.)

—— 1961. The association between young whiting, *Gadus merlangus*, and the jellyfish *Cyanea capillata*. *Sarsia*, no. 3, pp. 47–55, fig. 1.

DAHLGREN, Ulric, 1916. The production of light by animals. Porifera and Coelenterata. *J. Franklin Inst.* **181**, 243–61, figs. 1–8.

DALYELL, J. G., 1834. On the propagation of Scottish zoöphytes. *Edinb. New Philos. Journ.* **17**, 411; also *Rep. Brit. Assoc. Adv. Sci.* p. 598. [*Misspelt Dalzell in *Edinb. New Philos. J.*]

—— 1836. Further illustrations of the propagation of Scottish zoöphytes. *Edinb. New Philos. Journ.* **21**, 88.

—— 1847–8. *Rare and remarkable animals of Scotland, represented from living subjects: with practical observations on their nature*, **1** (1847), i–xii, 1–268, pl. I–LIII; **2** (1848), i–iv, 1–322, pl. I–LVI. London: J. van Voorst.

DAMAS, D., 1909a. Review of Norwegian fishery and marine investigations 1900–1908. Plankton. *Rep. Norw. Fishery mar. Invest.* **2**, no. 1, pp. 93–107.

—— 1909b. Contribution à la biologie des Gadides. *Rapp. P.-v. Réun. Cons. perm. int. Explor. Mer*, **10**, 3, 1–277, text-figs. 1–25, pl. I–XXI.

DANTAN, J.-L., 1925. Contribution à l'Étude du *Tetraplatia volitans*. *Annls Inst. océanogr., Monaco*, n.s. **2**, 429–59, text-figs. I–XII, pl. I–II.

DAUMAS, Raoul & CECCALDI, Hubert J., 1965. Contribution à l'étude biochimique d'organismes marins. I. Acid amines libres et protéiques chez *Beroë ovata* (Eschscholtz), *Ciona intestinalis* (L.), *Cymbulia peroni* (De Blainville) et *Rhizostoma pulmo* (Agassiz). *Recl Trav. Stn mar. Endoume*, bull. 38, fasc. 54, pp. 3–14.

DAVIDSON, Viola M. & HUNTSMAN, A. G., 1926. The causation of diatom maxima. *Trans. R. Soc. Can.* **20**, sect. V, pp. 119–25, figs. 1–3.

DELAGE, Y. & HÉROUARD, E., 1901. *Traité de zoologie concrète*, **2**, Coelentérés. 848 pp. 72 pl.

DELAP, Maude J., 1901. Notes on the rearing of *Chrysaora isosceles* in an aquarium. *Ir. Nat.* **10**, 25–8, pl. 1, 2.

—— 1905. Notes on the rearing, in an aquarium, of *Cyanea lamarcki*, Péron & Lesueur. *Rep. Sea inld Fish. Ire.* (1902–3), pt. II, Sci. Invest. pp. 20–2, pl. I–II.

—— 1907. Notes on the rearing, in an aquarium, of *Aurelia aurita* L. and *Pelagia perla* (Slabber). *Rep. Sea inld Fish. Ire.* (1905), pt. II, Sci. Invest. pp. 160–4, pl. I–II.

DELAP, M. & C., 1907. Notes on the plankton of Valencia harbour. *Rep. Sea inld Fish. Ire.* (1905), pt. II, Sci. Invest. pp. 141–59.

DENTON, E. J. & SHAW, T. I., 1962. The buoyancy of gelatinous marine animals. *J. Physiol. Lond.* **161**, 14–15P, fig. 1.

DERBÈS, A., 1850. Note sur les organes reproducteurs et l'embryogénie de *Cyanea chrysaora*. *Annls Sci. nat.* sér. 3, **13**, 377–82, pl. I, figs. 1–12.

DIGBY, Peter S. B., 1967. Pressure sensitivity and its mechanism in the shallow water environment. *Symp. zool. Soc. Lond.* no. 19, pp. 159–88, figs. 1–15.

DOYLE, Sir Arthur Conan, 1927. *The Case-book of Sherlock Holmes*. IX. The adventure of the lion's mane.

DUBOIS, Raphael, 1914. *La vie et la lumière*. 338 pp. 48 figs. Paris.

DUNCKER, Georg, 1894. Ueber ein abnormes Exemplar von *Aurelia aurita* L. *Arch. Naturgesch.* **60**, 7–9, pl. I, figs. 11, 12.

EBBECKE, U., 1935. Über die Wirkungen hoher Drucke auf marine Lebewesen. *Pfluegers Arch. ges. Physiol.* **236**, 648–57.

EHRENBERG, C. G., 1837. Über die Akalephen des Rothen Meeres und den Organismus der Medusen der Ostsee. *Abh. preuss. Akad. Wiss. Berlin* (1835), pp. 181–260, pl. I–VIII.

EIMER, Theodor, 1878. *Die Medusen. Physiologisch und morphologisch auf ihr Nervensystem*. 277 pp. 13 pl. Tübingen.

EL-DUWEINI, A. KHALAF, 1945. Cannibalism in *Aurelia*. *Nature, Lond.* **155**, 337.

ESCHSCHOLTZ, Fr., 1829. *System der Acalephen. Eine ausführliche Beschreibung aller medusenartigen Strahlthiere*. 190 pp. 16 pl. Berlin: Ferdinand Dümmler.

EVANS, H. Muir, 1943. *Sting-fish and seafarer*. 180 pp. 31 figs. London.

EVANS, William, 1916. *Haliclystus auricula* (Rathke), and other medusae in the Firth of Forth. *Scott. Nat.* pp. 283–6.

EVANS, Wm & ASHWORTH, J. H., 1909. Some medusae and ctenophores from the Firth of Forth. *Proc. R. phys. Soc. Edinb.* **17**, no. 6, pp. 300–11.

EYSENHARDT, F. W., 1821. Zur Anatomie und Naturgeschichte der Quallen. I. Von dem *Rhizostoma Cuvierii* Lam. *Nova Acta Acad. Caesar. Leop. Carol.* **10**, 377–410, pl. XXXIV.

FABER, F., 1829. *Naturgeschichte der Fische Islands. Mit einem Anhange von den Isländischen Medusen und Strahlthieren*. 206 pp. Frankfurt am Main.

FABRICIUS, O., 1780. *Fauna Groenlandica, systematice sistens Animalia Groenlandiae occidentalis, hactenus indagata. . . .* Hafniae et Lipsiae.

FEWKES, J. Walter, 1881. Studies of the jellyfishes of Narragansett Bay. *Bull. Mus. comp. Zool. Harv.* **8**, no. 8, pp. 141–82, pl. I.

—— 1883. The affinities of *Tetraplatia volitans*. *Am. Nat.* **17**, 426.

—— 1886. Report on the medusae collected by the U.S.F.C. Steamer 'Albatross', in the region of the Gulf Stream, in 1883–84. *U.S. Comm. Fish & Fisheries*, pt. XII, rep. for 1884, pp. 927–77, pl. I–X.

FISH, Charles J., 1925. Seasonal distribution of the plankton of the Woods Hole Region. *Bull. Bur. Fish. Washington*, **41**, doc. no. 975, pp. 91–179, figs. 1–81.

FORBES, Edward, 1848. A monograph of the British naked-eyed medusae: with figures of all the species. *Ray Society*, pp. 1–104, pl. I–XIII.

FORSKÅL, Petrus, 1775. *Descriptiones animalium, avium, amphibiorum, piscium, insectorum, vermium; quae in itinere orientali observavit Petrus Forskål*. Post mortem auctoris edidit Carsten Niebuhr. 164 pp. Hauniae.

FOX, D. L. & MILLOTT, N., 1954. The pigmentation of the jellyfish, *Pelagia noctiluca* (Forskål) var. *panopyra* Péron & Lesueur. *Proc. R. Soc.* B, **142**, 392–408, figs. 1–11.

FOX, D. L. & PANTIN, C. F. A., 1944. Pigments in the Coelenterata. *Biol. Rev.* **19**, 121–34.

FRÄNKEL, Gottfried, 1925. Der statische Sinn der Medusen. *Z. vergl. Physiol.* **2**, 658–90, figs. 1–17.

FRANZEN, Åke, 1956. On spermiogenesis, morphology of the spermatozoon, and biology of fertilization among invertebrates. *Zool. Bidr. Upps.* **31**, 355–482, text-figs. 1–223, pl. 1–6.

FRASER, J. H., 1954. North Sea Plankton. *Ann. Biol.* **10** (1953), 100.

—— 1961. An uncommon jellyfish. *Scottish Fisheries Bull.* no. 14, p. 23, fig. 11.

—— 1969. Observations on the experimental feeding of some medusae and Chaetognatha. *J. Fish. Res. Bd Canada*, **26**, in press.

FRIEDEMANN, Otto, 1902. Untersuchungen über die postembryonale Entwicklung von *Aurelia aurita*. *Z. wiss. Zool.* **71**, 227–67, pl. XII–XIII, text-figs. 1–3.

GAEDE, H. M., 1816. *Beiträge zur Anatomie und Physiologie der Medusen, nebst einem Versuch einer Einleitung über das, was den ältern Naturforschern in Hinsicht dieser Thiere bekannt war*. 23 pp. pl. I–II. Berlin.

GEMMILL, J. F., 1921. Notes on food-capture and ciliation in the ephyrae of *Aurelia*. *Proc. R. phys. Soc. Edinb.* **20**, 222–5, figs. 1 & 2.

GILCHRIST, Francis G., 1937. Budding and locomotion in the Scyphistomas of *Aurelia*. *Biol. Bull. mar. biol. Lab., Woods Hole*, **72**, 99–124, figs. 1–9.

GOETTE, A., 1887. *Abhandlungen zur Entwicklungsgeschichte der Tiere*. Viertes Heft. *Entwicklungsgeschichte der* Aurelia aurita *und* Cotylorhiza tuberculata. 79 pp. 16 text-figs. pl. I–IX. Hamburg & Leipzig.

—— 1892. Über die Entwickelung von *Pelagia noctiluca*. *Math.-naturw. Mitt.* **7**, 413–21.

—— 1893. Vergleichende Entwicklungsgeschichte von *Pelagia noctiluca* Pér. *Z. wiss. Zool.* **55**, 645–712, pl. XXVIII–XXXI.

BIBLIOGRAPHY

GOHAR, H. A. F. & EISAWY, A. M., 1960. The development of *Cassiopea andromeda* (Scyphomedusae)· *Publs mar. biol. Stn Ghardaqa*, no. 11, pp. 147–90, figs. 1–19.

GOLDFARB, A. J., 1914. Changes in salinity and their effects upon regeneration of *Cassiopea xamachana*. *Pap. Tortugas Lab.* **6**, 85–94, figs. 1–4.

GOODEY, T., 1908. On the presence of gonadial grooves in a medusa, *Aurelia aurita*. *Proc. zool. Soc. Lond.* pp. 55–9, pl. I.

—— 1909. A further note on the gonadial grooves of a medusa, *Aurelia aurita*. *Proc. zool. Soc. Lond.* pp. 78–81, pl. XXIV.

GOSSE, Philip Henry, 1853. *A Naturalist's Rambles on the Devonshire Coast.* 451 pp. pl. I–XXVIII. London: J. van Voorst.

—— 1856. *Tenby: a sea side holiday.* 400 pp. pl. I–XXIII. London: J. van Voorst.

—— 1863. The blue *Cyanea. The Intellectual Observer*, **4**, no. III, pp. 149–56, 1 pl.

GRAEFFE, Ed., 1884. Uebersicht der Seethierfauna des Golfes von Triest nebst Notizen über Vorkommen, Lebensweise, Erscheinungs- und Fortpflanzungszeit der einzelnen Arten. *Arb. zool. Inst. Univ. Wien*, **5**, 333–62.

GRIFFITHS, A.-B. & PLATT, C., 1895. Sur la composition de la pélagéine. *C. r. hebd. Séanc. Acad. Sci., Paris*, **121**, 451–2.

HAAHTELA, I. & LASSIG, J., 1967. Records of *Cyanea capillata* (Scyphozoa) and *Hyperia galba* (Amphipoda) from the Gulf of Finland and the Northern Baltic. *Annls Zool. Fennici*, **4**, 469–71, fig. 1.

HADŽI, Jovan, 1909. Einige Kapitel aus der Entwicklungsgeschichte von *Chrysaora*. *Arb. zool. Inst. Univ. Wien*, **17**, 17–44, pl. III–IV, text-figs. 1–15.

HAECKEL, Ernst, 1880. *System der Acraspeden. Zweite Hälfte des Systems der Medusen.* pp. 361–672, pl. XXI–XL. Jena.

—— 1881 a. Report on the deep-sea medusae dredged by H.M.S. 'Challenger' during the years 1873–1876. *Rep. Challenger Exped., Zool.* **4** (1882), 1–154, text-figs. A–Q, pl. 1–32.

—— 1881 b. *Metagenesis und Hypogenesis von* Aurelia aurita. *Ein Beitrag zur Entwickelungsgeschichte und zur Teratologie der Medusen.* 36 pp, pl. I–II. Jena.

HAGMEIER, A., 1933. *Die Züchtung verschiedener wirbelloser Meerestiere. Abderhalden's Handbuch der biologischen Arbeitsmethoden*, Abt. IX, Teil 5, pp. 465–598, figs. 135–174.

HALISCH, W., 1933. Beobachtungen an Scyphopolypen. *Zool. Anz.* **104**, 296–304, figs. 1–10.

HALSTEAD, B. W., 1965. *Poisonous and venomous marine animals of the World.* **1**, Invertebrates, 994 pp. Washington.

HAMANN, Otto, 1881. Die Mundarme der Rhizostomen und ihre Anhangsorgane. *Jena Z. Naturw.* **15**, 243–85, pl. IX–XI.

—— 1883. Beiträge zur Kenntnis der Medusen. *Z. wiss. Zool.* **38**, 419–28, pl. XXIII.

—— 1890. Ueber die Entstehung der Keimblätter. Ein Erklärungsversuch. *Int. Mschr. Anat. Physiol.* **7**, 255–67, 295–311, pl. XII.

HAND, Cadet, 1955. A study of the structure, affinities, and distribution of *Tetraplatia volitans* Busch (Coelenterata; Hydrozoa: Pteromedusae). *Pacif. Sci.* IX, 332–48, figs. 1–8.

HARDY, Alister C., 1956. *The open sea. Its natural history: the world of plankton.* 335 pp. 48 pl. 105 text-figs. London.

HARGITT, C. W., 1902. Notes on the coelenterate fauna of Woods Hole. *Am. Nat.* **36**, 549–60, 4 text-figs.

—— 1904. Regeneration in *Rhizostoma pulmo. J. exp. Zool.* **1**, 73–94, figs. 1–6.

—— 1905. Variations among Scyphomedusae. *J. exp. Zool.* **2**, 547–82, pl. 1, text-figs. 1–17.

HARGITT, C. W. & HARGITT, G. T., 1910. Studies on the development of Scyphomedusae. *J. Morph.* **21**, 217–62, figs. 1–48.

HARTING, P., 1875. Notices zoologiques faites pendant un séjour à Schéveningen, 1874. Œufs de *Cyanea*. Otolithes de *Cyanea* et de *Chrysaora. Niederl. Arch. Zool.* **2**, pt. 3, pp. 1–25, pl. 1.

HARTLAUB, Clemens, 1894. Die Coelenteraten Helgolands. *Wiss. Meeresunters. Abt. Helgoland*, N.F., **1**, Heft 1, IV, pp. 161–206.

—— 1909. Méduses. *Crois. Océanogr. 'Belgica'* (Mer du Grönland 1905), pp. 471–8, pl. LXXVI & LXXVII.

HARVEY, E. Newton, 1952. *Bioluminescence.* New York.

HATAI, S., 1917. On the composition of *Cassiopea xamachana* and the changes in it after starvation. *Pap. Tortugas Lab.* **11**, 95–109, fig. 1.

HAUROWITZ, Felix, 1920. Untersuchung des Fetts der Gonaden von *Rhizostoma Cuvieri. Hoppe-Seyler's Z. physiol. Chem.* **112** (1921), 28–37.

HEIN, W., 1900. Untersuchungen über die Entwicklung von *Aurelia aurita*. *Z. wiss. Zool.* **68**, 401–38, pl. XXIV–XXV, text-figs. 1–5.

HELA, Ilmo, 1951. On the occurrence of the jellyfish, *Aurelia aurita* L., on the south coast of Finland. *Suomal. eläin-ja kasvit. Seur. van Tiedon.* **6**, 71–8, fig. 1.

HEMMING, Francis (ed.), 1958. *Official Index of rejected and invalid generic names in zoology*. First instalment: names 1–1169, pp. 1–132. (*Aurelia*, p. 117, No. 1157.)

HENSCHEL, J., 1935. Untersuchungen über den chemischen Sinn der Scyphomedusen *Aurelia aurita* und *Cyanea capillata* und der Hydromeduse *Sarsia tubulosa*. *Wiss. Meeresunters. Abt. Kiel*, N.F., **22**, 25–42, 4 text-figs.

HERDMAN, William Abbot, 1880. On the invertebrate fauna of Lamlash Bay. *Proc. R. phys. Soc. Edinb.* **5**, 193–218, pl. IV.

—— 1894. Variation of *Aurelia*. *Nature, Lond.* **50**, 426.

HERIC, Mat., 1909. Zur Kenntnis der polydisken Strobilation von *Chrysaora*. *Arb. zool. Inst. Univ. Wien.* **17**, 95–108, pl. IX, text-fig. 1.

HÉROUARD, E., 1907. Existence de statoblastes chez le Scyphistome. *C. r. hebd. Séanc. Acad. Sci. Paris*, **145**, 601–3.

—— 1908. Sur un Acraspède sans Méduse: *Taeniolhydra roscoffensis*. *C. r. hebd. Séanc. Acad. Sci. Paris*, **147**, 1336–7.

—— 1909 a. Sur les cycles évolutifs d'un Scyphistome. *C. r. hebd. Séanc. Acad. Sci. Paris*, **148**, 320–3.

—— 1909 b. Sur les entéroïdes des Acraspèdes. *C. r. hebd. Séanc. Acad. Sci. Paris*, **148**, 1225–7.

—— 1911 a. Sur le mode de fixation au sol des Scyphistomes par des tonofibrilles. *Bull. Soc. zool. Fr.* **36**, 15–19, figs. 1–3.

—— 1911 b. Le pharynx des Scyphistomes. *Zool. Anz.* **38**, 231–3.

—— 1911 c. Sur la progénèse parthénogénésique à longue échéance de *Chrysaora*. *C. r. hebd. Séanc. Acad. Sci. Paris*, **153**, 1094–5.

—— 1912. Histoire du kyste pédieux de *Chrysaora* et sa signification. *Archs Zool. exp. gén. Paris*, 5° sér., **10**, Notes et Revue, II, pp. xi–xxv, figs. 1–6.

—— 1920 a. Existence d'une bistrobilisation; sa signification et ses conséquences. *Bull. Soc. zool. Fr.* **45**, 162–9, figs. A–E.

—— 1920 b. Les monstres doubles du Scyphistome. *C. r. hebd. Séanc. Acad. Sci. Paris*, **170**, 295–8.

—— 1921. Rétablissement de l'équilibre de corrélation par lacération chez le Scyphistome. *Bull. Soc. zool. Fr.* **46**, 68–72, figs. 1–7.

HERTWIG, Oscar & HERTWIG, Richard, 1878. *Das Nervensystem und die Sinnesorgane der Medusen*, 189 pp. pl. I–X, Leipzig.

—— 1879. *Studien zur Blättertheorie. Heft I. Die Actinien. Anatomisch und Histologisch mit besonderer Berücksichtigung des Nervenmuskelsystems*. Jena. 224 pp. pl. I–X.

HESSE, R., 1895. Über das Nervensystem und die Sinnesorgane von *Rhizostoma Cuvieri*. *Z. wiss. Zool.* **60**, 411–57, text-figs. I–III, pl. XX–XXII.

HEYMANS, C. & MOORE, A. R., 1923. Action des ions sur la luminescence et les pulsations de *Pelagia noctiluca*. *C.R. Soc. biol. Paris*, **89**, 430–2.

—— 1924. Luminescence in *Pelagia noctiluca*. *J. gen. Physiol.* **6**, 273–80.

HILLIS, J. P., 1967. *Cyanea lamarckii* Péron et Lesueur off the coast of Co. Dublin. *Ir. Nat. J.* **15**, 362–3.

HJORT, Johan & DAHL, Knut, 1900. Fishing experiments in Norwegian fiords. *Rep. Norw. Fishery mar. Invest.* **1**, no. 1, pp. 1–215, text-figs. 1–32, pl. I–III.

HÖGBERG, Bertil, SÜDOW, Göran, THON, Inga-Lisa & UVNÄS, Börje, 1957. The inhibitory action of a compound obtained from hip seeds (HSC) on the release of histamine and the disruption of mast cells produced by compound 48/80 and extracts from jellyfish (*Cyanea capillata*) and eelworm of swine (*Ascaris lumbricoides*). *Acta physiol. scand.* **38**, 265–74, figs. 1–3.

HÖGBERG, Bertil, THUFVESSON, Gunnar & UVNÄS, Börje, 1956. Histamine liberation produced in the perfused paw of the cat by 48/80 and extracts from jellyfish (*Cyanea capillata*) and eelworm (*Ascaris lumbricoides*) from swine. *Acta physiol. scand.* **38**, 135–44, figs. 1–3.

HOLLOWDAY, Eric D., 1947. On the commensal relationship between the amphipod *Hyperia galba* (Mont.) and the scyphomedusa *Rhizostoma pulmo* Agassiz, var. *octopus* Oken. *J. Quekett microsc. Club*, ser. 4, **2**, 187–90, pl. 25–6.

—— 1951. Planula, hydratuba and ephyra stages of the common jellyfish. *The Microscope*, **8**, 193–8, figs. 1–11.

HORRIDGE, G. A., 1953. An action potential from the motor nerves of the jellyfish *Aurellia aurita* Lamarck. *Nature, Lond.* **171**, 400, 1 fig.

BIBLIOGRAPHY

HORRIDGE, G. A., 1954a. Observations on the nerve fibres of *Aurellia aurita*. *Q. Jl microsc. Sci.* **95**, 85–92, figs. 3 & 4.

—— 1954b. The nerves and muscles of medusae. I. Conduction in the nervous system of *Aurellia aurita*. *J. exp. Biol.* **31**, 594–600, figs. 1–4.

—— 1955. The nerves and muscles of medusae. III. A decrease in the refractory period following repeated stimulation of the muscle of *Rhizostoma pulmo*. *J. exp. Biol.* **32**, no. 4, pp. 636–41, figs. 1–4.

—— 1956a. The nerves and muscles of medusae. V. Double innervation in Scyphozoa. *J. exp. Biol.* **33**, 366–83, figs. 1–9.

—— 1956b. The nervous system of the ephyra larva of *Aurellia aurita*. *Q. Jl microsc. Sci.* **97**, pt. 1, pp. 59–74, figs. 1–9.

—— 1959. The nerves and muscles of medusae. VI. The rhythm. *J. exp. Biol.* **36**, 72–91, figs. 1–9.

—— 1969. Statocysts of medusae and evolution of stereocilia. *Tissue & Cell*, **1**, 341–53, figs. 1–10.

HORRIDGE, G. A. & MACKAY, Bruce, 1962. Naked axons and symmetrical synapses in coelenterates. *Q. Jl microsc. Sci.* **103**, 531–51, figs. 1–6.

HORSTMANN, Ernst, 1934a. Untersuchungen zur Physiologie der Schwimmbewegungen der Scyphomedusen. *Pflügers Arch. ges. Physiol.* **234**, 406–20, figs. 1–6.

—— 1934b. Nerven- und muskelphysiologische Studien zur Schwimmbewegung der Scyphomedusen. *Pflügers Arch. ges. Physiol.* **234**, 421–31, figs. 1–6.

HUNT, O. D., 1952. Occurrence of *Pelagia* in the River Yealm Estuary, South Devon. *Nature, Lond.* **169**, 934.

HÜSING, Johannes Otto, 1956. Zur Frage der Ernährung der Ohrenqualle *Aurelia aurita* L. (Scyphozoa, Semaeostomae). *Wiss. Z. Martin-Luther Univ. Halle-Wittenb.* (V. Jahrg.), Heft 3, pp. 479–81.

HUXLEY, Thomas Henry, 1849. On the anatomy and the affinities of the family of the Medusae. *Phil. Trans. R. Soc.* pt. II, pp. 413–34, pl. XXXVII–XXXIX.

HYDE, Ida H., 1894. Entwicklungsgeschichte einiger Scyphomedusen. *Z. wiss. Zool.* **58**, 531–65, pl. XXXII–XXXVII, text-figs. A–D.

HYKES, O. V., 1928. Contribution à la physiologie de la luminescence et de la motilité des Coelentérés. *C.R. Soc. biol. Paris*, **98**, 259–61.

HYMAN, L. H., 1938. The water content of medusae. *Science N.Y.* **87**, 166–7.

—— 1940. Observations and experiments on the physiology of medusae. *Biol. Bull. mar. biol. Lab., Woods Hole*, **79**, 282–96, figs. 1–8.

—— 1943. Water content of medusae: sexuality in a planarian. *Nature, Lond.* **151**, 140.

IWANZOFF, N., 1896. Über den Bau, die Wirkungsweise und die Entwickelung der Nessel-Kapseln der Coelenteraten. *Byull. mosk. Obshch. Ispȳt. Prir.* (*Bull. Soc. nat. Moscau*), n.s. **10**, 95–161 & 323–55, 6 pl.

JOHNSON, M. E. & SNOOK, H. J., 1927. *Seashore animals of the Pacific coast* (pl. II). New York: Macmillan & Co.

JOHNSTONE, James, 1908. *Conditions of life in the sea*. Cambridge Univ. Press. 332 pp.

KAKINUMA, Yoshiko, 1965. On some factors for the differentiations of *Cladonema uchidai* and of *Aurelia aurita*. *Bull. biol. Stn Asamushi*, **11**, 81–5, fig. 1.

KÄNDLER, Rudolf, 1950. Jahreszeitliches Vorkommen und unperiodisches Auftreten von Fischbrut, Medusen und Dekapodenlarven im Fehmarnbelt in den Jahren 1934–1943. *Ber. dt. wiss. Kommn Meeresforsch.* **12**, Heft 1, pp. 49–85, figs. 1–6.

—— 1961. Über das Vorkommen von Fischbrut, Decapodenlarven und Medusen in der Kieler Förde. *Kieler Meeresforsch.* **17**, Heft 1, pp. 48–64.

KAUFMAN, S. Z. 1957. The dependence of regeneration in the scyphistoma of the scyphomedusa *Cyanea capillata* on its stage of development [in Russian]. *Dokl. Akad. Nauk SSSR*, **114**, 1317–19, fig. 1.

KOIZUMI, Tatsuo & HOSOI, Keizô, 1936. On the inorganic composition of the medusae: *Aequorea coerulescens* (Brandt), *Dactylometra pacifica* Goette and *Cyanea capillata* Eschscholtz. *Sci. Rep. Tohoku Univ.* ser. IV, **10**, 709–19.

KÖLLIKER, A., 1865. *Icones histiologicae oder Atlas der vergleichender Gewebelehre*. Abt. II. *Der feinere Bau der höheren Thiere*. Heft 1. *Die Bindesubstanz der Coelenteraten*. pp. 98–172, 13 text-figs. 10 pl.

KOMAI, Taku, 1935. On *Stephanoscyphus* and *Nausithoë*. *Mem. Coll. Sci. Kyoto Univ.* B, **10**, no. 5, art. 14, pp. 289–339, text-figs. 1–43, pl. XXI–XXII.

—— 1939. On the enigmatic coelenterate *Tetraplatia*. *Jap. J. Zool.* **8**, 231–50, text-figs. 1–11, pl. XXXI.

—— 1942. The nervous system in some coelenterate types. 2. Ephyra and Scyphula. *Annotnes zool. jap.* **21**, 25–9, figs. 1–4.

KORN, H., 1966. Zur ontogenetischen Differenzierung der Coelenteratengewebe (Polypstadium) unter besonderer Berücksichtigung des Nervensystems. *Z. Morph. Ökol. Tiere, Berlin*, **57**, 1–118, figs. 1–64.

KRAMP, Paul L., 1913. Medusae collected by the 'Tjalfe' Expedition. *Vidensk. Meddr dansk naturh. Foren.* **65**, 257–86, figs. 1–4.

—— 1924. Medusae. *Rep. Danish Oceanogr. Exped. 1908–10 Medit. adjac. Seas.* **2**, Biology. H. 1. pp. 1–67, 40 text-figs. text-charts I–XII.

—— 1934. An exceptional occurrence of *Rhizostoma octopus* and *Chrysaora hysoscella* in the Danish waters in 1933. *J. Cons. perm. int. Explor. Mer,* **9**, no. 2, pp. 211–21, figs. 1–2.

—— 1937. Polypdyr (Coelenterata) II. Gopler. *Danm. Fauna,* **43**, 1–223, figs. 1–90.

—— 1939. Medusae, Siphonophora and Ctenophora. *Zoology Iceland,* **2**, pt. 5*b*, pp. 1–37, 2 figs.

—— 1942. The Godthaab Expedition 1928. Medusae. *Meddr Grønland,* **81**, nr. 1, pp. 1–168, 38 text-figs.

—— 1947. Medusae. Part III. Trachylina and Scyphozoa, with zoogeographical remarks on all the medusae of the Northern Atlantic. *Dan. Ingolf-Exped.* **5**, pt. 14, pp. 1–66, 20 text-figs. pl. I–VI.

—— 1948. Trachymedusae and Narcomedusae from the 'Michael Sars' North Atlantic Deep-Sea Expedition 1910 with additions on Anthomedusae, Leptomedusae and Scyphomedusae. *Rep. 'Sars' N. Atl. Deep Sea Exped.* **5**, no. 9, 23 pp. 1 pl. text-figs. 7.

—— 1951*a*. Medusae collected by the Lund University Chile Expedition 1948–49. *Acta Univ. lund.,* N.F. Avd. 2, **47**, no. 7, pp. 1–19, figs. 1–8 (*Kungl. Fysiograf. Sälesk. Handl.* N.F. **62**, no. 7).

—— 1951*b*. Hydrozoa and Scyphozoa. *Rep. Swedish Deep Sea Exped.* **2** (Zool.), no. 10, pp. 121–7, 1 pl. text-fig. 1.

—— 1955*a*. The Medusae of the tropical west coast of Africa. *Atlantide Rep.* no. 3, pp. 239–324, text-figs. 1–14, pl. I–III.

—— 1955*b*. A revision of Ernst Haeckel's determinations of a collection of Medusae belonging to the Zoological Museum of Copenhagen. *Deep-Sea Res.,* supplement to Vol. 3, pp. 149–68.

—— 1956*a*. Medusae collected in the eastern tropical Pacific by Cyril Crossland in 1924–1925. *Vidensk. Meddr dansk. naturh. Foren.* **118**, 1–6, fig. 1.

—— 1956*b*. Medusae of the Iranian Gulf. *Vidensk. Meddr dansk. naturh. Foren.* **118**, 235–42, 1 chart.

—— 1957. Medusae. *Br. Aust. N.Z. Antarct. Res. Exped. Rep.* ser. B, **6**, pt. 8, pp. 151–64.

—— 1959*a*. Medusae. Mainly from the West Coast of Africa. *Expédit. Océanogr. Belge eaux côtières Afric. Atlant. Sud, Rés. Sci.* **3**, fasc. 6, 33 pp. figs. 1–5.

—— 1959*b*. *Stephanoscyphus* (Scyphozoa). *Galathea Rep.* **1**, 173–85, text-figs. 1–12, pl. 1.

—— 1961. Synopsis of the Medusae of the World. *J. mar. biol. Ass. U.K.* **40**, 1–469.

—— 1965. Some medusae (mainly Scyphomedusae) from Australian coastal waters. *Trans. R. Soc. S. Aust.* **89**, 257–78, pl. 1–3.

—— 1968. The Scyphomedusae collected by the Galathea Expedition 1950–52. *Vidensk. Meddr dansk naturh. Foren.* **131**, 67–98, figs. 1–2.

KRAMP, P. L. & DAMAS, D., 1925. Les Méduses de la Norvège. Introduction et partie spéciale I. *Vidensk. Meddr dansk naturh. Foren.* **80**, 217–323, 33 text-figs. pl. XXXV.

KRASIŃSKA, Sophie, 1914. Beiträge zur Histologie der Medusen. *Z. wiss. Zool.* **109**, Heft 2, pp. 256–348, text-figs. 1–5, pl. VII & VIII.

KROHN, A., 1855. Ueber die frühesten Entwicklungsstufen der *Pelagia noctiluca. Arch. Anat. Physiol.* pp. 491–7, pl. XX.

—— 1865. Ueber *Tetraplatia volitans. Arch. Naturgesch.* Jahrg. 31, **1**, 337–41, pl. XIV, text-figs. 1–3.

KRÜGER, F., 1968. Stoffwechsel und Wachstum bei Scyphomedusen. *Helgoländer wiss. Meeresunter.* **18**, 367–83, figs. 1–7.

KRUKENBERG, C. Fr. W., 1880. Über den Wassergehalt der Medusen. *Zool. Anz.* (III. Jahrg.), p. 306.

—— 1882. Ueber das Cyaniën und das Asterocyanin. *Vergl. physiol. Studien* (2 Reihe, 3 Abt.), pp. 1–115.

—— 1886. Vergleichend-physiologische Vorträge. **1**, 1–517.

KRUMBACH, Thilo, 1925. Scyphozoa. *Handb. Zool.* **1**, 522–686, figs. 512–80.

—— 1930. *Die Tierwelt der Nord- und Ostsee.* Lief. XVII, Teil III*d*: Scyphozoa. pp. 1–88, figs. 1–54.

KÜHL, H., 1964. Die Scyphomedusen der Elbmündung. *Veröff. Inst. Meeresforsch. Bremerh.* **9**, 84–94.

—— 1966. Der Abfluss der Elbe im Jahre 1965 und seine Wirkung auf Salzgehalt, Plankton und Bewuchsbildung bei Cuxhaven. *Veröff. Inst. Meeresforsch. Bremerh.* **10**, 61–70, figs. 1–4.

—— 1967. Scheibenquallen an unseren Küsten. *Mikrokosmos,* Heft 6, pp. 169–74, figs. 1–6.

KÜNNE, Clemens, 1948. Medusen als Transportmittel für Actinien-Larven. *Natur Volk,* **78**, 174–6, figs. 1–4.

—— 1952. Untersuchungen über das Grossplankton in der Deutschen Bucht und im Nordsylter Wattenmeer. I. Über das Grossplankton in der Deutschen Bucht. *Helgoländer wiss. Meeresunters.* **4**, Heft 1, pp. 1–54, figs. 1–3.

BIBLIOGRAPHY

LAMARCK, J. B. P. A. de M., 1816. *Histoire naturelle des Animaux sans Vertèbres*, **2**.

LAMBERT, F. J., 1936*a*. Observations on the Scyphomedusae of the Thames Estuary and their meta-morphoses. *Trav. Stn zool. Wimereux*, **12**, Mém. no. 3, pp. 281–307, figs. 1–6.

—— 1936*b*. Jellyfish. The difficulties of the study of their life history and other problems. *Essex Nat.* **25**, 70–86, pl. III–IV.

LAWLESS, E., 1877. On a fish-sheltering medusa. *Nature, Lond.* **16**, 227.

LEBOUR, Marie V., 1922. The food of plankton organisms. *J. mar. biol. Ass. U.K.* **12**, 644–77, figs. 1–3.

—— 1923. The food of plankton organisms. II. *J. mar. biol. Ass. U.K.* **13**, 70–92, figs. 1–12.

LE DANOIS, Ed., 1913. Coelentérés du plankton recueillis pendant la Croisière d'été 1913 par le Yacht 'Pourquoi pas?'. *Bull. Soc. zool. Fr.* **38**, 282–8.

LEGENDRE, R., 1940. La faune pélagique de l'Atlantique au large du Golfe de Gascogne, recueillie dans des estomacs de Germons. *Annls Inst. océanogr. Monaco*, **20**, 127–310, figs. 1–71.

LEHMANN, Conrad, 1923. Untersuchungen über die Sinnesorgane der Medusen. *Zool. Jb.* (Allg. Zool.), **39**, Heft 3, pp. 321–94, 18 text-figs.

LELOUP, E., 1935. Contribution à la Répartition de *Tetraplatia volitans* (Busch). *Bull. Mus. r. Hist. nat. Belg.* **11**, no. 4, pp. 1–7, 1 chart.

—— 1937. Hydropolypes et Scyphopolypes recueillis pas C. Dawidoff sur les côtes de l'Indochine française. *Mém. Mus. Hist. nat. Belg.* sér. 2, fasc. 12, pp. 1–73, figs. 1–43.

LELOUP, E. & MILLER, O., 1940. La Flore et la Faune du Bassin de Chasses d'Ostende (1937–1938). *Mém. Mus. Hist. nat. Belg.* no. 94, pp. 1–122, text-figs. 1–11, pl. I–III.

LENDENFELD, R. von, 1882. Über Coelenteraten der Südsee. I. Mittheilung. *Cyanea annaskala* nov. sp. *Z. wiss. Zool.* **37**, 465–552, pl. XXVII–XXXIII, 1 text-fig.

LESSON, René-Primevère, 1843. *Histoire Naturelle des Zoophytes. Acalèphes.* 596 pp. pl. 1–12, Paris.

LEUCKART, R., 1866. Bericht über die wissenschaftlichen Leistungen in der Naturgeschichte der niederen Tiere während der Jahre 1864 u. 1865. *Arch. Naturgesch.* p. 170.

LINNAEUS, C. 1746. *Fauna suecica.* Nr. 1287.

—— 1747. *Westgöta Resa.* Tab. III, fig. 2.

—— 1758. *Systema naturae.* 10th ed.

LITTLEFORD, Robert A., 1939. The life cycle of *Dactylometra quinquecirrha* L. Agassiz in the Chesapeake Bay. *Biol. Bull. mar. biol. Lab., Woods Hole*, **77**, 368–81, pl. I–III.

LO BIANCO, Salvatore. 1888. Notizie biologiche riguardanti specialmente il periodo di maturità sessuale degli animali del golfo di Napoli. *Mitt. zool. Stn Neapel*, **8**, 385–440.

—— 1903. Le pesche abissali eseguite da F. A. Krupp col Yacht 'Puritan' nelle adiacenze di Capri ed in altre località del Mediterraneo. *Mitt. zool. Stn Neapel*, **16**, 109–278, pl. 7–9.

LOHMANN, H., 1908. Untersuchungen zur Feststellung des vollständingen Gehaltes des Meeres an Plankton. *Wiss. Meeresunters. Abt. Kiel*, N.F., **10**, 131–370, text-figs. 1–22, pl. IX–XVII.

LOOMIS, W. F., 1961. In *The biology of* Hydra, pp. 335 and 336. (Univ. Miami Press; ed. H. M. Lenhoff & W. F. Loomis.)

LOW, James W., 1921. Variation in ephyrae of *Aurelia aurita. Proc. R. phys. Soc. Edinb.* **20**, 226–35, figs. 1–13.

LOWNDES, A. G., 1942. Percentage of water in jellyfish. *Nature, Lond.* **150**, 234–5.

LYTLE, Charles F., 1961. In *The biology of* Hydra, p. 336. (Univ. Miami Press; ed. H. M. Lenhoff & W. F. Loomis.)

MAADEN, H.v.d., 1939. Über das Sinnesgrübchen von *Aurelia aurita* Linné. *Zool. Anz.* **125**, 29–35, fig. 1.

—— 1942*a*. Beobachtungen über Medusen am Strande von Katwijk aan Zee (Holland) in den Jahren 1933–1937. *Archs. néerl. Zool.* **6**, 347–62, figs. 1–7.

—— 1942*b*. Über *Cyanea capillata* Eschscholtz und die sog. var. *lamarcki* Péron & Lesueur. *Zool. Anz.* **137**, 63–70.

MAAS, Otto, 1897. Reports on an exploration off the west coasts of Mexico, central and South America, and off the Galapagos Islands. XXI. Die Medusen. *Mem. Mus. comp. Zool. Harv.* **23**, no. 1, pp. 9–92, pl. I–XV.

—— 1903. Die Scyphomedusen der Siboga-Expedition. *Siboga-Exped. Mon.* **11**, 1–91, pl. I–XII.

—— 1904. Méduses provenant des Campagnes des Yachts *Hirondelle* et *Princesse-Alice* (1886–1903). *Résult. Camp. scient. Prince Albert I*, fasc. XXVIII, pp. 1–72, pl. I–VI.

MACALLUM, A. B., 1903. On the inorganic composition of the medusae, *Aurelia flavidula* and *Cyanea arctica. J. Physiol., Lond.* **29**, 213–41.

MacGINITIE, G. E. & MacGINITIE, N., 1949. *Natural History of Marine Animals*, 1st ed. (text-fig. 21, p. 122). New York: McGraw-Hill.

MACRI, Saverio, 1778. *Nuove osservazioni intorno la storia naturale del Polmone marino degli antichi*. Naples.

MANSUETI, Romeo, 1963. Symbiotic behaviour between small fishes and jellyfishes, with new data on that between the stromateid, *Peprilus alepidotus*, and the Scyphomedusa, *Chrysaora quinquecirrha*. *Copeia*, no. 1, pp. 40–80, figs. 1–5.

MARSHALL, C. F., 1888. Observations on the structure and distribution of striped and unstriped muscle in the animal kingdom, and a theory of muscular contraction. *Q. Jl microsc. Sci.* **28**, 75–107, pl. VI.

MAYER, Alfred Goldsborough, 1906 *a*. Medusae of the Hawaiian Islands collected by the steamer 'Albatross' in 1902. *U.S. Fish Comm. Bull. for 1903*, pt. III, pp. 1131–43, pl. I–III.

—— 1906 *b*. Rhythmical pulsation of Scyphomedusae. *Yearbk Carneg. Instn*, no. 4 (for 1905), p. 120.

—— 1910. *Medusae of the World*. **3**. *The Scyphomedusae*, pp. 499–735, pls. LVI–LXXVI. Washington.

—— 1914 *a*. The effects of temperature upon tropical marine animals. *Publs Carnegie Instn*, no. 183, pp. 1–24, figs. 1–8 (Pap. Tortugas Lab. **6**).

—— 1914 *b*. The law governing the loss of weight in starving *Cassiopea*. *Pap. Tortugas Lab.* **6**, 57–82, text-figs. 1–21, pl. I.

M'ANDREW, Robert & FORBES, Edward, 1847. XLIII. Notices of new or rare British animals observed during cruises in 1845 and 1846. II. On the occurrence of a species of *Pelagia* in the British Seas. *Ann. Mag. nat. Hist.* **19**, ser, 1, no. CXXVIII, pp. 390–2, pl. IX, fig. 5.

M'INTOSH, D. C., 1910. Variation in *Aurelia aurita*. *Proc. R. phys. Soc. Edinb.* **18**, 125–43, figs. 1–4.

—— 1911. Note on variation in the jellyfish *Aurelia aurita*. *Ann. Scot. nat. Hist.* no. 77 (January), pp. 25–9.

M'INTOSH, W. C., 1885. Notes from the St Andrews Marine Laboratory. I. On the British species of *Cyanea*, and the reproduction of *Mytilus edulis*. *Ann. Mag. nat. Hist.* ser. 5, **15**, 148–52.

—— 1926. Additions to the marine fauna of St Andrews since 1874. *Ann. Mag. nat. Hist.* ser. 9, **18**, pp. 241–66, figs. 1–16.

—— 1927. Notes from the Gatty marine laboratory, St Andrews. *Ann. Mag. nat. Hist.* **20**, 9th ser. no. 115, pp. 1–22, pl. I.

McMURRICH, J. Playfair, 1891. The development of *Cyanea arctica*. *Amer. Nat.* **25**, pp. 287–9.

METSCHNIKOFF, Elias, 1880. Über die intracelluläre Verdauung bei Coelenteraten. *Zool. Anz.* (III. Jahrg.), pp. 261–2.

—— 1886. *Embryologische Studien an Medusen. Ein Beitrag zur Genealogie der Primitivorgane*, pp. 1–159, Atlas, Taf. I–XII. Wien.

—— 1892. *Leçons sur la pathologie comparée de l'inflammation*. 239 pp. 65 text-figs. 3 pl. Paris: Masson.

MIELCK, W. & KÜNNE, C., 1935. Fischbrut- und Plankton-Untersuchungen auf dem Reichsforschungs-dampfer 'Poseidon' in der Ostsee, Mai-Juni 1931. *Wiss. Meeresunters. Abt. Helgoland*, N.F., **19**, nr. 7, pp. 1–120, text-figs. 1–12, pl. 1–8.

MILLOTT, Norman & FOX, Denis L., 1954. Pigmentation of the jellyfish, *Pelagia noctiluca*. *Nature, Lond.* **173**, 169.

MINCHIN, Edward A., 1889. Note on the mode of attachment of the embryos to the oral arms of *Aurelia aurita*. *Proc. zool. Soc. Lond.* **59**, 583–5, pl. 67, 68.

MITCHELL, John H., 1962. Eye injuries due to jellyfish (*Cyanea annaskala*). *Med. J. Aust.* **3**, 25 August, pp. 303–5, figs. I–V.

MÖBIUS, Karl, 1866. Ueber den Bau, den Mechanismus und die Entwicklung der Nesselkapseln einiger Polypen und Quallen. *Abh. Verh. naturw. Ver. Hamburg*, **5**, 1–22, pl. I–II.

—— 1880. Medusen werden durch Frost getödtet. *Zool. Anz.* (III. Jahrg.), pp. 67–8.

—— 1882. Wassergehalt der Medusen. *Zool. Anz.* (v. Jahrg.), pp. 586–7.

MONNÉ, L., 1957. On the influence of hot trichloracetic acid on the staining properties of the exoplasms of various animals, particularly parasitic worms. *Ark. Zool.* **11**, 1–19.

MONTICELLI, Fr. Sav., 1897. Adelotacta Zoologica. I. *Pemmatodiscus socialis* n. gen. n. sp. *Mitt. zool. Stn Neapel*, **12**, Heft 3 (1896), pp. 432–62, pl. 19.

MOORE, A. R., 1926. Galvanic stimulation of luminescence in *Pelagia noctiluca*. *J. gen. Physiol.* **9**, 375–91.

MORSE, Max, 1910. Pulsations in Scyphomedusae deprived of their marginal organs. *Science, N.Y.* **31**, 544–5.

MÜLLER, Fritz, 1859. Die Magenfäden der Quallen. *Z. wiss. Zool.* **9** (1858), 542–3 (also in English in *Ann. Mag. nat. Hist.* 3rd ser. **3**, 446–7).

MÜLLER, O. F., 1780. *Zoologiae Danicae seu animalium Daniae et Norvegiae rariorum ac minus notorum icones*. Fasc. II, Havniae.

—— 1784. *Zoologia Danica seu animalium Daniae et Norvegiae rariorum ac minus notorum descriptiones et historia*. **2**. Lipsiae.

BIBLIOGRAPHY

NAGABHUSHANAM, A. K., 1959. Studies on the biology of the commoner gadoids in the Manx area, with special reference to their food and feeding habits. Thesis for Ph.D., University of Liverpool.

NATHORST, A. G., 1881. Om aftryck af Medusor i sveriges Kambriska Lager. *K. svenska VetenskAkad. Handl.* **19**, 1–34, pl. 1–6.

NAUMOV, D. V., 1961. Scyphoid medusae of the seas of U.S.S.R. *Acad. Sci. Keys to the Fauna of U.S.S.R.* no. 75, pp. 1–98, text-figs. 1–72, pl. I–III. (In Russian.)

NAYLOR, E., 1965. Biological effects of a heated effluent in docks at Swansea, S. Wales. *Proc. zool. Soc. Lond.* **144**, 253–68, figs. 1–4.

NEPPI, Valeria, 1931. Osservazioni sulle Scifomeduse. *Boll. Zool.* **2**, 143–9, figs. 1–4.

NETCHAEFF, A. & NEU, W., 1940. *Aurelia aurita* L. und *Pilema (Rhizostoma) pulmo* Heack. im Schwarzen Meer und im Bosporus. *Zool. Anz.* **129**, 61–3.

NEWELL, G. E., 1954. The marine fauna of Whitstable. *Ann. Mag. nat. Hist.* ser. XII, **7**, 321–50, 1 chart.

NICOL, J. A. C., 1955. Physiological control of luminescence in animals. *Amer. Assoc. Advanc. Sci.* pp. 299–319, figs. 1–10.

—— 1958. Observations on luminescence in pelagic animals. *J. mar. biol. Ass. U.K.* **37**, 705–52, text-figs. 1–19, pl. I.

NORMAN, Alfred Merle, 1867. On *Cyanea imporcata*, an undescribed medusa taken off the Northumberland coast. *Nat. Hist. Trans. Northumb.* **1**, 58–60, pl. XI.

NOWIKOFF, M., 1912. Studien über das Knorpelgewebe von Wirbellosen. *Z. wiss. Zool.* **103**, 661–717, text-figs. 1–13, pl. XV–XVII.

OKADA, Yô. K., 1927. Sur l'Origine de l'Endoderme des Discoméduses. (La nature des Cellules isolées dans la cavité de segmentation et leur destinée.) *Bull. Biol. Fr. Belg.* **61**, 250–62, text-figs. I–VI, pl. V, figs. 1–12.

ORTON, J. H., 1922. The mode of feeding of the jellyfish *Aurelia aurita* on the smaller organisms of the plankton. *Nature, Lond.* **110**, 178–9.

OSTROUMOV, A., 1896. Scientific results of the 'Atmanay' Expedition. *Bull. Acad. Imp. Sci. St. Pétersbourg,* **4**, 389–408, pl. I. (In Russian.)

ÖSTERGREN, Hjalmar, 1909. *Cyanea palmstruchii* (Swartz) eine verkannte Qualle aus dem Skagerrak. *Zool. Anz.* **34**, 464–74.

PALMSTRUCH, J. W., 1809. Text by O. Swartz. *Svensk Zoologi,* **2**, Stockholm.

PANCERI, P., 1872. Études sur la phosphorescence des animaux marins. *Annls Sci. nat. Paris,* Zool. sér. 5, **13**, 45–51.

PANTIN, C. F. A. & DIAS, M. Vianna, 1952. Rhythm and after-discharge in medusae. *Anais Acad. bras. Cienc.* **24**, 351–64, figs. 1–14.

PAPENFUSS, E. J., 1936. The utility of the nematocysts in the classification of certain Scyphomedusae. *Acta Univ. Lund.,* N.F. 2, **31**, no. 11, 1–26, figs. 1–22.

PASPALEV, G. V., 1938. Über die Entwicklung von *Rhizostoma pulmo* Agass. *Arb. biol. Meeresst. Varna,* no. 7, pp. 1–17, figs. 1–25.

PASSANO, L. M., 1958. Intermittent conduction in scyphozoan nerve nets. *Anat. Rec.* **132**, 486.

—— 1965. Pacemakers and activity patterns in medusae: homage to Romanes. *Am. Zoologist,* **5**, 465–81, figs. 1–14.

PASSANO, L. M. & McCULLOUGH, Coyla B., 1960. Nervous activity and spontaneous beating in Scyphomedusae. *Anat. Rec.* **137**, 387.

—— 1961. Pacemaker activity in jellyfish ganglia. *Fed. Proc.* **20**, 338.

PENNANT, T., 1812. *British Zoology.* New (5th) edition, **4**. London.

PERCIVAL, E., 1923. On the strobilization of *Aurelia. Q. Jl microsc. Sci.* **67**, no. 265, pp. 85–100, text-figs. 1–3, pl. 6.

—— 1929. A report on the fauna of the estuaries of the River Tamar and the River Lynher. *J. mar. biol. Ass. U.K.* **16**, 81–108, fig. 1, 1 chart.

PÉRÈS, J.-M., 1958. Trois plongées dans le canyon du Cap Sicié, effectuées avec le bathyscaphe F.N.R.S. III de la Marine Nationale. *Bull. Inst. océanogr., Monaco,* no. 1115, pp. 1–21, pl. I–III.

PÉREZ, Charles, 1920*a*. Un élevage de Scyphistomes de *Cyanea capillata. Bull. biol. Fr. Belg.* **54**, 168–78, pl. III–IV.

—— 1920*b*. Processus de multiplications par bourgeonnement chez un Scyphistome. *Bull Soc. zool. Fr.* XLV, pp. 260–1.

—— 1922. Observations sur la multiplication gemmipare d'un Scyphistome. *Bull biol. Fr. Belg.* **56**, 244–74, figs. 1–34.

PÉRON, Fr. & LESUEUR, C. A., 1809. Tableau des caractères génériques et spécifiques de toutes les espèces de Méduses connues jusqu'à ce jour. *Annls Mus. Hist. nat. Paris*, **14**, 325–66 (pl. I–XCVI not published).

—— 1807–16. *Voyages de découvertes aux Terres Australes...pendant les années 1800–1804.* 2 tom. & Atlas 38 pls. 15 maps. Paris.

PETERSEN, K. W., 1957. On some medusae from the North Atlantic. *Vidensk. Meddr dansk naturh. Foren.* **119**, 25–45, figs. 1–4.

POPE, Elizabeth C., 1949. A large medusa in Sydney Harbour. *Aust. Mus. Mag.* **10**, 14–16, 3 figs.

—— 1951. Te Baïtari—An edible jellyfish from Tarawa. *Aust. Mus. Mag.* **10**, 270–2, 2 pl.

PURCHON, R. D., 1937. Studies on the biology of the Bristol Channel. II. An ecological study of the beach and the dock at Portishead. *Proc. Bristol Nat. Soc.* (4), **8**, 311–29.

QUOY, J. R. C. & GAIMARD, J. P., 1827. Observations zoologiques faites à bord de l'*Astrolabe* en Mai 1826, dans le détroit de Gibraltar. *Annls Sci. nat.* **10**, 1–21, & pp. 172–93, pl. 1–2 & 4–9. (Text 8vo; pls. 4to).

RALPH, Patricia M., 1959. Notes on the species of the pteromedusan genus *Tetraplatia* Busch, 1851. *J. mar. biol. Ass. U.K.* **38**, 369–79, figs. 1–3.

—— 1960. *Tetraplatia*, a coronate scyphomedusan. *Proc. R. Soc. B*, **152**, 263–81, text-figs. 1–7, pl. 19.

RANSON, Gilbert, 1924. Méduses du plankton recueilli par La Tanche pendant sa première croisière de 1923 (avec deux cartes de répartition des *Pelagia* et des *Rhopalonema*). *Bull. Mus. Hist. nat. Paris*, no. 1, pp. 88–92.

—— 1936. Observations morphologiques, systématiques et biogéographiques sur une Scyphoméduse rare, *Paraphyllina intermedia* O. Maas 1903, trouvée sur la Plage de Biarritz. *Bull. Mus. Hist. nat. Paris*, sér. 2, **8**, no. 3, pp. 269–76, 1 fig.

—— 1945. Scyphoméduses provenant des campagnes du Prince Albert Ier de Monaco. *Résult. Camp. scient. Prince Albert I*, fasc. CVI, pp. 1–92, pl. I–II.

—— 1949. Résultats scientifiques des Croisières du Navire-École Belge 'Mercator'. II. Méduses. *Mém. Inst. r. Sci. nat. Belg.* sér. 2, fasc. XXXIII, pp. 123–58.

RATHKE, H. & ZADDACH, E. G. 1849. *Preussische Provincial-Blätter. Vierter Bericht des Vereins für die Fauna der Provinz Preussen im März 1849.* Königsberg.

RAYMONT, J. E. G., KRISHNASWAMY, S. & TUNDISI, J., 1967. Biochemical studies on marine zooplankton. IV. Investigations on succinic dehydrogenase activity in zooplankton with special reference to *Neomysis integer*. *J. Cons. perm. int. Explor. Mer*, **31**, 164–9, figs. 1–2.

REES, W. J., 1957. Proposed validation under the plenary powers of the generic name 'Aurelia' Lamarck 1816 (class Scyphozoa). *Bull. zool. Nom.* **13** (1957–58), 26–8; support on pp. 103, 153, 199.

—— 1966. *Cyanea lamarcki* Péron & Lesueur (Scyphozoa) and its association with young *Gadus merlangus* L. (Pisces). *Ann. Mag. nat. Hist.* sér. 13, **9**, 285–7.

REES, William J. & WHITE, Ernest, 1957. New observations on the aberrant medusa *Tetraplatia volitans* Busch. '*Discovery*' *Rep.* **29**, 129–40, figs. 1–7.

REID, J., 1848. Observations on the development of the medusae (*Aurelia aurita?*). *Ann. Mag. nat. Hist.* **1**, 25, pl. V & VI.

RENTON, Rachel M., 1930. On the budding of a Scyphistoma. *Proc. zool. Soc. Lond.* pp. 893–6, text-fig. 1, pl. 1.

REPELIN, R., 1962a. Une nouvelle Scyphoméduse bathypélagique: *Atolla russelli*, n.sp. *Bull. Inst. fr. Afr. noire*, **24**, sér. A, no. 3, pp. 664–76, figs. 1–5.

—— 1962b. Scyphoméduses de la famille des Atollidae dans le Bassin de l'Angola. *Bull. Inst. Res. scient. Congo*, **1**, 89–99, text-figs. 1–3, pl. 1, figs. 4–5.

—— 1964. Scyphoméduses de la famille des Atollidae dans le Golfe de Guinée. *Cah. O.R.S.T.O.M. océanogr.* **2**, 13–30, text-figs. 1–3, pl. I–V.

—— 1965. Le méduse *Paraphyllina ransoni* dans la Vallée du Trou sans Fond. *Cah. O.R.S.T.O.M. océanogr.* **3**, 81–6, figs. 1–6.

—— 1966. Scyphoméduses Atollidae du Bassin de Guinée. *Cah. O.R.S.T.O.M. Océanogr.* **4**, 21–33, figs. 1–5.

—— 1967. *Stygiomedusa stauchi* n.sp. Scyphoméduse géante des profondeurs. *Cah. O.R.S.T.O.M. océanogr.* **5**, 23–8, figs. 1–6.

RETZIUS, Gustav, 1904. Zur Kenntniss der Spermien der Evertebraten. *Biol. Unters.* N.F., **11**, no. 1, pp. 1–32, pl. I–XIII.

RICE, A. L., 1964. Observations on the effects of changes of hydrostatic pressure on the behaviour of some marine animals. *J. mar. biol. Ass. U.K.* **44**, 163–75, fig. 1.

RIES, Julius v. & RIES, Marie v., 1924. Farbe und Wärme. Eine lichtbiologische Studie. *Umschau* (28 Jahrg.), pp. 666–70, figs. 1–6.

ROBERTSON, James D., 1949. Ionic regulation in some marine invertebrates. *J. exp. Biol.* **26**, 182–200.

BIBLIOGRAPHY

ROMANES, George J., 1876a. An account of some new species, varieties, and monstrous forms of medusae. *J. Linn. Soc., Zool.* **12**, no. 64, pp. 524–31.

—— 1876b. Preliminary observations on the locomotor system of medusae. *Phil. Trans. R. Soc.* **166**, pt. 1, pp. 269–313, pl. 32 & 33. (The Croonian Lecture.)

—— 1877a. An account of some new species, varieties, and monstrous forms of medusae. II. *J. Linn. Soc., Zool.* **13**, no. 68, pp. 190–4, pl. XV & XVI.

—— 1877b. Further observations on the locomotor system of medusae. *Phil. Trans. R. Soc.* **167**, pt. II, pp. 659–752, pl. 30 & 31.

—— 1877c. The fish-sheltering medusa. *Nature, Lond.* **16**, 248.

—— 1880. Concluding observations on the locomotor system of medusae. *Phil. Trans. R. Soc.* **170**, pt. 1, pp. 161–202, figs. 1–10.

RUSSELL, Findley E., 1965. Marine toxins and venomous and poisonous marine animals. *Adv. mar. Biol.* **3**, 255–384, figs. 1–20.

RUSSELL, F. S., 1927. The vertical distribution of marine macroplankton. V. The distribution of animals caught in the ring-trawl in the daytime in the Plymouth area. *J. mar. biol. Ass. U.K.* **14**, pp. 557–608, figs. 1–11.

—— 1928. The vertical distribution of marine macroplankton. VIII. Further observations on the diurnal behaviour of the pelagic young of teleostean fishes in the Plymouth area. *J. mar. biol. Ass. U.K.* **15**, 829–50, figs. 1–6.

—— 1930. The vertical distribution of marine macroplankton. IX. The distribution of the pelagic young of teleostean fishes in the daytime in the Plymouth area. *J. mar. biol. Ass. U.K.* **16**, 639–76, figs. 1–7.

—— 1931. Notes on *Cyanea* caught in the ring-trawl in the Plymouth area during the years 1925 to 1930. *J. mar. biol. Ass. U.K.* **16**, pp. 573–6.

—— 1937. The seasonal abundance of the pelagic young of teleostean fishes in the Plymouth area. IV. The year 1936, with notes on the conditions shown by the occurrence of plankton indicators. *J. mar. biol. Ass. U.K.* **21**, 679–86, figs. 1–4.

—— 1938. The Plymouth offshore medusa fauna. *J. mar. biol. Ass. U.K.* **22**, pp. 411–39, figs. 1–5.

—— 1956a. On a new scyphomedusa, *Paraphyllina ransoni* n.sp. *J. mar. biol. Ass. U.K.* **35**, 105–11, pl. I–II, text-figs. 1–3.

—— 1956b. On the scyphomedusae *Nausithoë atlantica* Broch and *Nausithoë globifera* Broch. *J. mar. biol. Ass. U.K.* **35**, 363–70, pl. I, text-figs. 1–6.

—— 1957. On a new species of scyphomedusa, *Atolla vanhöffeni* n.sp. *J. mar. biol. Ass. U.K.* **36**, 275–9, pl. I, 1 text-fig.

—— 1958. A new species of *Atolla. Nature, Lond.* **181**, 1811.

—— 1959. Some observations on the scyphomedusa *Atolla. J. mar. biol. Ass. U.K.* **38**, 33–40, text-figs. 1–3.

—— 1962. On the scyphomedusa *Poralia rufescens* Vanhöffen. *J. mar. biol. Ass. U.K.* **42**, 387–90, text-figs. 1–2, pl. I.

—— 1964. On Scyphomedusae of the genus *Pelagia. J. mar. biol. Ass. U.K.* **44**, 133–6.

—— 1967. On the occurrence of the scyphomedusan *Pelagia noctiluca* in the English Channel in 1966. *J. mar. biol. Ass. U.K.* **47**, 363–6, fig. 1.

RUSSELL, F. S. & REES, W. J., 1960. The viviparous scyphomedusa *Stygiomedusa fabulosa* Russell. *J. mar. biol. Ass. U.K.* **39**, 303–17, text-figs. 1–7, pl. I–IV.

RUSTAD, Ditlef, 1952. Zoological notes from the biological station. *Univ. Bergen Årb. 1951, naturv. rekke,* nr. 4, pp. 1–11.

SARS, M., 1829. *Bidrag til Söedyrenes Naturhistorie af M. Sars,* Cand. Theol. Förste-Haefte, med sex illuminerede Steentryktafler, 8vo. I. Halfte, pp. 17–26. Bergen.

—— 1835. *Beskrivelser og Jagttagelser over nogle maerkelige eller nye i Havet ved den Bergenske kyst levende Dyr af Polypernes, Acalephernes, Radiaternes, Annelidernes og Molluskernes Classer, etc.* Bergen, 4to, with 15 pl.

—— 1837. *Wiegmann's Archiv für Naturgeschichte,* **1**, 406.

—— 1841. Ueber die Entwickelung der *Medusa aurita* und der *Cyanea capillata. Arch. Naturgesch.* (VII. Jahrg.), **1**, 9–34, pl. I–IV.

SCHÄFER, Edward Albert, 1878. Observations on the nervous system of *Aurelia aurita. Phil. Trans. R. Soc.* **169**, II, pp. 563–75, pl. 50–1.

SCHAEFER, J. Georg, 1921. Untersuchungen an Medusen. I. Teil. *Pflügers Arch. ges. Physiol.* **188**, 49–59, figs. 1–2.

SCHEURING, Ludwig, 1915. Beobachtungen über den Parasitismus pelagischer Jungfische. *Biol. Zbl.* **35**, 181–90.

SCHEWIAKOFF, Wladimir, 1889. Beiträge zur Kenntnis der Acalephenauges. *Morph. Jb.* **15**, 21–60, pl. I–III.

SCHNEIDER, A., 1870. Zur Entwickelungsgeschichte der *Aurelia aurita*. *Arch. mikrosk. Anat. EntwMech.* **6**, 363, pl. 19.

SCHODDYN, M., 1926. Observations faites dans la baie d'Ambleteuse (Pas de Calais). *Bull. Inst. océanogr. Monaco*, no. 482, pp. 1–64.

SCHULTZE, M., 1856. Ueber den Bau der Gallertscheibe bei den Medusen. *Müller's Arch.* p. 314.

SCHULZE, F. E., 1877. *Spongicola fistularis*, ein in Spongien wohnenden Hydrozoon. *Arch. mikrosk. Anat. EntwMech.* **13**, 795–817, 3 pl.

SEGERSTRÅLE, Sven G., 1951. The recent increase in salinity off the coasts of Finland and its influence upon the fauna. *J. Cons. perm. int. Explor. Mer*, **17**, 103–10, figs. 1–2.

—— 1965. On the salinity conditions off the south coast of Finland since 1950, with comments on some remarkable hydrographical and biological phenomena in the Baltic area during this period. *Commentat. biol.* **28**, nr. 7, pp. 1–28, figs. 1–6.

SHOJIMA, Yoichi, 1963. Scyllarid phyllosomas' habit of accompanying the jelly-fish. *Bull. Jap. Soc. scient. Fish.* **29**, 349–53, pl. I–II.

SIEBOLD, Carl Theodor von, 1839. *Beiträge zur Naturgeschichte der Wirbellosen Thiere. Ueber* Medusa, Cyclops, Loligo, Gregarina *und* Xenos. 94 pp., 3 pl.

SIPOS, J. C. & ACKMAN, R. G., 1968. Jellyfish (*Cyanea capillata*) lipids: fatty acid composition. *J. Fish. Res. Bd Can.* **25**, 1561–9.

SJÖGREN, Lennart, 1962. Scyphistoma-stage of the jellyfish *Aurelia aurita*, on *Fucus* at Tvärminne Zoological Station in Finland. *Memo. Soc. Fauna Flora fenn.* **37** (1960–1), 3–4, figs. 1–2.

SKRAMLIK, Emil von, 1945. Beobachtungen an Medusen. *Zool. Jb. Allg. Zool. Physiol.* **61** (1948), 296–336, figs. 1–4.

SLABBER, Mart., 1781. *Physikalische Belustigungen oder mikroscopische. Wahrnehmungen von 43 in- u. ausländ. Wasser- u. Landthierchen,...Aus dem Holländischen übersezt.* 99 pp. 18 pl. Nürnberg.

SLONIMSKI, Pierre, 1926. Un nouveau procédé pour la mise en évidence du système gastro-vasculaire chez les méduses. *C. r. Séanc. Soc. Biol.* **95**, 926–8, 1 fig.

SMITH, Frank, 1891. The gastrulation of *Aurelia flavidula*, Pér. & Les. *Mem. Mus. comp. Zool. Harv.* **22**, 115–25, pl. I–II.

SMITH, H. G., 1936. Contribution to the anatomy and physiology of *Cassiopea frondosa*. *Publs. Carnegie Instn*, no. 475, pp. 17–52, text-figs. 1–20, pl. 1.

SORBY, H. C., 1894. Symmetry of *Aurelia aurita*. *Nature, Lond.* **50**, 476.

SOUTHWARD, A. J., 1949. Ciliary mechanisms in *Aurelia aurita*. *Nature, Lond.* **163**, 536.

—— 1955. Observations on the ciliary currents of the jellyfish *Aurelia aurita* L. *J. mar. biol. Ass. U.K.* **34**, 201–16, figs. 1–7.

SPANGENBERG, Dorothy Breslin, 1964. New observations on *Aurelia*. *Trans. Am. microsc. Soc.* **83**, 448–55, pl. I–II.

—— 1965a. Cultivation of the life stages of *Aurelia aurita* under controlled conditions. *J. exp. Zool.* **159**, 303–18, pl. 1–8.

—— 1965b. A study of strobilation in *Aurelia aurita* under controlled conditions. *J. exp. Zool.* **160**, 1–10, figs. 1–6.

—— 1967. Iodine induction of metamorphosis in *Aurelia*. *J. exp. Zool.* **165**, 441–9, fig. 1.

—— 1968. Statolith differentiation in *Aurelia aurita*. *J. exp. Zool.* **169**, 487–99, figs. 1–6.

SPANGENBERG, D. B. & BECK, C. W., 1968. Calcium sulphate dihydrate in *Aurelia*. *Trans. Am. microsc. Soc.* **87**, 329–35, figs. 1–3.

STADEL, Otto, 1965. Über die Nesselwirkung der Quallen auf den Menschen und ihre medizinische Behandlung. *Abh. Verh. naturw. Ver. Hamburg*, N.F. (1964), 61–80.

STAMMER, H. J., 1928. Die Fauna der Ryckmündung, eine Brackwasserstudie. *Z. Morph. Ökol. Tiere*, **11**, 36–114, figs. 1–15.

—— 1932. Die Fauna des Timavo. Ein Beitrag zur Kenntnis der Höhlengewässer, des Süss- und Brackwassers im Karst. *Zool. Jb.* (Syst.), **63**, 521–656, figs. 1–16.

STEENSTRUP, J. J. Sm., 1837. 'Acta Musei Hafniensis'—according to Kramp (1947), p. 40, these Acta were never printed.

STEINBERG, Sonia N., 1963. The regeneration of whole polyps from ectodermal fragments of scyphistoma larvae of *Aurelia aurita*. *Biol. Bull. mar. biol. Lab., Woods Hole*, **124**, 337–43, figs. 1–4.

BIBLIOGRAPHY

STEINER, Gerolf, 1934. Der Verlust der Glockenautomatie bei randorganlos aufgezogenen Ohrenquallen (*Aurelia aurita*). *Biol. Zbl.* **54**, 102–5, figs. 1–5.

—— 1935. Über eine eigenartige Degeneration bei Ephyren und Medusen von *Aurelia aurita*. *Zool. Anz.* **109**, 176–81, figs. 1–5.

STEPHENS, Jane, 1905. A list of Irish Coelenterata, including the Ctenophora. *Proc. R. Ir. Acad.* **25** (B), pp. 25–92.

STIASNY, Gustav, 1914. Zwei neue Pelagien aus der Adria. *Zool. Anz.* **44**, nr. 12, pp. 529–33, figs. 1–4.

—— 1921*a*. Results of Dr E. Mjöberg's Swedish Scientific Expeditions to Australia 1910–13. XXX. Scyphomedusen. *K. svenska VetenskAkad. Handl.* **62**, no. 2, pp. 1–13, 1 fig.

—— 1921*b*. Studien über Rhizostomeen mit besonderer Berücksichtigung der Fauna des Malaiischen Archipels nebst einer Revision des Systems. *Capita zool.* **1**, Afl. 2, pp. 1–177, text-figs. 1–17, pl. I–V.

—— 1922. Die Scyphomedusen-Sammlung von Dr Th. Mortensen nebst anderen Medusen aus dem zoologischen Museum der Universität in Kopenhagen. *Vidensk. Meddr dansk naturh. Foren.* **73**, 513–58, figs. 1–14.

—— 1923. Das Gastrovascularsystem als Grundlage für ein neues System der Rhizostomeen. *Zool. Anz.* **17**, 241–7, figs. 1–17.

—— 1924. Ueber einige von Dr C. J. van der Horst bei Curaçao gesammelte Medusen. *Bijdr. Dierk.* **23**, 83–91.

—— 1927. Ueber Variation der Zeichnung und Färbung bei *Chrysaora hysoscella* Eschscholtz. *Zoöl. Meded., Leiden*, **10**, 73–86, pl. I–III.

—— 1928. Mitteilungen über Scyphomedusen. II. 1. Über einige Entwicklungstadien von *Rhizostoma octopus* Linn. 2. Das Gefässystem der Mundarme von *Rhizostoma octopus* Linn. 3. Ueber die Anhänge an den Mundarmen von *Rhizostoma octopus* Linn. *Zoöl. Meded., Leiden*, **11**, 7, 177–98, figs. 1–10.

—— 1929. Ueber Anomalien des Gastrovascularsystems von *Rhizostoma octopus* L. und ihre Bedeutung für die Phylogenie. *Zoöl. Meded., Leiden*, **12**, 2, 4–15, pl. I–IX.

—— 1930. Die Scyphomedusen-Sammlung des 'Musée royal d'Histoire Naturelle de Belgique' in Brüssel. *Mém. Mus. r. Hist. nat. Belg.* no. 42, pp. 1–32, pl. I–II.

—— 1931. Die Rhizostomeen-Sammlung des British Museum (Natural History) in London. *Zool. Meded., Leiden*, **14**, 9, 137–78, figs. 1–9.

—— 1934. Scyphomedusae. '*Discovery*' *Rep.* **8**, 329–96, pl. XIV–XV, text-figs. 1–12.

—— 1935. Die Scyphomedusen der Snellius-Expedition. *Verh. K. ned. Akad. Wet.* (Afd. Natuurk.) Tweede Sectie, **34**, no. 6, pp. 1–44, pl. I, text-figs. 1–9.

—— 1937. Scyphomedusae. *John Murray Exped. 1933–34. Sci. Rep.* **4** (Zoology, no. 7), 203–42, pl. I, text-figs. 1–14, 2 maps.

—— 1938. Die Scyphomedusen des Roten Meeres. *Verh. K. ned. Akad. Wet.* (Afd. Natuurk.) Tweede Sectie, **37**, 1–35, text-figs. A–F, pl. I–II.

—— 1940. Die Scyphomedusen. '*Dana*' *Rep.*, no. 18, pp. 1–28, pl. I–II, text-figs. A–C, Karte 1–2.

STIASNY, G. & MAADEN, H. van der, 1943. Über Scyphomedusen aus dem Ochotskischen und Kamtschatka Meer nebst einer Kritik der Genera *Cyanea* und *Desmonema*. *Zool. Jb.* (Syst.), **76**, 227–66, figs. 1–15.

STOCKARD, Charles R., 1908. Studies on tissue growth. I. An experimental study of the rate of regeneration in *Cassiopea xamachana* (Bigelow). *Pap. Tortugas Lab.* **2**, 63–102, figs. 1–29.

SUGIURA, Yasuo, 1963. On the life-history of rhizostome medusae. I. *Mastigias papua* L. Agassiz. *Annotnes zool. jap.* **36**, 194–202, figs. 1–12.

—— 1964. On the life-history of rhizostome medusae. II. Indispensability of *Zooxanthellae* for strobilation in *Mastigias papua*. *Embryologia*, **8**, 223–33, figs. 1–3.

—— 1965. On the life-history of rhizostome medusae. III. On the effects of temperature on the strobilation of *Mastigias papua*. *Biol. Bull. mar. biol. Lab., Woods Hole*, **128**, 493–6, fig. 1.

—— 1966. On the life-history of rhizostome medusae. IV. *Cephea cephea*. *Embryologia*, **9**, 105–22, text-figs. 1–4, pl. 6, 7.

SWARTZ, O., 1791. In Modeer, A. (1791), Slagret Sjokalf, *Medusa. Kongl. Vetensk. Acad. nya Handlingar*, **12**, 188–90, pl. V.

TATTERSALL, W. M., 1907. The marine fauna of the coast of Ireland. VIII. Pelagic Amphipoda of the Irish Atlantic Slope. *Rep. Sea inld Fish. Ire.* (1905), pt. II. *Sci. Invest.* 63–99, pl. I–V.

TCHÉOU-TAI-CHUIN, 1928. Absence de strobilisation et persistance du bourgeonnement pendant l'hiver chez des Scyphistomes alimentés artificiellement. *C. r. hebd. Séanc. Acad. Sci. Paris*, **186**, 790–1.

—— 1929*a*. Histolyse musculaire et phagocytose dans le strobile de *Chrysaora isosceles*. *C. r. Séanc. Soc. Biol.* **101**, 1005–7, figs. 1–6.

Tchéou-Tai-Chuin, 1929*b*. Sur les nématocystes du scyphistome de *Chrysaora isosceles*. *Bull. Soc. zool. Fr.* **54**, 531–6, figs. I–IV.

—— 1929*c*. La digestion chez le Scyphistome de *Chrysaora*. *C. r. Séanc. Soc. Biol.*, **102**, 520–2.

—— 1929*d*. Les phénomènes cytologiques au cours de la digestion intracellulaire chez le Scyphistome de *Chrysaora*. *C. r. Séanc. Soc. Biol.* **102**, 557–8.

—— 1929*e*. Sur les vacuoles digestives chez le Scyphistome. *C. r. Séanc. Soc. Biol.* **102**, 825–7.

—— 1930*a*. Contribution à l'étude de la formation des grains de sécrétion chez le Scyphistome. *C. r. Séanc. Soc. Biol.* **103**, 315–17.

—— 1930*b*. Le cycle évolutif de Scyphistome de *Chrysaora*: Étude histophysiologique. *Trav. Stn biol. Roscoff*, fasc. 8, pp. 1–179, text-figs. 1–50, pl. I–IV.

Teissier, Georges, 1926. Sur la teneur en eau et en substances organiques de *Chrysaora hysoscella* (L.) aux différents stades de son ontogénèse. *Bull. Soc. Zool. Fr.* **51**, 266–73.

—— 1929. La croissance embryonnaire de *Chrysaora hysoscella* (L.). *Arch. Zool. exp. gén.* **69**, 137–78, text-figs. 1–10.

—— 1932. Sur la composition chimique des planulas de *Chrysaora hysoscella* (L.). *Bull. Soc. zool. Fr.* **57**, 160–2.

Thiel, Hjalmar, 1962. Untersuchungen über die Strobilisation von *Aurelia aurita* Lam. an einer Population der Kieler Förde. *Kieler Meeresforsch.* **18**, 198–230, figs. 1–20.

—— 1963*a*. Untersuchungen über die Entstehung abnormer Scyphistomae, Strobilae und Ephyrae von *Aurelia aurita* Lam. und ihre theoretische Bedeutung. *Zool. Jb.* (Anat.), **81**, 311–58, figs. 1–23.

—— 1963*b*. Teil- und Spiralephyren von *Aurelia aurita* und ihre Regulation. *Zool. Anz.* **171**, 303–27, figs. 1–10.

—— 1966. The evolution of Scyphozoa. A review in 'The Cnidaria and their evolution'. *Symp. Zool. Soc. Lond.* no. 16, pp. 77–117, figs. 1–24.

Thiel, Max Egon, 1935. Über die Wirkung des Nesselgiftes der Quallen auf den Menschen. *Ergebn. Fortschr. Zool.* **8**, 1–35, figs. 1–9.

—— 1936*a* & *b*. Scyphomedusae. Coronatae. *Bronn's Kl. Ordn. Tierreichs*, **2**, Abt. II, Buch 2, Lief. 2, pp. 308–20, figs. 157–64; Lief. 3, pp. 321–480, figs. 165–224.

—— 1938*a* & *b*. Scyphomedusae. *Bronn's Kl. Ordn. Tierreichs*, **2**, Abt. II, Buch 2, Semaeostomae. Morphologie und Histologie, Lief. 4, pp. 481–672, figs. 225–325; Ontogenie, Lief. 5, pp. 673–848, figs. 326–429.

—— 1938*c*. Über eine Scyphistomabildung durch Strobilisation und ihre phylogenetische und eidonomische Bedeutung. *Zool. Anz.* **121**, 97–107.

—— 1959*a*. Beiträge zur Kenntnis der Wachstums- und Fortpflanzungsverhältnisse von *Aurelia aurita* L. *Abh. Verh. naturw. Ver. Hamburg*, N.F. **3** (1958), 13–26, figs. 1–4.

—— 1959*b*. Scyphomedusae. Semaeostomae, Physiologie. *Bronn's Kl. Ordn. Tierreichs*, **2**, Abt. II, Buch 2, Lief. 6, pp. 849–1072, figs. 430–95.

—— 1960. Beobachtungen über Wachstum, Variationen und Abnormitäten bei *Cyanea capillata* der Ostsee. *Abh. Verh. naturw. Ver. Hamburg*, N.F. **4** (1959), 89–108, figs. 1–16.

—— 1962*a*. Scyphomedusae. Semaeostomae. Ökologie, Geschichte ihrer Kenntnis, ihre Stammesgeschichte und Klassifikation. *Bronn's Kl. Ordn. Tierreichs*, **2**, Abt. II, Buch 2, Lief. 7, pp. 1073–1308, figs. 496–549.

—— 1962*b*. Untersuchungen zur Artfrage von *Cyanea lamarckii* Pér. et Les. und *Cyanea capillata* L. *Abh. Verh. naturw. Ver. Hamburg*, N.F. **6** (1961), 277–93, figs. 1 & 2.

—— 1964. Untersuchungen über die Ernährungsweise und den Nahrungs-Kreislauf bei *Rhizostoma octopus* L. Ag. *Mitt. zool. St. Inst. Hamb.* Kosswig-Festschr., pp. 247–69, pl. VI–VIII.

—— 1965. Untersuchungen zur Systematik der Gattung *Rhizostoma*. *Abh. Verh. naturw. Ver. Hamburg*, N.F. **9** (1964), 37–53, text-figs. 1–4, pl. I–II.

—— 1966*a*. Untersuchungen über die Herkunft, das Auftreten, das Wachstum und die Fortpflanzung von *Rhizostoma octopus* L. Ag. im Elbmündungsgebiet. *Abh. Verh. naturw. Ver. Hamburg*, N.F. **10** (1965), 59–88, figs. 1–10.

—— 1966*b*. Beitrag zur Kenntnis westafrikanischer Hydro- und Scyphomedusen. *Mitt. hamb. zool. Mus. Inst.* **63**, 3–27, text-figs. 1–4, pl. I–III.

Thill, Hans, 1937. Beiträge zur Kenntnis der *Aurelia aurita* (L.). *Z. wiss. Zool.* **150**, 51–96, text-figs. 1–18.

Thomas, Len R., 1963. Phyllosoma larvae associated with medusae. *Nature, Lond.* **198**, 208.

Thompson, W., 1840. Additions to the fauna of Ireland. *Ann. Mag. nat. Hist.* **5**, 245–57.

Tilesius, W. G. 1831. Beiträge zur Naturgeschichte der Medusen. I. Cassiopeae. *Nov. Act. acad. Leopold*, **15** (2), 249–88, pl. LXX–LXXIII.

223

BIBLIOGRAPHY

TOPPE, Otto, 1910. Untersuchungen über Bau und Funktion der Nesselzellen der Cnidarier. Teil I. Der feinere Bau der Nesselzellen sowie systematische Beiträge zur Kenntnis des Genus *Hydra. Zool. Jb.* (Anat.), **29**, 191–280, pl. 13–16.

TRETYAKOFF, D. K., 1937. Mesogloea of Black Sea Scyphomedusae. *Russk. Arkh. Anat. Gistol. Émbriol.* **17**, 279. (Russian, with English summary.)

TSUKAGUCHI, R., 1914. Über die feinere Struktur des Ovarialeies von *Aurelia aurita* L. *Arch. mikrosk. Anat. EntwMech.* **85**, 114–23, pl. IX.

UCHIDA, Tohru, 1926. The anatomy and development of a rhizostome medusa, *Mastigias papua* L. Agassiz, with observations on the phylogeny of Rhizostomeae. *J. Fac. Sci. Tokyo Univ.* (sect. IV, Zool.), **1**, 45–95, text-figs. 1–50, pl. VI.

—— 1934a. Metamorphosis of a Scyphomedusa (*Pelagia panopyra*). *Proc. imp. Acad. Japan*, **10**, no. 7, pp. 428–30, figs. 1–4.

—— 1934b. A semaeostome medusa with some characters of Rhizostomae. *Proc. imp. Acad. Japan*, **10**, 698–700, figs. 1–2.

—— 1935. Remarks on the Scyphomedusan family Pelagidae. *Trans. Sapporo nat. Hist. Soc.* **14**, pt. 1, pp. 42–5, figs. 1–2.

—— 1938. Medusae in the vicinity of the Amakusa Marine Biological Station. *Bull. Biogeogr. Soc. Tokyo*, **8**, no. 10, pp. 143–9, 1 fig.

—— 1947. Medusae in the vicinity of Shimoda. *Journ. Fac. Sci., Hokkaido Univ.* ser. VI (Zool.), **9**, no. 4, pp. 331–43, figs. 1–3.

—— 1954. Distribution of Scyphomedusae in Japanese and its adjacent waters. *J. Fac. Sci. Hokkaido Univ.* ser. VI (Zool.), **12**, 209–19, figs. 1–2.

—— 1955. Scyphomedusae from the Loochoo Islands and Formosa. *Bull. Biogeogr. Soc. Japan*, **16–19**, (Recent conceptions of Japanese Fauna), 14–16, figs. 1–2.

—— 1963a. Two phylogenetic lines of Coelenterates from the viewpoint of symmetry. *J. Fac. Sci. Hokkaido Univ.* ser. VI (Zool.), **15**, no. 2, pp. 276–82, figs. 1–4.

—— 1963b. On the interrelationships of the Coelenterata, with remarks on their symmetry, in *The Lower Metazoa: comparative biology and phylogeny. Univ. Calif. Press*, **12**, 169–77, 4 figs.

—— 1964. Some medusae from New Caledonia. *Publ. Seto. mar. biol. Lab.* **12**, no. 1, pp. 109–12, figs. 1–3.

UCHIDA, Tohru & NAGAO, Zen, 1963. The metamorphosis of the Scyphomedusa, *Aurelia limbata* (Brandt). *Annotnes zool. jap.* **36**, 83–91, figs. 1–18.

UEXKÜLL, J. v., 1901. Die Schwimmbewegungen von *Rhizostoma pulmo. Mitt. zool. Stn. Neapel*, **14**, 620–6.

UNTHANK, H. W., 1894. Under 'Notes', *Nature, Lond.* **50**, 413.

USSING, Hj., 1927. Bidrag til *Aurelia aurita*'s Biologi i Mariagerfjord. *Vidensk. Meddr dansk. naturh. Foren.* **84**, 91–106, figs. 1–6.

—— 1945. Brâendegoplen (*Cyanea capillata*). Biologiske Bidrag. *Flora og Fauna*, **51**, 46–8, 1 text-fig.

UVNÄS, Börje, 1960. Mechanism of action of a histamine-liberating principle in jellyfish (*Cyanea capillata*). *Ann. N.Y. Acad. Sci.* **90**, 751–9, figs. 1–7.

VALLENTIN, Rupert, 1900. Notes on the fauna of Falmouth. *Jl R. Instn Cornwall*, no. XLVI, 14 pp.

—— 1907. The fauna of St. Ives Bay. *Jl R. Instn Cornwall*, no. LIII, 29 pp.

VANHÖFFEN, Ernst, 1888. Untersuchungen über Semaeostome und Rhizostome Medusen. *Zoologica, Cassel*, Heft 3, pp. 1–54, pl. I–VI, 1 chart.

—— 1892. Die Akalephen der Plankton-Expedition. *Ergebn. Plankton Exped.* **2**, K. d., pp. 1–30, pl. I–IV.

—— 1896. Schwarmbildung im Meere. *Zool. Anz.* **19**, 523–6.

—— 1900. Über Tiefseemedusen und ihre Sinnesorgane. *Zool. Anz.* **23**, 277–9.

—— 1902. Die acraspeden Medusen der deutschen Tiefsee-Expedition 1898–1899. *Wiss. Ergebn. dt. Tiefsee-Exped. 'Valdivia'*, **3**, 3–86, pl. I–XII.

—— 1906. Acraspedae, *Nordisches Plankton*, **6**, XI, pp. 40–64, figs. 1–37.

—— 1908. Die Lucernariden und Scyphomedusen der Deutschen Südpolar-Expedition 1901–1903. *Deutsche Südpolar-Exped.* **10** (Zool.), **2**, Heft 1, pp. 25–50, pl. II–III, text-figs. 1–11.

VANNUCCI, M., 1957. Double monster ephyrae of *Aurelia aurita. Nature, Lond.* **179**, 326–7.

VANNUCCI, M. & TUNDISI, J., 1962. Las medusas existentes en los Museos de la Plata y Buenos Aires. *Comun. Mus. argent. Cienc. nat. Bernardino Rivadavia, Ciencias Zoológicas*, **3**, 203–15.

VERESS, E., 1911. Sur les mouvements des Méduses. *Arch. int. Physiol.* **10**, 253–89, 19 figs.

—— 1938. Studien über die rhythmischen Bewegungen der Medusen. *Allatorv. Közlem*, **35** (3–4), 153–70.

VERNON, H. M., 1895. The respiratory exchange of the lower marine invertebrates. *J. Physiol., Lond.* **19**, 18–70, figs. 1–13.

VERWEY, J., 1942. Die Periodizität im Auftreten und die activen und passiven Bewegungen der Quallen. *Archs néerl. Zool.* **6**, livr. 4, pp. 363–468, figs. 1–2.

VERWEY, J., 1964. Annual report of the Netherlands Institute for Sea Research for the year 1962. *Neth. J. mar.* **2**, no. 2, pp. 293–318.

—— 1965. Annual report of the Netherlands Institute for Sea Research for the year 1963. *Neth. J. mar. Res.* **2**, no. 4, pp. 615–37.

—— 1966. The role of some external factors in the vertical migration of marine animals. *Neth. J. mar. Res.* **3**, 245–66, figs. 1–3.

VIGUIER, Camille, 1890. Études sur les animaux inférieurs de la Baie d'Alger. IV. Le Tétraptère (*Tetraplatia volitans*, Busch.). *Archs Zool. exp. gén. Paris*, sér. 2, **8**, 101–40, pl. VII–IX.

WAGNER, Rudolph, 1835. Entdeckung männlicher Geschlechtstheile bei den Actinien. *Arch. Naturgesch.* (1 Jahrg.), **2**, 215–19, pl. III.

—— 1836. *Prodromus historiae generationis hominis atque animalium*, 15 pp. pl. 1–2.

—— 1841. *Ueber den Bau der* Pelagia noctiluca *und die Organisation der Medusen zugleich als Prodromus seines zootomischen Handatlasses.* 4 pp. pl. XXXIII. Leipzig.

WALCOTT, Charles Doolittle, 1898. Fossil Medusae. *Monogr. U.S. Geol. Survey*, **30**, 4, pl. 30, 31.

WEILL, Robert, 1934. Contribution à l'étude des cnidaires et de leurs nématocystes. I. Recherches sur les nématocystes (Morphologie–Physiologie–Développement). *Trav. Stn zool. Wimereux*, **10**, 1–347, figs. 1–208; II. Valeur taxonomique du cnidome. *Trav. Stn zool. Wimereux*, **11**, 349–701, figs. 209–432.

WELSH, John H., 1955. On the nature and action of coelenterate toxins. *Deep-sea Res.* **3** (suppl.), 287–97.

WERNER, Bernhard, 1966. *Stephanoscyphus* (Scyphozoa, Coronatae) und seine direkte Abstammung von den fossilen Conulata. *Helgoländer wiss. Meeresunters.* **13**, 317–47, figs. 1–15.

—— 1967. Morphologie, Systematik und Lebensgeschichte von *Stephanoscyphus* (Scyphozoa, Coronatae) sowie seine Bedeutung für die Evolution der Scyphozoa. *Zool. Anz.* Suppl. **30**, 297–319, figs. 1–16.

WETOCHIN, J. A., 1926 *a*. Ueber die Arbeit des Flimmerepitels des Gastrovascularsystem der Meduse *Aurelia aurita* (L.) Lam. *Rab. Murmansk. biol. Sta.* **2**, 107–20, 3 figs. (Russian with German résumé.)

—— 1926 *b*. Über die Erregungsprocesse im Schirm der Qualle *Aurelia aurita* und über die Regulation der Bewegungen dieses Tieres im Meerwasser. *Russk. fiz. Zh.* **9**, 518–36, figs. 1–5. (Russian with German summary.)

—— 1930. Über das Ergreifen der Nahrung und die Ernährung bei den Medusen *Aurelia aurita* L. im Zusammenhang mit der Arbeit des Flimmerpithels des Manubriums. *Russk. fiz. Zh.* **13**, 585–601, figs. 1–4. (Russian with German summary.)

WIDERSTEN, Bernt, 1965. Genital organs and fertilization in some Scyphozoa. *Zool. Bidr. Upps.* **37**, 45–58, text-figs. 1–11, pl. I–III.

—— 1967. On the development of septal muscles, supporting fibrils and periderm in the scyphistoma of semaeostome Scyphozoa. *Ark. Zool.* ser. 2, **18**, 567–74, text-figs. 1–7, pl. I.

—— 1968. On the morphology and development in some cnidarian larvae. *Zool. Bidr. Upps.* **37**, 139–82, text-figs. 1–16, pl. 1–3.

WIDMARK, Erik M. P., 1911. Über die Gastrovascularströmungen bei *Aurelia aurita* L. und *Cyanea capillata* Eschz. *Zool. Anz.* **38**, 378–82, figs. 1–3.

—— 1913. Über die Wasserströmungen in dem Gastrovascularapparat von *Aurelia aurita* L. *Z. allg. Physiol.* **15**, 33–48, figs. 1–4.

WIERSBITZKY, Siegfried & SCHEIBE, Ernst, 1966. Über das Vorkommen blutgruppenaktiver Stoffe in *Aurelia aurita*. *Z. ImmunForsch. Allerg. klin. Immun.* **130**, 301–5.

WIKSTRÖM, D. A., 1921. Iakttagelser om öronmaneten (*Aurelia aurita*). *Meddn Soc. Fauna Flora fenn.* **47**, 169–73.

—— 1925 *a*. Vidare jakttagelser om öronmaneten (*Aurelia aurita*). *Meddn Soc. Fauna Flora fenn.* **48**, 244–6.

—— 1925 *b*. *Aurelia aurita* i Hitis och Houtskär. *Meddn Soc. Fauna Flora fenn.* **49**, 209–10.

—— 1932–3. Beobachtungen über die Ohrenqualle (*Aurelia aurita* L.) in den Schären SW-Finnlands. *Memo. Soc. Fauna Flora fenn.* **8** (1931–2), 14–17.

WILL, L., 1927. Ueber Hungerreduktion bei *Aurelia aurita*. *Sber. Abh. naturf. Ges. Rostock*, ser. III, **1** (1925/6), 66–9.

WOOLLARD, H. H. & HARPMAN, J. A., 1939. Discontinuity in the nervous system of coelenterates. *J. Anat.* **73**, 559–62, pl. I.

WRIGHT, T. Strethill, 1861. On hermaphrodite reproduction in *Chrysaora hysoscella*. *Ann. Mag. Nat. Hist.* ser. 3, **7**, XL, 357–9, pl. XVIII.

WU, Hsien Wen, 1927. Preliminary observations on the sense organs and the adjacent structures of two Scyphomedusae at young stage. *Contr. biol. Lab. Sci. Soc. China*, **3**, no. 4, pp. 1–5, figs. 1–3.

YAMASHITA, T., 1957 *a*. Über den Statolithen in den Sinneskörpen der Meduse *Aurelia aurita*. *Z. Biol.* **109**, 111–15, figs. 1–5.

BIBLIOGRAPHY

YAMASHITA, T., 1957*b*. Das Aktionspotential der Sinneskörper (Randkörper) der Meduse *Aurelia aurita*. *Z. Biol.* **109**, 116–22, figs. 1–3.

ZEYNEK, R. v., 1912. Chemische Studien über *Rhizostoma Cuvieri*. *Sber. Akad. Wiss. Wien, math. naturw. kl.* **121**, Heft x, abt. 11*b*, pp. 1539–79.

PLATES

PLATE I

FIG. I

Nausithoë atlantica Broch, × 2.

FIG. 2

Nausithoë globifera Broch, × 3.

FIG. 3

Atolla wyvillei Haeckel, × 1.

FIG. 4

Paraphyllina ransoni Russell, × 1·5.

FIG. 5

Periphylla periphylla (Pér. & Les.), × ½.

Plate I

1

2

3

4

5

Del. F.S.R.

PLATE II

Pelagia noctiluca (Forskål), × 1.

Plate II

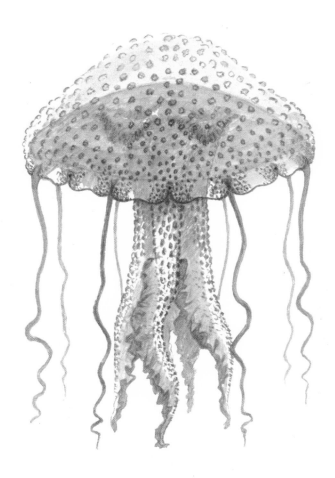

PLATE III

Chrysaora hysoscella (L.), × $\frac{2}{3}$.

Plate III

PLATE IV

Cyanea capillata (L.), × $\frac{1}{3}$.
The full numbers of tentacles have not been drawn,
but they are shown as about twice as many
as those of *C. lamarckii* in Pl. V.

Plate IV

Cyanea lamarckii (Pér. & Les.), × 1.
The full numbers of tentacles have not been drawn,
but they are shown as about half the number
in *C. capillata* in Pl. IV.

Plate V

PLATE VI

Cyanea lamarckii (Pér. & Les.).
Colour photographs taken by C. I. D. Moriarty
of specimens from the Irish Sea to show colour
range from blue to yellow.

Plate VI

PLATE VII

Plate VII

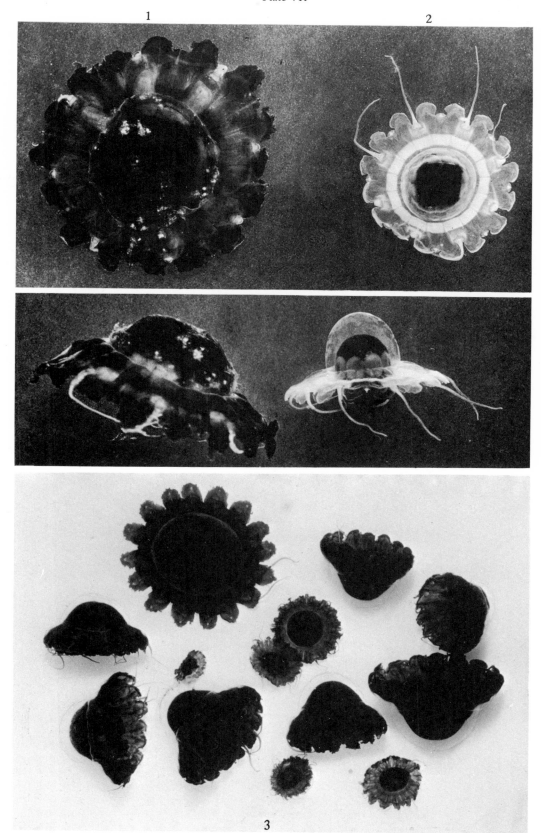

1   2

3

PLATE VIII

Plate VIII

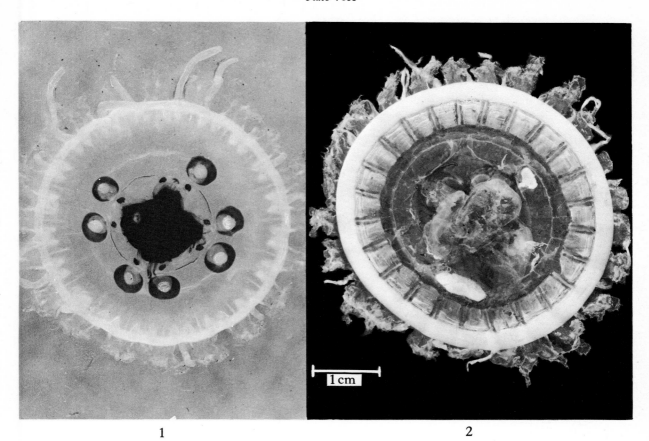

1

2

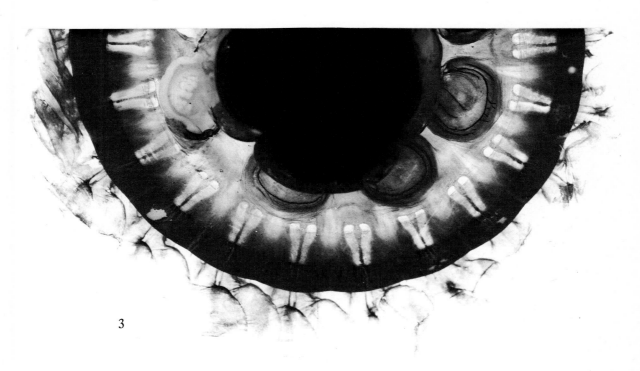

3

PLATE IX

Periphylla periphylla (Pér. & Les.), × 1.

Plate IX

PLATE X

Pelagia noctiluca (Forskål).

Plate X

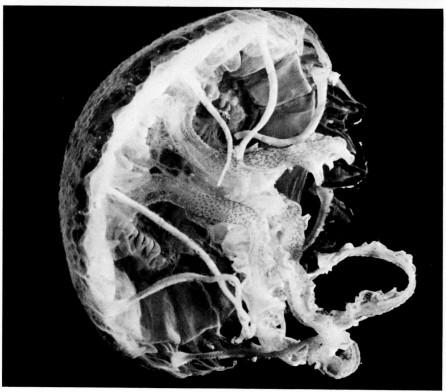

Plate XI

160 mm.

1

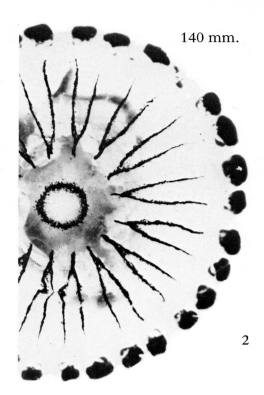

140 mm.

2

160 mm.

3

120 mm.

4

PLATE XII

Cyanea capillata (L.).
Specimen from Gruinard Bay, Ross-shire, *ca.* 140 mm. in diameter,
September 1959. The manubrium, gonads and marginal tentacles
have been removed to show the gastrovascular canal system.
In the lower photograph an abnormal confluence of canals can be
seen in one lappet.

Plate XII

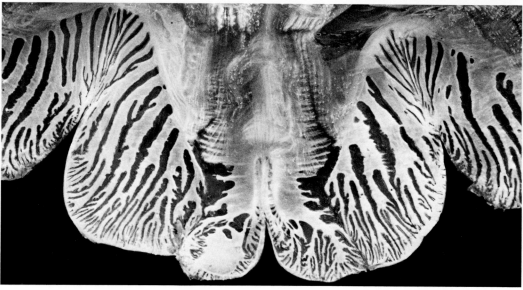

Above: Cyanea capillata (L.).
Specimen from Millport *ca.* 60 mm. in diameter.
Below: Cyanea lamarckii (Pér. & Les.).
Specimen from English Channel *ca.* 60 mm. in diameter.
Note: pits into coronal muscle folds in *C. capillata*
and radial furrows only in *C. lamarckii*.
Higher number of marginal tentacles in *C. capillata*
and greater development of gonad in *C. lamarckii*.

Plate XIII

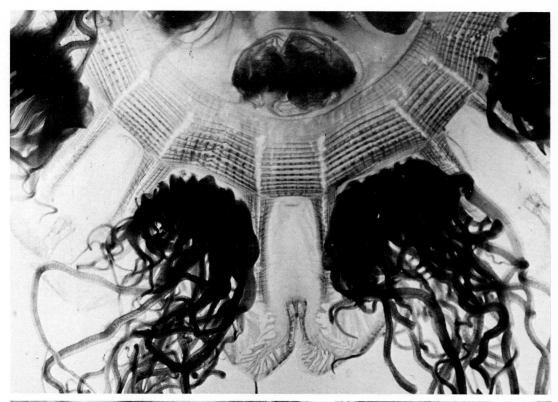

Aurelia aurita (L.).
Above: specimen from Loch Melfort, June 1967.
Below: exumbrellar view of excised
manubrium showing brood pouches.

Plate XIV

Rhizostoma octopus (L.).
Above left: young specimen *ca.* 20 mm. in diameter;
Above right: specimen *ca.* 45 mm. in diameter with terminal clubs.
Below: enlarged portion of gastrovascular canal system of specimen
ca. 45 mm. in diameter. (Collected by T. G. Skinner
from Solway Firth, July 1966.)

Plate XV

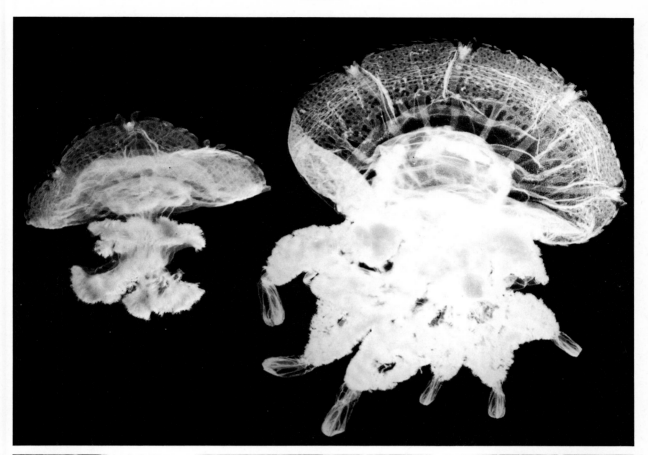

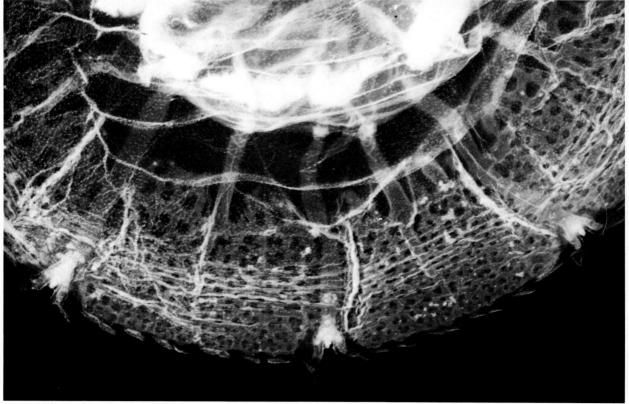

SUPPLEMENT
TO THE FIRST VOLUME ON
THE MEDUSAE OF THE BRITISH ISLES
PUBLISHED IN 1953

ERRATA FOR 1953 VOLUME

Page 6, line 29: For *Melicertum* read *Liriope*.

Page 6, line 32: Delete (Pl. XIII).

Page 187, four lines from bottom: under characters of *Leuckartiara nobilis* 'with abaxial exumbrellar spur' should read 'without abaxial exumbrellar spur'.

Page 392, line 12, for *Sabellaria* read *Sabella*.

Page 392, Text-fig. 256, for *Sabellaria* read *Sabella*.

SUPPLEMENT TO
THE MEDUSAE OF THE BRITISH ISLES
DEALING WITH HYDROMEDUSAE,
PUBLISHED IN 1953

Since the publication of my first volume on British medusae in 1953 sufficient advances in our knowledge have been made to justify a section in the present monograph to bring the reader up to date.

There have been numerous records extending the known distribution of some species, but the most significant additions to our knowledge have been due to the researches of C. Edwards and B. Werner, to both of whom I am indebted for information and drawings not yet published.

Edwards, working at Millport, has been very successful in rearing hydroids and medusae and has filled in many of the gaps in our knowledge of the identity of the hydroids and their medusae.

Werner has also filled a number of gaps in the linking of medusae to their hydroids, but in addition he has made most interesting observations and experiments explaining the distribution and ecology of certain species. Among these observations the discovery of the existence of two kinds of eggs is remarkable. He has also studied the effects of temperature on sexual and asexual reproduction.

With the opportunity to examine collections from deep water off the mouth of the English Channel I have myself discovered a number of new species which can be added to the British fauna.

Kramp (1959, 1961, 1968) has done a signal service to zoologists by his three great summarizing works: the 'Dana' Reports on 'The Hydromedusae of the Atlantic Ocean and adjacent waters', 'The Hydromedusae of the Pacific and Indian Oceans', and his 'Synopsis of the Medusae of the World'.

Mention should also be made of the following faunistic reports: Vannucci (1956, Clyde sea area); Edwards (1958, Clyde sea area); Werner (1959a, Port Erin); Hamond (1957, 1963, Norfolk coast). An important faunistic contribution now in the press is that of Brinckmann–Voss (1969) on the Anthomedusae/Capitata of the Mediterranean. Edwards (1968) discussed the distribution of Hydromedusae in British waters in relation to water movements.

In the following pages I have added, under the British species concerned, records that have extended their known distribution, and all significant work on their systematics, morphology and general biology.

ORDER

ANTHOMEDUSAE

FAMILY CORYNIDAE

Genus **Sarsia**

Sarsia eximia (Allman)

Distribution includes: Norfolk (Hamond, 1957); Port Erin (Werner, 1959*a*); scarce in Clyde sea area (Edwards, unpublished); Helsingør (Allwein, 1968); Mediterranean (Kramp, 1957); Brazil (Vannucci, 1957); West and South Africa (Millard, 1959, 1966); Chile (Kramp, 1966); east coast of North America (Petersen, 1964).

Rice (1964) found that pressure decrease caused decreased activity with consequent passive sinking away from light, while pressure increase induced renewed activity and movement towards light. A detailed study of the development and histology of the planula was made by Bodo and Bouillon (1968).

Sarsia prolifera Forbes

Distribution includes Clyde sea area (Vannucci, 1956; Edwards, unpublished).

Sarsia tubulosa (M. Sars)

Distribution includes Azores (Rees & White, 1966).

Kühl (1962) suggested that *Coryne lovéni* was possibly a degenerate form of *C. sarsi* due to estuarine conditions. Werner (1963) found that, if the temperature was lowered from 14° to 2° C., after two to three weeks medusa buds began to develop, always below the tentacle zone. If the temperature was then increased again to 14° C. medusa budding ceased. Above 6–8° C. there was excessive growth of the manubrium of the medusa bud or the formation of a polyp bud. Kakinuma (1966) stated that lowering the temperature to 5° C. induced medusa bud formation.

Rice (1964) found similar effects of pressure changes to those recorded above for *Sarsia eximia*.

Mackie, Passano & Ceccatty (1967), Passano, Mackie & Ceccatty (1967) and Mackie & Passano (1968) record aneural conduction and spontaneous activity.

Sarsia gemmifera Forbes

Distribution includes Port Erin (Southward, 1954).

Genus **Stauridiosarsia**

Stauridiosarsia producta (Wright)

Distribution includes Clyde sea area (Vannucci, 1956).

Genus **Dipurena**

Dipurena ophiogaster Haeckel

Distribution includes Chile (Kramp, 1966).

Family TUBULARIIDAE

Genus **Ectopleura**

Ectopleura dumortierii (van Beneden)

Distribution includes: Elbe estuary (Kühl, 1962); Helsingør (Allwein, 1968); Brazil (Vannucci, 1957, 1963); Ems estuary (Kühl, 1967).

Werner & Aurich (1955) studied the development of the egg into a free-swimming actinula with ten to thirteen long aboral tentacles and four to five short oral tentacles. The actinula settled 4–5 days after the egg was laid and started feeding before settlement. The eggs were a little over 0·3 mm. in diameter; they were few in number and developed at the expense of other oocytes. Widersten (1968) described the planula.

Regeneration and healing in the pro-actinula was studied by Kuhl & Kuhl (1967).

Kühl (1965) studied the occurrence of the medusa in relation to the tide at Cuxhaven.

Genus **Hybocodon**

Hybocodon prolifer L. Agassiz

Distribution includes Elbe estuary (Kühl, 1962).

Werner (1959a) at Port Erin found that actinulae started to feed on the first day after detachment from the medusa and settled in the culture dish on the second day, indicating that the planktonic phase is short.

Genus **Steenstrupia**

Steenstrupia nutans (M. Sars)

Werner (1959a) at Port Erin did culture experiments. At 12° to 15° C. some polyps appeared on the third day, but the majority on the fourth and fifth days after shedding of the egg by the parent medusa.

Some only of the eggs developed at once into polyps, the remainder being winter-resting eggs developing into polyps the following spring.

Genus **Euphysa**

Euphysa aurata Forbes

Distribution includes Chile (Kramp, 1966).

Werner (1959a) at Port Erin reared the hydroid. The spherical fertilized eggs sank to the bottom and small polyps developed about 10 days later at 12°–15° C. Some only of the eggs developed at once into polyps, the remainder being winter-resting eggs developing into polyps the following spring.

Genus **Plotocnide**

Plotocnide borealis Wagner

Added to the British list, Clyde sea area (Edwards, 1958); Oslofjord, Beyer (1955).

Genus **Eucodonium**

Eucodonium brownei Hartlaub

Distribution includes Brazil (Vannucci, 1957, 1963).

233

Family MARGELOPSIDAE

Genus **Margelopsis**

Margelopsis haeckeli Hartlaub

Distribution includes: Norfolk (Hamond, 1963); Solway Firth, August 1966 (T. G. Skinner, personal communication); Elbe estuary (Kühl, 1962); Ems estuary (Kühl, 1967).

Werner (1954a, 1956a) could find no male sexual organs and assumed that *Margelopsis* could reproduce parthenogenetically. In only one medusa out of 250 reared in cultures were sperm cells seen.

This was the first species in which Werner (1954b) discovered that there were two kinds of eggs, some which developed at once into polyps and others which did not produce polyps until the following spring. The former eggs were small, averaging 0·125 mm. in diameter. The overwintering eggs were larger, with an average diameter of 0·187 mm. The egg divides on the parent medusa to form a mass of endoderm cells rich in yolk surrounded by ectoderm cells, a 'sterroblastula'. At this stage further development ceases, and the egg falls to the bottom to which it attaches. Here it forms a lens-shaped object with a protective periderm. It may rest in this state for six months, when a small polyp slips out of it. Werner found that at lower temperatures normal quick-developing eggs were produced and that these gave way to the production of overwintering or resting eggs at higher temperatures.

Kühl (1965) studied the occurrence of the medusa in relation to the tide at Cuxhaven.

Family ZANCLEIDAE

Genus **Zanclea**

Zanclea costata Gegenbauer

Distribution includes: Norfolk (Hamond, 1957); Brazil (Vannucci, 1957); Japan (Yamada, 1959).

Martin & Brinckmann (1963) found *Zanclea* medusae infected by larval stages of the mollusc *Phyllirhoë*. The snail feeds by sucking at the manubrium and ring- and radial canals; when it is larger than the medusa and able to swim it devours the tentacles and manubrium of the medusa. The authors thus showed that *Mnestra parasites* Krohn is in fact *Zanclea costata*.

They found that all medusae budded from the polyp had two marginal tentacles and that they reached the stage with four marginal tentacles after 3–4 weeks. The authors reared the medusae successfully in the laboratory, specimens 4·5 mm. high and 3·0 mm. in diameter being produced with good feeding in 2–3 weeks at 20° C.

Rees & Roa (1966) record a specimen from Venezuela with four polyps bearing medusa buds growing from the stomach wall. They identify the specimen as *Z. gemmosa* McCrady which they regard as identical with *Z. implexa* (Alder). Picard (1957) regards the Mediterranean species as different from the north European and North American species.

Family CLADONEMATIDAE

Genus **Cladonema**

Cladonema radiatum Dujardin

Brinckmann & Petersen (1960) described characters by which the hydroid can be distinguished from that of *Dipurena reesi* Vannucci. Bouillon (1968) described the adhesive tentacles.

Family ELEUTHERIIDAE

Eleutheria dichotoma Quatrefages

Hauenschild (1946, 1947 *a*, *b*) studied asexual clones and migration of nematocysts. Bouillon (1968) described the adhesive tentacles.

Family CLAVIDAE

Genus **Turritopsis**

Turritopsis nutricula McCrady

Distribution includes Brazil (Vannucci, 1957, 1963).

Family HYDRACTINIIDAE

Genus **Podocoryne**

Podocoryne carnea M. Sars

Bénard-Boirard (1962) studied the embryonic development at Roscoff. Eggs were laid at any time of day or night, and embryos were seen developing in the subumbrellar cavity. The eggs, 90–120 μ in diameter, were surrounded with a mucous envelope. Segmentation was total and equal. The gastroblastula was 300 × 50 μ. Gastrulation was by unipolar migration or by thickening of the cell layer of the posterior third. The planula was 200–350 μ long.

The whole cycle from planula to production of medusa buds by the hydroid took about 4 weeks. It was thought that the life of the medusa is probably not more than a day or two.

Widersten (1968) also described the planula and Bodo & Bouillon (1968) made a detailed study of its development and histology.

Braverman (1968) studied the replacement of digestive gland cells from a region at the base of the hypostomial ridges.

Brinckmann & Vannucci (1965) examined the nematocysts and found desmonemes and microbasic euryteles in which the terminal ampulla of the butt may be absent.

Rice (1964) found no reactions to pressure changes.

Podocoryne borealis (Mayer)

Edwards (unpublished) has found the hydroid to be common in the Clyde sea area and has reared the medusa.

Podocoryne hartlaubi Neppi & Stiasny

Distribution includes: Clyde sea area (Edwards, unpublished); Villefranche, Mediterranean (Kramp, 1957).

Edwards (unpublished) has identified the hydroid as *Podocoryne areolata* (Alder), which name must now take precedence. Hydroid and development stages will be figured.

Podocoryne minima (Trinci)

Distribution includes: Brazil (Vannucci, 1957); Beaufort, North Carolina (Allwein, 1967).

FAMILY RATHKEIDAE

Genus Rathkea

Rathkea octopunctata (M. Sars)

Werner (1956*b*, 1958, 1962) studied the relation between temperature and sexual and asexual reproduction by budding of medusae. He found that *Rathkea* reproduces asexually below about 6–7° C. and sexually between about 8° and 12° C. A survey of distribution in relation to temperature indicated that *Rathkea* is a boreal arctic species. In the laboratory he made female medusae with ripe gonads reverse to asexual reproduction by reducing the temperature. Werner (1956) gives a photograph of the hydroid and (1958) gives results of culture experiments and a photograph of medusa buds produced on the hydroid stolon. Medusa bud formation could be induced by a lowering of temperature. The primary medusae produced from the hydroid have a different ratio of height to width from that of the secondary medusae produced by asexual reproduction of the medusa.

Bouillon (1961) confirmed earlier observations that the formation of medusa buds from the manubrium of the medusa is entirely ectodermal and this was further confirmed by Bouillon & Werner (1965), who showed that medusa bud formation on the hydroid involves both ectoderm and endoderm.

Kühl (1962) found the hydroid in the Elbe estuary producing medusa buds. Kühl (1965) studied the occurrence of the medusa in relation to the tide at Cuxhaven.

FAMILY BOUGAINVILLIIDAE

Genus Lizzia

Lizzia blondina Forbes

Distribution includes Beaufort, North Carolina, and Woods Hole (Allwein, 1967); first record from western Atlantic.

Genus Bougainvillia

A review of the genus *Bougainvillia* was made by Vannucci & Rees (1961); the British species were reviewed by Edwards (1966).

Bougainvillia ramosa (van Beneden)

Distribution includes: Norfolk (Hamond, 1957); Elbe estuary (Kühl, 1962); Brazil (Vannucci, 1957, 1963).

Bougainvillia britannica Forbes

Edwards (1964*a*) discovered the hydroid (Text-fig. 1 s) in the Firth of Clyde on shells of living *Aporrhais pespelecani* and medusae were given off by it in the laboratory. He identified it as *Atractylis linearis* Alder and *Bougainvillia flavida* Hartlaub. Edwards reared the medusae to a size of over 6 mm. in diameter (Text-figs. 2 s).

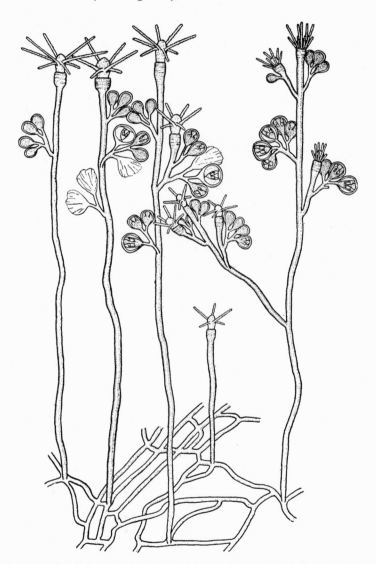

Text-fig. 1 s. *Bougainvillia britannica*. The hydroid growing on the shell of *Aporrhais pespelecani*, Firth of Clyde (from Edwards, 1964*a*, fig. 1).

237

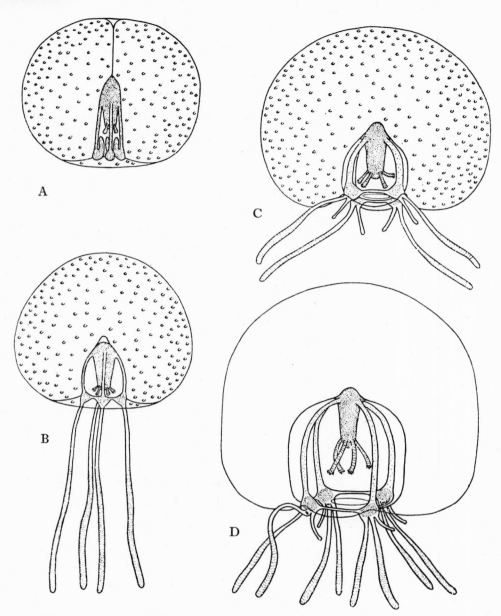

Text-fig. 2s. *Bougainvillia britannica*. Young stages reared from the hydroid. A, 0·96 mm. in diameter; B, 0·98 mm. in diameter; C, 1·26 mm. in diameter; D, 1·85 mm. in diameter (from Edwards, 1964a, fig. 2).

Bougainvillia principis (Steenstrup)

Distribution includes: Clyde sea area (Vannucci, 1956); Loch Creran, May 1968 (personal observation).

Edwards (1966) discovered the hitherto unknown hydroid in the Firth of Clyde growing on pieces of coal and clinker. Medusae were released from it in the laboratory.

DESCRIPTION OF HYDROID (Text-fig. 3s)

Colonies of small sessile sparsely distributed single hydranths rising from a creeping stolon. Perisarc of hydranth often wrinkled. Hydranths up to 1·1 mm. in height, with single whirl of five to eight filiform tentacles. Medusa buds borne singly on short stalks arising from stolon.

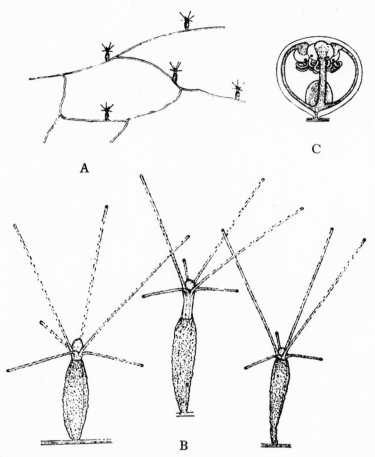

Text-fig. 3s. *Bougainvillia principis*. A, hydroid on a piece of coal, Firth of Clyde; B, polyps; C, gonophore (from Edwards, 1966, fig. 1).

DESCRIPTION OF YOUNG MEDUSA

Umbrella deep bell-shaped, 1·26 mm. high and 1·08 mm. in diameter. Jelly fairly thick. Abundant nematocysts on exumbrella. Four moderately deep interradial longitudinal furrows on exumbrella and four less deep perradial furrows. Stomach small and tubular with quadrate base. Four perradial oral tentacles each once or twice dichotomously branched. Four radial canals smooth and moderately wide; ring-canal narrow. Four marginal bulbs somewhat triangular, rounded distally, each bearing three marginal tentacles and three adaxial round black ocelli situated on bulb above bases of tentacles. Velum broad. Colour of bulbs and stomach brick-red; radial canals pink.

239

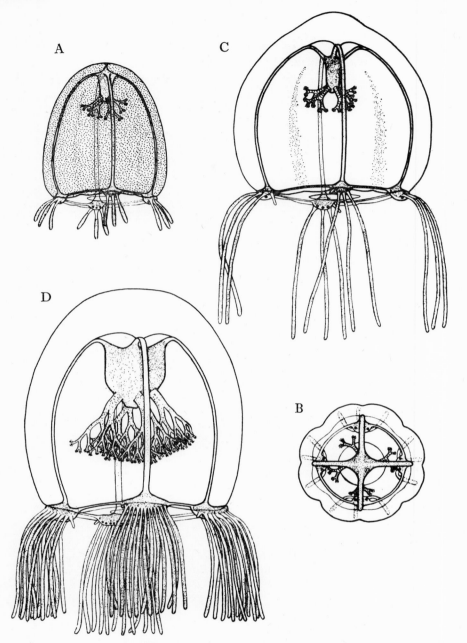

Text-fig. 4s. *Bougainvillia principis*. Medusae reared from the hydroid. A, 1·26 mm. high; B, aboral view of A; C, 1·66 mm. high; D, 4·18 mm. high (from Edwards, 1966, fig. 2).

Edwards reared the medusa to a size of over 6 mm. in diameter and described developmental stages (Text-fig. 4s).

Edwards regards *Medusa duodecilia* Dalyell and *Hippocrene simplex* Forbes & Goodsir as possibly synonyms.

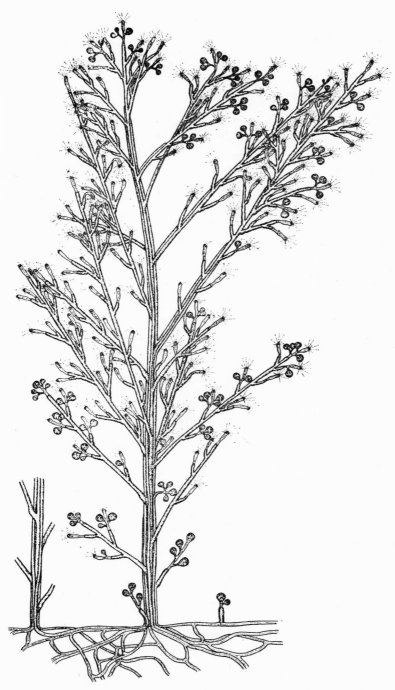

Text-fig. 5 s. *Bougainvillia pyramidata*. Portion of colony of hydroid growing on the test of *Ascidiella aspersa*, Firth of Clyde (from Edwards, 1964 b, fig. 1).

Bougainvillia pyramidata (Forbes & Goodsir)

Edwards (1964*b*) discovered the hitherto unknown hydroid in the Firth of Clyde growing on stems of *Virgularia mirabilis* and *Tubularia indivisa*, and on the test of *Ascidiella aspersa*. He reared the medusae from the hydroid.

DESCRIPTION OF HYDROID (Text-fig. 5s)

Hydrocaulus rising erect about 40 mm., much branched, main stem composed of aggregated tubes or polysiphonic; arising from creeping stolon. Hydranths up to 0·66 mm. in height each with ten to thirteen filiform tentacles in single whorl. Medusa buds borne below hydranths on stalks carrying one to three buds. Perisarc thins out as delicate transparent pseudohydrotheca over base of hydranth. Colour of hydranth and medusa buds pink.

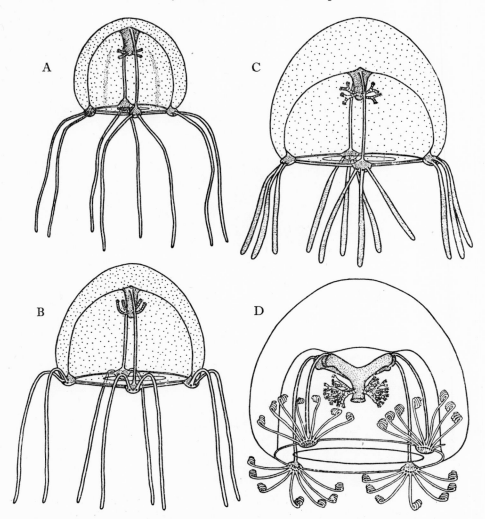

Text-fig. 6s. *Bougainvillia pyramidata*. Medusae reared from the hydroid. A, 0·63 mm. high; B, 1·25 mm. high; C, 1·44 mm. high; D, 3·64 mm. high (from Edwards, 1964*b*, fig. 3).

DESCRIPTION OF YOUNG MEDUSA (Text-fig. 6s)

Umbrella subglobular, 0·63 mm. high and 0·72 mm. in diameter. Jelly moderately thick. Scattered nematocysts on exumbrella. Four shallow interradial longitudinal furrows on exumbrella. Stomach small. Four short perradial unbranched oral tentacles. Four radial canals smooth and rather narrow; ring-canal narrow. Four perradial marginal bulbs round and rather small, each bearing two equal-sized marginal tentacles and two adaxial round ocelli situated on bulb opposite base of tentacle. Velum broad. Colour of stomach and marginal bulbs pinkish yellow; ocelli black.

Edwards reared the medusa to a size of 5 mm. in diameter. When the medusa was 1·69 mm. in diameter and 1·44 mm. high the oral tentacles had each branched once and there were three tentacles on each marginal bulb. The base of the stomach was beginning to change its shape and the gonads were developing in a specimen 2·3 mm. high and 2·6 mm. in diameter, which had four marginal tentacles on each bulb and in which the oral tentacles were dichotomously branched three times.

Edwards (1966) regards *Medusa ocilia* Dalyell as a possible synonym.

Bougainvillia superciliaris (L. Agassiz)

Distribution includes: Norfolk (Hamond, 1964); Bembridge, Isle of Wight (Edwards, 1966); Clyde sea area (Edwards, 1968); Thames estuary near Sheerness (specimens collected by D. W. Lilley, winter 1963–4).

The hydroid was found in Akkeshi Bay, Japan, by Uchida & Nagao (1960) and again by Nagao (1964) when medusae were liberated in the laboratory.

Werner (1961) made a detailed investigation of the hydroid and medusa by rearing. He found that the production of medusa buds by the hydroid was restricted to falling temperatures below an upper limit of 7–5° C. A survey of its distribution and temperature conditions confirmed it as an Arctic boreal species.

It has two types of nematocyst: desmonemes, $5–6·2 \times 3·1–3·7 \mu$; microbasic euryteles $7·5–9·4 \times 3·1–5·0 \mu$. The chromosomes numbered $2n = 30$.

Bougainvillia macloviana Lesson

Distribution includes: Clyde sea area and off Mull of Kintyre (Edwards, 1958, 1966); Cape Town (Millard, 1959); west coast of S. Africa (Millard, 1966).

Bougainvillia muscoides (M. Sars)

Medusa and hydroid recorded for the first time in the British fauna by Edwards (1958, 1964b) in the Clyde sea area.

Edwards reared medusae in the laboratory from the hydroid.

DESCRIPTION OF YOUNG MEDUSA (Text-fig. 7s)

Umbrella bell-shaped, 0·74 mm. high and 0·68 mm. in diameter. Jelly thin. Scattered nematocysts on exumbrella. Four perradial longitudinal furrows on exumbrella. Stomach short and

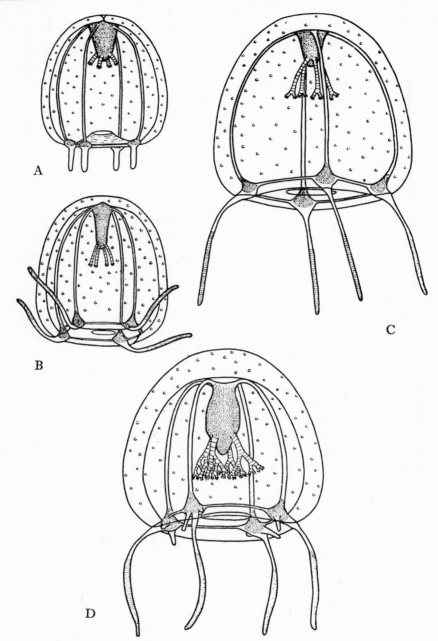

Text-fig. 7s. *Bougainvillia muscoides*. Medusae reared from the hydroid. A, 0·74 mm. high; B, 1·08 mm. high; C, 1·39 mm. high; D, 1·98 mm. high (from Edwards, 1964b, fig. 5).

conical with broad square base. Four perradial unbranched oral tentacles. Four radial canals and ring-canal narrow. Four perradial marginal bulbs large and rounded, each bearing one marginal tentacle; without ocelli. Velum broad. Colour of stomach reddish-brown; marginal bulbs orange.

Edwards reared the medusa to a size of over 3 mm.

244

The adult medusa was originally described by Browne (1903) as *Margelis nordgaardi*.

Edwards examined specimens described by me (1953, p. 150) as *Thamnostoma* sp. in E. T. Browne's collection in the British Museum and concluded that they were typical young *Bougainvillia muscoides*.

Widersten (1968) described the planula.

Genus **Nemopsis**

Nemopsis bachei L. Agassiz

Distribution includes: Gironde estuary (Tiffon, 1957); Gulf of Mexico (Moore, 1962).

In my earlier monograph (1953, p. 152) I drew attention to the uncertainty of inclusion of *Nemopsis bachei* in the British fauna. This is further discussed by Edwards (1966, p. 142) who associates this species with *Bougainvillia charcoti* Le Danois from the Little Minch, Hebrides; and by M. E. Thiel (1968) who reviews all the literature on the occurrence of *Nemopsis* in European waters and suggests that while not native on the north-west coasts of the British Isles, it may one day be found in the Thames or Humber estuaries.

Kühl (1965) studied its occurrence in relation to the tide at Cuxhaven.

FAMILY PANDEIDAE

Genus **Amphinema**

Amphinema rugosum (Mayer)

Distribution includes Clyde sea area (Edwards, unpublished).

Amphinema dinema (Péron & Lesueur)

Distribution includes Brazil (Vannucci, 1957, 1963).

Chapman (in Chapman, Pantin & Robson, 1962) examined the fine structure of the muscle with the electron microscope.

Amphinema krampi Russell

A new deep-water species described (Russell, 1956*a*, 1958) from specimens caught west of the mouth of the English Channel (Text-fig. 8s).

DESCRIPTION OF MEDUSA

Umbrella bell-shaped, higher than wide, with small apical process. Jelly moderately thick. Stomach cross-like in section, about two-thirds length of umbrella cavity, without peduncle. Mouth with four simple perradial lips, without crenulation. Four radial canals and ring-canal moderately broad, with smooth margins. Radial canals attached to stomach over half its length to form 'mesenteries'. Up to seventeen strands of tissue running from each radial canal to umbrella surface. Four cushion-like gonads one on each interradial wall of stomach. Two opposite perradial marginal tentacles with swollen elongated basal bulbs; eight marginal tentacula. No ocelli. Height of umbrella 6 mm. Colour of stomach and gonads rich reddish brown; ring-canal brownish.

Nematocysts; ? microbasic euryteles.

Genus **Merga**

Merga reesi Russell

A new deep-water species described (Russell, 1956*b*) from a specimen caught west of the mouth of the English Channel (Text-fig. 9s).

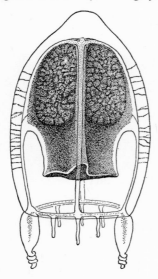
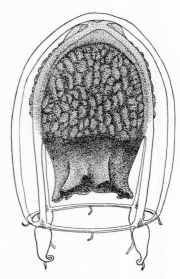

Text-fig. 8s. *Amphinema krampi*. 7 mm. high. (from Russell, 1958, fig. 1).

Text-fig. 9s. *Merga reesi*. Reconstruction from damaged specimen (from Russell, 1956*b*, fig. 1).

DESCRIPTION OF MEDUSA

Umbrella bell-shaped, higher than wide, no apical process. Stomach flask-shaped, not extending beyond umbrella margin, with broad base. Mouth with four slightly crenulated lips. Four radial canals and ring-canal broad, with smooth outlines. Radial canals attached to stomach over more than half its length to form 'mesenteries'. Gonads interradial, irregularly folded to form numerous small raised corrugations. Four perradial marginal tentacles with swollen elongated basal bulbs, without exumbrellar spurs. Four small tentaculae, one in each interradius. No ocelli. Height 10 mm. Colour of stomach and gonads dark chocolate-red; basal bulbs of marginal tentacles colourless except for slight internal pigmentation at junction of radial canal with ring-canal.

Genus **Leuckartiara**

Leuckartiara octona (Fleming)

Distribution includes: Chile (Kramp, 1966); south coast of South Africa (Millard, 1966).
Latham (1963) recorded the hydroid in the Clyde sea area on the polychaete worm *Aphrodite aculeata*. Widersten (1968) described the planula.

Leuckartiara breviconis (Murbach & Shearer)

Distribution includes Clyde sea area, rare (Edwards, unpublished).

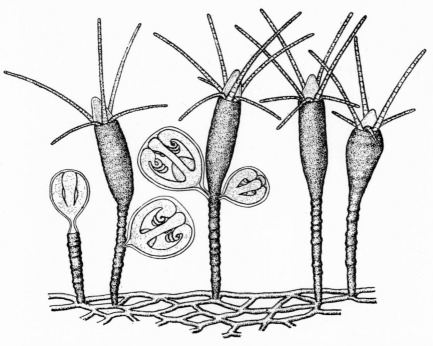

Text-fig. 10s. *Neoturris pileata*. Portion of colony of hydroid on shell of *Nucula sulcata*, Clyde sea area (from Edwards, 1965, fig. 1).

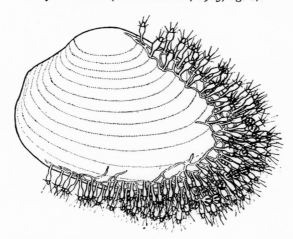

Text-fig. 11s. *Neoturris pileata*. Colony of hydroid on *Nucula sulcata* shell (from Edwards, 1965, fig. 2).

Genus **Neoturris**

Neoturris pileata (Forskål)

Distribution includes Clyde sea area (Edwards, 1965).

Edwards (1965) discovered the previously unknown hydroid from the Clyde sea area, and reared the medusae.

16-2

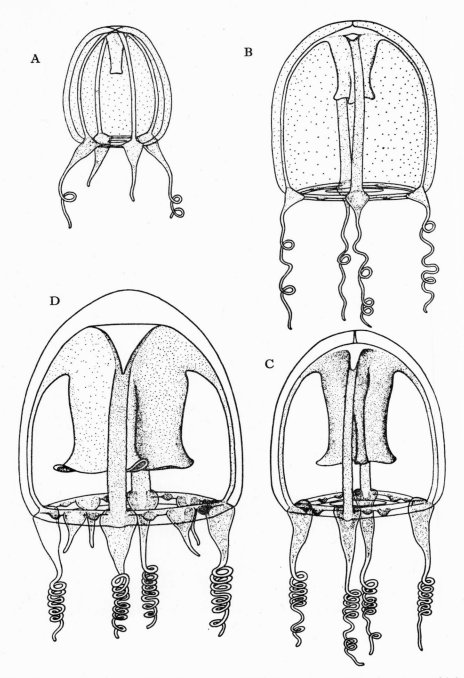

Text-fig. 12 s. *Neoturris pileata*. A, 1·2 mm. high; B, 1·75 mm. high; C, 1·8 mm. high; D, 2·3 mm. high (from Edwards, 1965, fig. 3).

DESCRIPTION OF HYDROID (Text-figs. 10s, 11s)

Hydrorhiza forming close network of anastomosing stolons. Hydranths borne on erect un-branched stems, with irregular or spirally coiled perisarc, up to 3·3 mm. in height. Hydranth spindle-shaped with prominent conical hypostome. Four to nine filiform tentacles in single whorl. Perisarc forms pseudohydrotheca at base of hydranth. Medusa buds on stems, or less commonly on stolon.

Hydroid attached to shells of *Nucula* living in deep water.

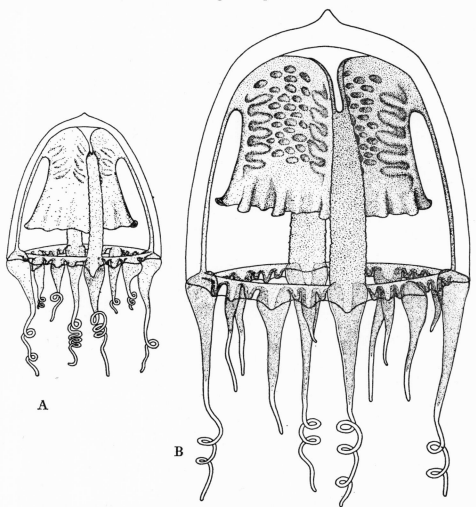

Text-fig. 13s. *Neoturris pileata.* A, 2·7 mm. high; B, 5 mm. high (from Edwards, 1965, fig. 4).

DESCRIPTION OF YOUNG MEDUSA (Text-figs. 12s, 13s)

Umbrella deep bell-shaped up to 1·25 mm. high and 1·1 mm. in diameter, without apical process. Jelly thin. Numerous nematocysts on exumbrella. Velum broad. Stomach quadrate in section with four simple perradial lips, one-quarter to one-third height of umbrella. Four radial canals moderately wide, smooth. Ring-canal moderately narrow. Four perradial tentacles each with large hollow conical base not laterally compressed; two opposite tentacles less developed than other two. No rudiments of tentacle bulbs present. Colour of stomach, marginal tentacle bulbs, and canals pale straw. No ocelli.

Edwards reared the medusae through successive stages. Interradial marginal tentacle bulbs appeared when the medusa was 1·75 mm. high. Interradial tentacles were forming and adradial bulbs appearing at a height of about 2·0 mm. The gonadial ridges were developing when the medusa was 2·7 mm. high. When it had reached a height of about 5·0 mm. there were five or six gonadial folds on each adradial wall of the stomach and numerous rounded pockets in the interradial areas. The lips were frilled and the edges of the radial canals were becoming irregular. The interradial marginal tentacles were well developed and the adradial tentacles were developing.

Edwards discusses in detail the distribution of *Neoturris pileata*.

Rice (1964) found that this medusa showed no reactions to changes in pressure.

Hydroid Genus **Perigonimus**

A revision of the Genus *Perigonimus* was made by Rees (1956).

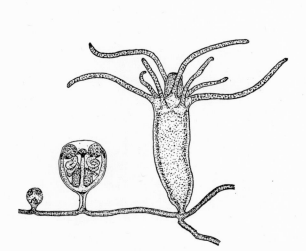

Text-fig. 14s. *Pandea conica*. Portion of colony of hydroid (from Picard, 1956, fig. 2).

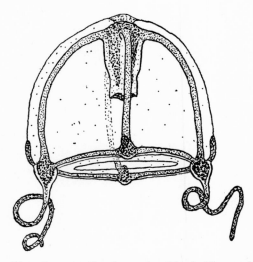

Text-fig. 15s. *Pandea conica*. Newly liberated medusa (from Picard, 1956, fig. 1).

Genus **Pandea**

Pandea conica (Quoy & Gaimard)

Picard (1956) reared young medusae from the hydroid *Campaniclava cleodorae* (Gegenbaur) living on the pteropod *Cleodora cuspidata* (Text-fig. 14s).

DESCRIPTION OF YOUNG MEDUSA (Text-fig. 15s)

Umbrella bell-shaped as high as wide. Jelly moderately developed. Stomach with broad base, about half height of umbrella in length, four simple lips. Four perradial canals broad with smooth edges; ring-canal moderately broad. Two opposite perradial marginal tentacles with large basal bulbs, two small opposite perradial marginal bulbs without tentacles. Four perradial elongated nematocyst clusters (cnidactines) immediately above each marginal bulb, the two larger being on tentacular bulbs. Colour of stomach and tentacular bulbs ochre-brown. No ocelli.

NEMATOCYSTS

Desmonemes $5 \times 3 \cdot 5\ \mu$ in hydroid only. Microbasic euryteles $6 \cdot 5$–$10\ \mu \times 3$–$5\ \mu$ in hydroid and medusa.

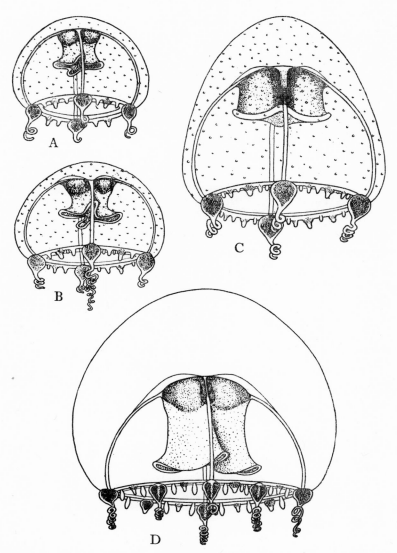

Text-fig. 16s. *Tiaranna rotunda*. Young specimens, Firth of Clyde. A, 1·0 mm. diameter; B, 1·1 mm. diameter; C, 1·5 mm. diameter; D, 2·6 mm. diameter (from Edwards, 1963, fig. 1).

Pandea rubra Bigelow

I have seen a young specimen 10 mm. high collected on 24 November 1965 off the Canaries (Discovery Station 5823). This had seven marginal tentacles of which three were perradial, but it had no developing tentacles.

Genus **Bythotiara**

Bythotiara murrayi Günther

Edwards (1967) caught a young medusa in the Clyde sea area and partly reared it. He considered it to be a young *B. murrayi*.

FAMILY TIARANNIDAE

Tiaranna rotunda (Quoy & Gaimard) = **Modeeria formosa** Forbes

Distribution includes Clyde sea area (Edwards, 1958, 1963).

A series of specimens (Text-fig. 16s) from the plankton in different stages of development were described by Edwards (1963). The youngest was about 1·0 mm. in diameter. Jelly thin; numerous scattered nematocysts on exumbrella. Velum fairly broad. Stomach relatively large, attached to subumbrella along perradii only. Mouth large, quadrate, with four simple lips. Four radial canals narrow and smooth; ring-canal broad. Four perradial marginal tentacles with large conical hollow basal bulbs, four interradial bulbs without tentacles, and eight adradial marginal swellings. Colour of stomach and perradial and interradial marginal bulbs reddish.

The marginal cordylus-like structures develop at an early stage when the medusa is a little over 1·0 mm. in diameter. When the medusa is 1·5 mm. in diameter the apical jelly of the umbrella is thickening. The pouches in which the gonadial folds will form are developing in a specimen 1·8 mm. in diameter.

In specimens 5–6 mm. in diameter there are four or five gonadial folds on the sides of each stomach pouch.

Edwards identifies *Tiaranna rotunda* with *Modeeria formosa* Forbes; and accordingly the medusa should now be called *Modeeria rotunda* (Quoy & Gaimard). The family name Tiarannidae remains unchanged.

ORDER

LEPTOMEDUSAE

FAMILY LAODICEIDAE

Genus **Laodicea**

Laodicea undulata (Forbes & Goodsir)

Distribution now includes Clyde sea area (Vannucci, 1956; Edwards, 1968).

Kramp (1957) made observations on numbers of marginal tentacles and ocelli on specimens from the Mediterranean.

Edwards (unpublished) has found the hydroid to be common in the Clyde sea area and has reared the medusa.

Genus **Staurophora**

Staurophora mertensi Brandt

Distribution includes: Norfolk (Hamond, 1957), St Andrews, Scotland (Rutherford, 1962), Clyde sea area, young medusae common but adults scarce (Edwards, 1968); Northumberland coast, May 1968 (young specimen collected by Frank Evans).

Naumov (1951) reared the primary polyp. The eggs were 0·13–0·14 mm. in diameter. Planulae developed in 24 hours and swam for 3–5 days before settling. The polyp was fully formed in 5–7 days. It was 0·5–0·6 mm. high with a maximum diameter of 0·12–0·13 mm., and developed up to 9 tentacles. Naumov identified it with *Cuspidella humilis* Hincks.

Rutherford (1962) found that there were no transitional stages between the marginal tentacles and the cordylus. She described two abnormal specimens and made observations on behaviour.

Genus **Krampella**

Krampella dubia Russell

A damaged specimen* of a medusa from deep water off the mouth of the English Channel which might have affinities with the Laodiceidae was described by Russell (1957).

DESCRIPTION OF MEDUSA (Text-fig. 17s)

Umbrella hemispherical; jelly moderately thick. Stomach possibly in form of open cross with mouth lips extending along each arm. Four radial canals broad. About sixteen fine strands of tissue connecting walls of radial canals with exumbrella surface. Gonads along nearly whole length of radial canals, divided longitudinally. Four perradial and four interradial marginal tentacles each with conical basal bulb. Three or four small marginal cirrus-like tentacles between each pair of marginal tentacles. Yellowish brown pigment on parts of gonad and on what appear to be mouth lips. Diameter of umbrella 3 mm.

* A complete specimen has now been recorded from the Azores (Baker 1967).

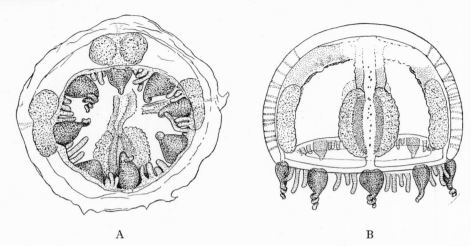

Text-fig. 17s. *Krampella dubia*. A, viewed from subumbrellar side; B, slightly idealized with suggested outlines of stomach (from Russell, 1957, fig. 1*b*).

Family DIPLEUROSOMATIDAE

Genus **Dipleurosoma**

Dipleurosoma typicum Boeck

Distribution includes Port Erin (Werner, 1959*a*).

Edwards (unpublished) has reared medusae from hydroids collected in the Clyde sea area. The medusa undergoes repeated transverse fission.

Family MITROCOMIDAE

Mitrocomella polydiademata (Romanes)

Edwards (unpublished) has reared the hydroid from the medusa in the Clyde sea area. It is a *Cuspidella*.

Genus **Tiaropsis**

Tiaropsis multicirrata (M. Sars)

Distribution includes Norfolk (Hamond, 1957, who recorded the hydroid and medusa).

Naumov (1951) reared the hydroid and confirmed the description and identification of Rees (1941). Naumov refers also to the rearing of the hydroid by G. F. Korsakova (*Dokl. Akad. Nauk SSSR*, **65**, no. 3, 1949) not previously cited by me.

Genus **Tiaropsidium**

Tiaropsidium atlanticum Russell

A new species of *Tiaropsidium* was described by Russell (1956*b*) from deep water off the mouth of the English Channel. The specimen was damaged.

DESCRIPTION OF MEDUSA (Text-fig. 18s)

Umbrella flatter than a hemisphere with fairly thick jelly. Velum narrow. Stomach small, about one-twelfth the diameter of the umbrella, attached to subumbrella along arms of perradial cross, leaving small triangular pouches between dorsal wall of stomach and subumbrella. Mouth with four short broad lips with slightly folded margins. Four straight radial canals and ring-canal narrow. Gonads along four radial canals, linear, with median division; along middle three-quarters of radial canal, not reaching to umbrella margin. Large hollow marginal tentacles, probably up to twenty-four in number, with elongated swollen bases. Small solid marginal tentacles, probably up to seventy-two in number, three in each space between adjacent large marginal tentacles. Probably up to forty-eight open marginal vesicles, two in each space between adjacent large marginal tentacles, each with *ca.* twelve to twenty, or more, concretions and one black ocellus at base. Diameter *ca.* 60 mm. Colour on interradial walls of stomach and large marginal tentacles black.

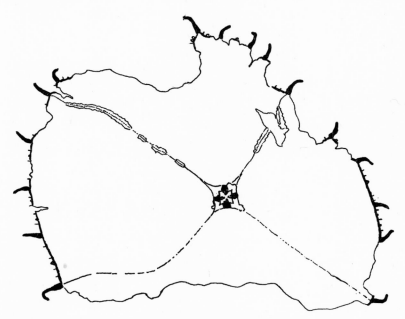

Text-fig. 18s. *Tiaropsidium altanticum* (from Russell, 1956*b*, fig. 2).

FAMILY CAMPANULARIIDAE

Genus **Obelia**

Chapman (1968) studied *Obelia* with the electron microscope and described its unusual type of musculature. He discussed this in relation to the swimming movements of *Obelia*. A study of the development and histology of the planula was made by Bodo & Bouillon (1968). The rate at which growth sequences proceed in relation to temperature and the resulting effect on stem form were examined by Ralph & Thomson (1968).

Genus **Phialidium**

Phialidium hemisphaericum (L.)

Distribution includes west and south coasts of South Africa (Millard, 1966).

Roosen-Runge (1962) and Roosen-Runge & Szollosi (1965) studied sexual reproduction in *Phialidium*. Most females have two ovulation periods, some hours after sunset and in morning before sunrise. Males release some sperm almost continuously, but the major periods of spermiation occur at about the times of ovulation. Females release 50–100 eggs in 24 hr. and males 3 million spermatozoa in the same period.

One hundred spermatozoa per c.c. were sufficient to give 100% normal fertilization. Spermatozoa remain fertile in laboratory for 11–12 hr. and eggs for at least 3–4 hr.

The time of ovulation can be advanced or retarded by altering hours of darkness.

Horridge & Mackay (1962) examined the marginal sense organ with the electron microscope, and Passano (1965) made electrophysiological observations.

Knight-Jones & Morgan (1966) found that the medusae pulsated more frequently and for longer periods after increase of pressure of 3 and 5 decibars.

Roosen-Runge (1967) studied the circulation in the radial canal and its dependence on a combination of flagellar and muscular action. A detailed study of the development and histology of the planula was made by Bodo & Bouillon (1968).

Kramp (1957*b*) recorded an abnormal specimen from the Mediterranean with six radial canals and a secondary stomach.

FAMILY LOVENELLIDAE

Genus **Lovenella**

Lovenella clausa Hincks

Distribution includes: Clyde sea area (Vannucci, 1956); Elbe estuary (Kühl, 1962).

Genus **Eucheilota**

Eucheilota maculata Hartlaub

Distribution includes: Thames estuary near Sheerness (specimens collected by D. W. Lilley, winter 1963–4); Elbe estuary (Kühl, 1962).

Kühl (1962) recorded that a medusa fed on copepods, *Oikopleura*, and a *Pleurobrachia* twice its own size. Werner (1963) reared the hydroid (Text-fig. 19s) and showed that its form of growth was dependent upon the space available on the substratum. With plenty of space it assumed a creeping form, but if space was restricted the colony was erect and branching. He also obtained newly liberated medusae (Text-fig. 20s).

Werner (1965) described a new type of nematocyst in the end clubs of the marginal cirri in the medusa, namely a merotrichous haploneme (see *Eutima gracilis*, p. 259); and (1968*a*) he gave a detailed description of the development of the hydroid and medusa. The chromosome count was $2n = 30$. He considered, in view of the structure of the hydroid, that this species belongs to the Campanulinidae.

Kühl (1965) studied the occurrence of the medusa in relation to the tide at Cuxhaven.

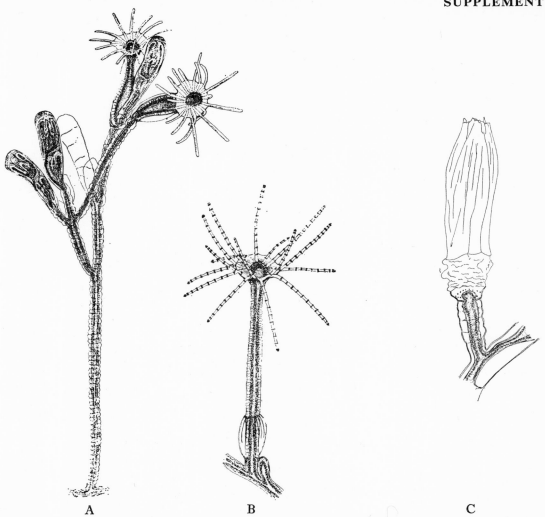

Text-fig. 19s. *Eucheilota maculata*. Hydroid. A, branched colony with gonangia; B, fully developed hydranth,; C, empty hydrotheca. (B and C from Werner, 1968a, figs. 12b, 13a; A, from original drawings.)

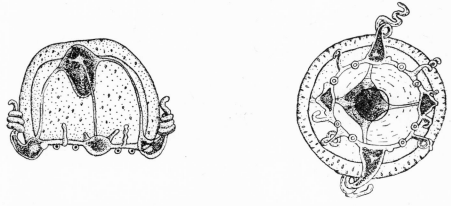

Text-fig. 20s. *Eucheilota maculata*. Newly liberated medusa, diameter *ca.* 0·5 mm. (from Werner, 1968a, fig. 19).

257

FAMILY PHIALELLIDAE

Genus **Phialella**

Phialella quadrata (Forbes)

Distribution includes: Norfolk (Hamond, 1957); Chile (Kramp, 1966).

FAMILY EIRENIDAE

Genus **Eirene**

Eirene viridula (Péron & Lesueur)

Distribution includes Gulf of Guinea (Repelin, 1965).

Kramp (1957) pointed out that the absence of traces of gonads in young specimens, before the development of a peduncle, and of interradial marginal tentacle bulbs were distinguishing characters for separation from young *Phialidium*.

Genus **Helgicirrha**

Helgicirrha schulzei (Hartlaub)

Distribution includes Clyde sea area (Edwards, unpublished).

FAMILY AEQUOREIDAE

Genus **Aequorea**

Aequorea forskalea Péron & Lesueur

This is now regarded as synonymous with *Aequorea aequorea* (Forskål), see, e.g. Kramp (1961).

Horridge (1955*b*) studied nervous inhibition. *Aequorea* contains a bioluminescent protein, called 'aequorin' by Shimomura, Johnson & Saiga (1962). A study of the quantum efficiency of this was made by Johnson *et al.* (1963). The protein emits light on the addition of calcium (Shimomura *et al.* 1963) and this reaction is so sensitive that it is very useful for detecting calcium in minute quantities in body fluids, see, e.g. Ashley & Ridgway (1968).

Aequorea vitrina Gosse

Distribution includes: Clyde sea area (Edwards, unpublished) and Norfolk (Hamond, 1957) where it was common in June and July, reaching 150 mm. in diameter. Hamond remarked on its violet-blue fluorescent appearance, and that it could be seen on calm days breaking the surface of the water or swimming just below it.

Uchida (1968) showed that this species is distinct from *A. coerulescens*.

Aequorea pensilis (Haeckel) (Modeer)

Aequorea macrodactyla (Brandt)

Kramp (1959, 1961) keeps these two as separate species on the shape of the marginal tentacle bulb with a spur in *A. macrodactyla*, without a spur and laterally extended in *A. pensilis*.

FAMILY EUTIMIDAE

Genus **Eutima**

Eutima gracilis (Forbes & Goodsir)

Distribution includes: Norfolk (Hamond, 1957); Villefranche, Mediterranean (Kramp, 1957).

Werner (1965) described a new type of nematocyst in the end clubs of the lateral cirri, a merotrichous haploneme. He comments on the possible relationship between the Eutimidae and the Lovenellidae in that *Eucheilota maculata* has similar nematocysts. He also reared the hydroid (Text-fig. 21s) and the newly liberated medusa (Text-fig. 22s).

Knight-Jones & Morgan (1966) recorded somewhat increased rates and periods of pulsation after pressure increases of 3 and 5 decibars.

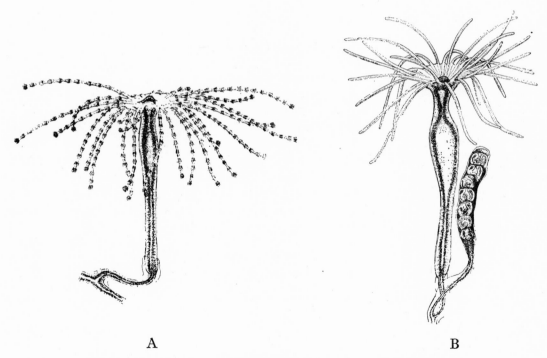

A   B

Text-fig. 21 s. *Eutima gracilis*. Hydroid. A, Fully developed hydranth; B, hydroid with gonangium (from Werner, unpublished.

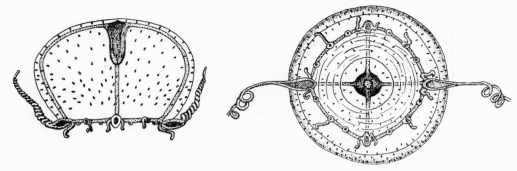

Text-fig. 22 s. *Eutima gracilis.* Newly liberated medusa, 0·85 mm. in diameter; 0·5 mm high (from Werner, unpublished).

Genus **Octorchis**

Regarded as a subgenus of *Eutima*.

Eutima (Octorchis) gegenbauri (Haeckel)

Distribution includes: Elbe estuary (Kühl, 1962); Beaufort, North Carolina (Allwein, 1967). Kramp (1957) examined a series of developmental stages in specimens from Villefranche, Mediterranean, and found that the rate of development was somewhat different from that

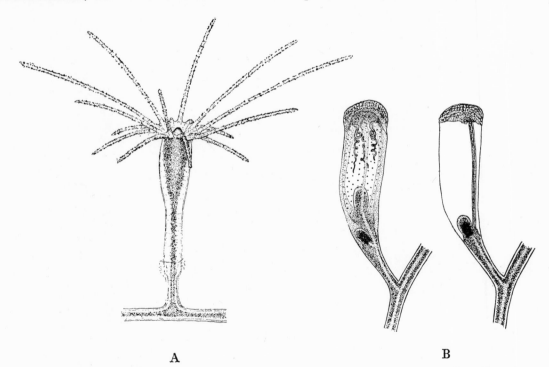

A   B

Text-fig. 23 s. *Eutima (Octorchis) gegenbauri.* A, hydroid: height from mouth to basal stolon, 0·7 mm. B, gonangium before and after liberation of medusa. Length 0·7 mm. (from Werner, unpublished).

recorded by previous authors. He concluded that different local conditions might be responsible.

Werner (1965) found in this species the new type of nematocyst, namely merotrichous haploneme, in the end clubs of the lateral cirri as in *Eucheilota maculata* and *Eutima gracilis*.

He also reared the hydroid (Text-fig. 23 s) and the newly liberated medusa (Text-fig. 24 s).

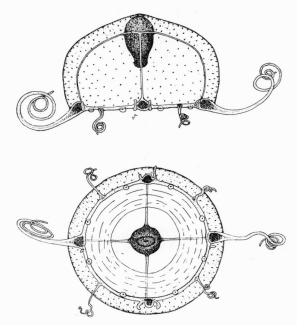

Text-fig. 24 s. *Eutima* (*Octorchis*) *gegenbauri*. Newly liberated medusa, 0·8 mm. in diameter, 0·5 mm. high (from Werner, unpublished).

Genus **Eutonina**

Eutonina indicans (Romanes)

Distribution includes: Clyde sea area (Vannucci, 1956; Edwards, 1968); Elbe estuary (Kühl, 1962).

Edwards (unpublished) has reared the hydroid from the medusa, as did Werner (1968 *b*) (Text-fig. 25 s) who also obtained the newly liberated medusa (Text-fig. 26 s).

Horridge (1966) found that when stimulated by a vibrating needle *Eutonina* would turn over from the upright position in a direction away from the stimulus, thus presenting the mouth to the source of the stimulus and by turning downwards avoiding disturbance of the water surface.

Kühl (1965) studied the occurrence of the medusa in relation to the tide at Cuxhaven.

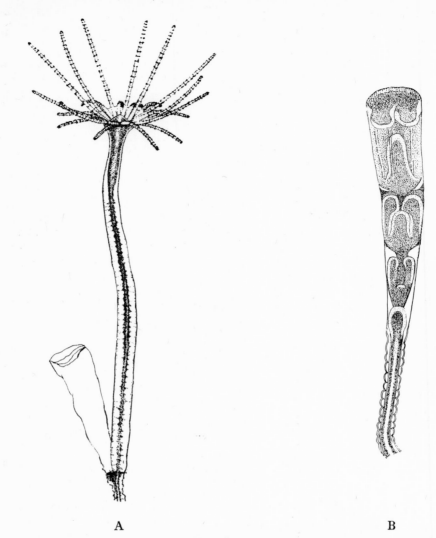

A   B

Text-fig. 25 s. *Eutonina indicans*. A, hydroid with empty gonangium; B, gonangium with medusa buds (from Werner, 1968 *b*, figs. 6, 10). (There should be a small hydrotheca in A.)

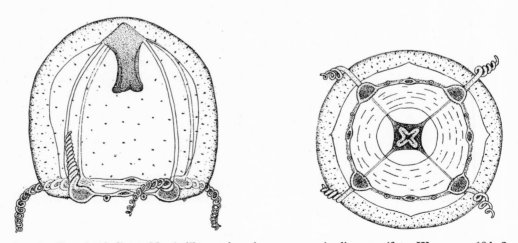

Text-fig. 26 s. *Eutonina indicans*. Newly liberated medusa, 1·1 mm. in diameter (from Werner, 1968 *b*, fig. 14).

Genus **Tima**

The genus *Tima* was reviewed by Petersen (1962).

Tima bairdii (Johnston)

Distribution includes Norfolk (Hamond, 1957).

Tima bairdii appeared in such quantities in the Skagerak in the winter 1966–7 as to make sorting of prawns from the trawl catches difficult or even not worth while. They may also have caused the prawn schools to disperse (Bernt, 1967).

ADDENDUM

An experimental study of the food and manner of feeding of a number of different medusae has recently been made by J. H. Fraser (1969, in Press) in Newfoundland. As well as the Scyphomedusae *Cyanea* and *Aurelia*, the following Hydromedusae were studied: *Bougainvillia superciliaris, Leuckartiara nobilis, Melicertum octocostatum, Laodicea undulata*, and *Staurophora mertensi*.

Fraser observed the composition of the food, and the numbers and weight of organisms consumed. He attempted to assess the effect that medusae might have on the recruitment of young fish in the sea.

ORDER

LIMNOMEDUSAE

Family OLINDIIDAE

Genus **Gonionemus**

Gonionemus vertens L. Agassiz

Distribution includes: Clyde sea area, one specimen (Edwards, unpublished); Aberdour, Fife-shire (Howe, 1959); and Cuckmere Haven, Sussex (R. C. Vernon and the late W. J. Rees, personal communication). (See also Tambs-Lyche, 1964 and Todd, Kier & Ebeling, 1966.)

A detailed study of the development and histology of the planula was made by Bodo & Bouillon (1968).

Genus **Gossea**

Gossea corynetes (Gosse)

Distribution includes Clyde sea area (Edwards, 1968).

Knight-Jones & Morgan (1966) recorded that the medusa pulsated more frequently and for longer periods after pressure increases of 3 and 5 decibars.

NOTE. The following additional references on fresh-water medusae are given in the Bibliography: Bouillon (1957, 1959); Kidd (1956): Kuhl, G. (1960); Lytle (1959, 1961); McClary (1959); Ramozzoti (1962); Reisinger (1957); Uchida (1951, 1955).

TRACHYMEDUSAE

The Trachymedusae are all oceanic species and as such their distribution is worldwide. Since 1953 there have been many additional records but it seems unnecessary to include them here as their distribution is as might be expected.

A new trachymedusan *Tesserogastria musculosa* was described by Beyer (1958) from the Oslo Fjord. It should be added to the list of species given in Appendix II on page 486 of the 1953 monograph.

As far as I am aware the only experimental research on a trachymedusan has been that of Passano (1965) on nervous activity in *Liriope tetraphylla*, and of Horridge (1969) who made electron micrographs of the statocyst of *Rhopalonema*.

ORDER

NARCOMEDUSAE

Like the Trachymedusae, the Narcomedusae are also oceanic and new records are not here included except for the following.

Solmaris corona (Keferstein & Ehlers)

Distribution includes: St Andrews Bay, south-east Scotland, based on photographs of a living specimen taken by M. S. Laverack in July or August 1966 (Pl. Is); coast of Northumberland, taken by Frank Evans in the autumn of 1968.

BIBLIOGRAPHY TO SUPPLEMENT

REFERENCES TO HYDROMEDUSAE

ALLWEIN, Joann, 1967. North American Hydromedusae from Beaufort, North Carolina. *Vidensk. Meddr dansk naturh Foren.* **130**, 117–36.

—— 1968. Seasonal occurrence of Hydromedusae at Helsingør, Denmark, 1966–67. *Ophelia*, **5**, 207–14.

ASHLEY, C. C. & RIDGWAY, E. B., 1968. Aspects of the relationship between membrane potential, calcium transient and tension in single barnacle muscle fibres. *Proc. Physiol. Soc.* 20–21 September, pp. 50–2.

AURICH, H. J., 1958. II. Verbreitung der Medusen und Actinulae von *Ectopleura dumortieri* (van Beneden) und *Hybocodon prolifer* L. Agassiz in der südlichen Nordsee. *Helgol. Wiss. Meeresunt.* **6**, Heft 2, pp. 207–28, figs. 1–11.

BAKER, I. H., 1967. Coelenterates, in Final Report: Chelsea College Azores Expedition, pp. 39–46.

BÉNARD-BOIRARD, Josette, 1962. Développement embryonnaire de *Podocoryne carnea* (Sars) de Roscoff, *Podocoryne carnea* (Sars) forma *exigua* (Haeckel). *Cah. Biol. mar.* **3**, 137–55, figs. 1–8.

BENEDEN, P.-J. van, 1866. Recherches sur la Faune Littorale de Belgique. *Mém. Acad. R. Belg.* **36**, 3–207, pl. I–XVIII.

BERNT, I. Dybern, 1967. The influence of the medusa, *Tima bairdii*, on the deep sea prawn fishery in the Skagerack in 1966–1967. *I.C.E.S.*, C.M. 1967/K:2, Shellfish Committee.

BEYER, Frederick, 1955. *Plotocnide borealis* Wagner in the Oslo Fjord. *Nytt Mag. Zool.* **3**, 94–8, fig. 1.

—— 1958. A new, bottom-living Trachymedusa from the Oslo Fjord. *Nytt Mag. Zool.* **6**, 121–43, pl. I–II.

BODO, France, & BOUILLON, JEAN, 1968. Étude histologique du développement embryonnaire de quelques Hydroméduses de Roscoff: *Phialidium hemisphaericum* (L.), *Obelia* sp. Péron et Lesueur, *Sarsia eximia* (Allman), *Podocoryne carnea* (Sars), *Gonionemus vertens* Agassiz. *Cah. Biol. Mar.* **9**, 69–104, figs. 1–5.

BOUILLON, J., 1957. Étude monographique du genre *Limnocnida* (Limnoméduse). *Annls Soc. r. zool. Belg.* **87** (1956–7), fasc. II, pp. 253–500, figs. 1–114.

—— 1959. *Limnocnida congoensis*, nouvelle espèce de Limnoméduse du Bassin du Congo. *Annls Mus. r. Congo belge* (Sér. in 8°, Sci. Zool.), **71**, 175–85, pl. XXVII–XXX, text-figs. A–D.

—— 1961. Sur le bourgeonnement médusaire manubrial de *Rathkea octopunctata*. *Annls. Soc. r. zool. Belg.* **92**, 7–25.

BOUILLON, J., CASTIAUX, P. & VANDERMEERSSCHE, G., 1958a. Ultrastructure des éléments basophiles de certaines cellules de Coelentérés. *Bull. Micr. Appl.* (2), **8**, no. 2, pp. 33–7, pl. I–III.

—— 1958b. Structure submicroscopique des cnidocils. *Bull. micr. Appl.* (2), **8**, no. 3, pp. 61–3, pl. I–IV.

—— 1958c. Musculature de la Méduse *Limnocnida tanganyicae* (Hydroméduse). *Bull. Micr. Appl.* (2), **8**, no. 4, pp. 81–7, figs. 1–5.

—— 1968. Sur la structure des tentacules adhésifs des Cladonematidae et Eleutheriidae (Anthomedusae). *Pubbl. Staz. zool. Napoli*, **36**, 471–504, 13 figs.

BOUILLON, Jean & WERNER, Bernhard, 1965. Production of medusae buds by the polyps of *Rathkea octopunctata* (M. Sars) (Hydroida Athecata). *Helgoland wiss. Meeresunters.* **12**, 137–48, figs. 1–7.

BRAVERMAN, Max, 1968. Studies on hydroid differentiation. III. The replacement of hypostomal gland cells of *Podocoryne carnea*. *J. Morph.* **126**, 95–106, figs. 1–8.

BRINCKMANN, Anita, 1959. Über den Generationswechsel von *Eucheilota cirrata* (Haeckel, 1879). *Pubbl. Staz. Zool. Napoli*, **31**, 82–9, figs. 1–3.

BRINCKMANN, Anita & PETERSEN, K. W., 1960. On some distinguishing characters of *Dipurena reesi* Vannucci 1956 and *Cladonema radiatum* Dujardin 1843. *Pubbl. Staz. zool. Napoli*, **31**, 386–92, figs. 1–6.

BRINCKMANN, Anita & VANNUCCI, Marta, 1965. On the life-cycle of *Proboscidactyla ornata* (Hydromedusae, Proboscidactylidae). *Pubbl. Staz. zool. Napoli*, **34**, 357–65, figs. 1–2.

BRINCKMANN-VOSS, Anita, 1969. Anthomedusae/Athecatae (Hydrozoa, Coelenterata) of the Mediterranean. Part I. Capitata. *Fauna e Flora del Golfo di Napoli.* 38 Monografia. (In press.)

BROWNE, Edward T., 1903. Report on some medusae from Norway and Spitzbergen. *Bergens Mus. Årb.* no. 4, pp. 1–36, pl. I–V.

BULLOCK, T. H., 1943. Neuromuscular facilitation in medusae. *J. Cell. Comp. Physiol.* **22**, 251–73, figs. 1–5.

BIBLIOGRAPHY TO SUPPLEMENT

CHAPMAN, David M., 1968. A new type of muscle cell from the subumbrella of the jellyfish, *Obelia*. *J. mar. biol. Ass. U.K.* **48**, 667–88, text-figs. 1–11, pl. I–VIII.

CHAPMAN, D. M., PANTIN, C. F. A., & ROBSON, E. A., 1962. Muscle in coelenterates. *Rev. canad. Biol.* **21**, 267–78, figs. 1–2.

DAVENPORT, D. & NICOL, J. A. C., 1955. Luminescence in Hydromedusae. *Proc. roy. Soc.* B, **144**, 399–411, figs. 1–6.

DENTON, E. J. & SHAW, T. I., 1962. The buoyancy of gelatinous marine animals. *J. Physiol. Lond.* **161**, 14–15P, fig. 1.

EDWARDS, C., 1958. Hydromedusae new to the British list from the Firth of Clyde. *Nature, Lond.* **182**, 1564–5.

—— 1963. On the Anthomedusae *Tiaranna rotunda* and *Modeeria formosa*. *J. mar. biol. Ass. U.K.* **43**, 457–67, fig. 1.

—— 1964a. The hydroid of the anthomedusa *Bougainvillia britannica*. *J. mar. biol. Ass. U.K.* **44**, 1–10, figs. 1, 2.

—— 1964b. On the hydroids and medusae *Bougainvillia pyramidata* and *B. muscoides*. *J. mar. biol. Ass. U.K.* **44**, 725–52, figs. 1–6.

—— 1965. The hydroid and the medusa *Neoturris pileata*. *J. mar. biol. Ass. U.K.* **45**, 443–68, figs. 1–4.

—— 1966. The hydroid and the medusa *Bougainvillia principis*, and a review of the British species of *Bougainvillia*. *J. mar. biol. Ass. U.K.* **46**, 129–52, figs. 1–2.

—— 1967. A calycopsid medusa from the Firth of Clyde. *J. mar. biol. Ass. U.K.* **47**, 367–72, figs. 1–2.

—— 1968. Water movements and the distribution of Hydromedusae in British and adjacent waters. *Sarsia*, no. 34, pp. 331–46, fig. 1.

FRASER, J. H., 1969. Observations on the experimental feeding of some medusae and Chaetognatha. *J. Fish. Res. Bd Canada*, **26**, in press.

HAMOND, Richard, 1957. Notes on the Hydrozoa of the Norfolk coast. *J. Linn. Soc.* (Zool.), **43**, 294–324, text-figs. 1–26, pl. VII.

—— 1963. A preliminary report on the marine fauna of the north Norfolk coast. *Trans. Norfolk Norwich Nat. Soc.* **20**, 1–31, figs. 1–4.

HAUENSCHILD, C., 1956. Experimentelle Untersuchungen über die Entstehung asexueller Klone bei der Hydromeduse *Eleutheria dichotoma*. *Z. Naturf.* **11**, B, 394–402.

—— 1957a. Ergänzende Mitteilung über die asexuellen Medusen klone bei *Eleutheria dichotoma*. *Z. Naturf.* **12**, B, 412–13.

—— 1957b. Versuche über die Wanderung der Nesselzellen bei der Meduse von *Eleutheria dichotoma*. *Z. Naturf.* **12**, B, 472–7.

HORRIDGE, G. A., 1955a. The nerves and muscles of medusae. II. *Geryonia proboscidalis* Eschscholtz. *J. Exp. Biol.* **32**, no. 3, pp. 555–68, figs. 1–5.

—— 1955b. The nerves and muscles of medusae. IV. Inhibition in *Aequorea forskalea*. *J. Exp. Biol.* **32**, no. 4, pp. 642–8, figs. 1–4.

—— 1966. Some recently discovered underwater vibration receptors in invertebrates. In *Some contemporary studies in marine science*, ed. Harold Barnes, pp. 395–405, figs. 1–6. London: George Allen & Unwin Ltd.

—— 1969. Statocysts of medusae and evolution of stereocilia. *Tissue & Cell*, **1**, 341–53, figs. 1–10.

HORRIDGE, G. A. & MACKAY, Bruce, 1962. Naked axons and symmetrical synapses in coelenterates. *Q. Jl microsc. Sci.* **103**, 531–41, figs. 1–6.

HOWE, D. C., 1959. A rare hydromedusa. *Nature, Lond.* **184**, 1963.

JOHNSON, Frank H., SHIMOMURA, Osamu, SAIGA, Yo, GERSHAM, Lewis, C., REYNOLDS, George T. & WATERS, John R., 1963. Quantum efficiency of *Cypridina* luminescence, with a note on that of *Aequorea*. *J. Cell. comp. Physiol.* **60**, 85–103, figs. 1–10.

KAKINUMA, Yoshiko, 1966. Life cycle of a hydrozoan, *Sarsia tubulosa* (Sars). *Bull. mar. biol. Stn Asamushi*, **12**, 207–10, figs. 1–5.

KIDD, L. N., 1956. *Craspedacusta sowerbyi* Lankester and its hydroid *Microhydra ryderi* Potts in Lancashire. *The Naturalist*, October/December, pp. 139–40.

KNIGHT-JONES, E. W. & MORGAN, E., 1966. Responses of marine animals to changes in hydrostatic pressure. *Oceanogr. mar. Biol. Ann. Rev.* **4**, 267–99.

KRAMP, P. L., 1955. Hydromedusae in the Indian Museum. *Rec. Indian Mus.* **53**, 339–76, figs. 1–5.

—— 1957a. Hydromedusae from the Discovery Collections. *Discovery Rep.* **29**, 1–128, pl. I–VII, text-figs. 1–19.

KRAMP, P. L., 1957*b*. Some Mediterranean Hydromedusae collected by A. K. Totton in 1954 and 1956. *Vidensk. Meddr dansk naturh. Foren.* **119**, 115–28, fig. 1.

—— 1966. A collection of medusae from the coast of Chile. *Vidensk. Meddr dansk naturh. Foren.* **129**, 1–38, figs. 1–4.

—— 1968. The Hydromedusae of the Pacific and Indian Oceans. Sections II and III. *'Dana' Rep.* no. 72, pp. 1–200, 367 text-figs.

KUHL, Gertrud, 1960. Über die Umbildung einer 'Meduse' von *Craspedacusta sowerbii* Lank. in eine Frustel. *Z. Morph. Ökol. Tiere,* **48**, 439–46, figs. 1–6.

KUHL, Willi & KUHL, Gertrud, 1967. Regenerations- und Heilungsversuche an der Proactinula von *Ectopleura dumortieri* (Athecatae–Anthomedusae) unter Anwendung der Zeittransformation. *Helgoländer wiss. Meeresunters.* **16**, 75–91, figs. 1–11.

KÜHL, Heinrich, 1962. Die Hydromedusen der Elbmündung. *Abh. verh. Naturwiss. Ver. Hamburg,* N.F. **6** (1961), 209–32, text-figs. 1–13, pl. V–VI.

—— 1965. Veränderung des Zooplanktons während einer Tide in der Elbmündung bei Cuxhaven. *Botanica Gothoburgensia,* **3**, 113–26, figs. 1–7.

—— 1967. Die Hydromedusen der Emsmündung. *Veröff. Inst. Meeresforsch. Bremerh.* **10**, 239–46, fig. 1.

LATHAM, E., 1963. The hydroid *Leuckartiara octona* (Fleming) and its association with the polychaete *Aphrodite aculeata* (L.). *Ann. Mag. nat. Hist.* ser. 13, **5**, 523–8, pl. XII.

LYTLE, Charles F., 1959. The records of freshwater medusae in Indiana. *Proc. Indiana Acad. Sci.* **67**, 304–8.

—— 1961. Patterns of budding in the freshwater hydroid *Craspedacusta*. In *The Biology of* Hydra, pp. 317–33, figs. 1–12. (Univ. Miami Press; ed. H. M. Lenhoff and W. F. Loomis.)

MACKIE, G. O. & PASSANO, L. M., 1968. Epithelial conduction in Hydromedusae. *J. gen. Physiol.* **52**, 600–21, figs. 1–11.

MACKIE, G. O., PASSANO, L. M. & CECCATTY, M. P. de, 1967. Physiologie du comportement de l'hydroméduse *Sarsia tubulosa* Sars. Les systèmes à conduction aneurale. *C. r. hebd. Séanc. Acad. Sci. Paris,* **264**, 466–9, 1 fig.

MARTIN, Rainer & BRINCKMANN, Anita, 1963. Zum Brutparasitismus von *Phyllirrhoë bucephala* Pér. & Les. (Gastropoda, Nudibranchia) auf der Meduse *Zanclea costata* Gegenb. (Hydrozoa, Anthomedusae). *Pub. stn zool. Neapel,* **33**, 206–23, figs. 1–15.

McCLARY, A., 1959. The effect of temperature on growth and reproduction in *Craspedacusta sowerbii*. *Ecology,* **40**, 158–62, figs. 1–5.

MILLARD, N. A. H., 1959. Hydrozoa from ships' hulls and experimental plates in Cape Town docks. *Ann. S. Afr. Mus.* **45**, 239–56, figs. 1–3.

—— 1966. The Hydrozoa of the south and west coasts of South Africa. Part III. The Gymnoblastea and small families of Calyptoblastea. *Ann. S. Afr. Mus.* **48**, 427–87, text-figs. 1–15, pl. I.

MOORE, Donald R., 1962. Occurrence and distribution of *Nemopsis bachei* Agassiz (Hydrozoa) in the northern Gulf of Mexico. *Bull. mar. Sci. Gulf Caribbean,* **12**, 399–402.

NAGAO, Zen, 1964. The life cycle of the Hydromedusa, *Nemopsis dofleini* Maas, with a supplementary note on the life-history of *Bougainvillia superciliaris* (L. Agassiz). *Annotnes zool. Jap.* **37**, 153–62.

NAUMOV, D. V., 1951. Some data about life cycles of metagenetic medusae. *Dokl. Akad. Nauk SSSR,* **76**, no. 5, pp. 747–50, figs. 1–2. (In Russian.)

PANTIN, C. F. A. & VIANNA DIAS, M., 1952. Rhythm and after-discharge in medusae. *Anais. Acad. bras. Cienc.* **24**, 351–64, figs. 1–14.

PASSANO, L. M., 1965. Pacemakers and activity patterns in medusae: homage to Romanes. *Am. Zoologist,* **5**, 465–81, figs. 1–14.

PASSANO, L. M., MACKIE, G. O. & CECCATTY, M. P. de, 1967. Physiologie du comportement de l'hydroméduse *Sarsia tubulosa* Sars. Les systèmes des activités spontanées. *C. r. hebd. Séanc. Acad. Sci. Paris,* **264**, 614–17, 1 fig.

PETERSEN, K. W., 1962. A discussion of the Genus *Tima* (Leptomedusae, Hydrozoa). *Vidensk. Meddr dansk naturh. Foren.* **124**, 191–13, figs. 1–10.

—— 1964. Some preliminary results of a taxonomic study of the Hydrozoa of the Cape Cod area. *Biol. Bull. mar. biol. Lab., Woods Hole,* **127**, 384–5.

PICARD, J., 1956. Le premier stade de l'Hydroméduse *Pandea conica*, issu de l'Hydropolype *Campaniclava cleodorae*. *Bull. Inst. océanogr. Monaco,* no. 1086, pp. 1–11, figs. 1–3.

—— 1957. Études sur les hydroïdes de la super-familie Pteronematoidea. *Bull. Inst. océanogr. Monaco,* no. 1106, pp. 1–12.

RALPH, Patricia M. & THOMSON, Helen G., 1968. Seasonal changes in growth of the erect stem of *Obelia geniculata* in Wellington Harbour, New Zealand. *Zool. Publ. Victoria Univ. Wellington,* **44**, 21 pp., figs. 1–11.

BIBLIOGRAPHY TO SUPPLEMENT

RAMOZZOTTI, Giuseppe, 1962. Ritrovamento della Medusa dulciacquicola *Craspedacusta sowerbyi* nella regione del Lago Maggiore. *Memorie Ist. ital. Idrobiol.* **15**, 178–81, fig. 1.

REES, William J., 1956. A revision of the hydroid Genus *Perigonimus* M. Sars, 1846. *Bull. Br. Mus. nat. Hist.* **3**, 337–50.

—— 1957. Evolutionary trends in the classification of capitate hydroids and medusae. *Bull. Br. Mus. nat. Hist.* **4**, 455–534, text-figs. 1–58, pl. 12–13.

REES, W. J. & ROA, E., 1966. Asexual reproduction in the medusa *Zanclea implexa* (Alder), *Vidensk. Meddr dansk naturh. Foren.* **129**, 39–41, fig. 1.

REES, W. J. & WHITE, E., 1966. New records and fauna list of hydroids from the Azores. *Ann. Mag. nat. Hist.* ser. 13, **9**, 271–84.

REISINGER, E., 1957. Zur Entwicklungsgeschichte und Entwicklungsmechanik von *Craspedacusta* (Hydrozoa: Limnotrachylina). *Z. Morph. Ökol. Tiere*, **45**, 656–98, figs. 1–26.

REPELIN, R., 1965. Quelques méduses de l'Ile Anno Bon (Golfe de Guinée). *O.R.S.T.O.M. océanogr.* **3**, 73–9, figs. 1–10.

RICE, A. L., 1964. Observations on the effects of changes of hydrostatic pressure on the behaviour of some marine animals. *J. mar. biol. Ass. U.K.* **44**, 163–75, fig. 1.

ROOSEN-RUNGE, Edward C., 1962. On the biology of sexual reproduction of Hydromedusae, genus *Phialidium* Leuckhart. *Pacif. Sci.* **16**, 15–24, figs. 1–5.

—— 1967. Gastrovascular system of small Hydromedusae: mechanisms of circulation. *Science, N.Y.* **156**, 74–6, figs. 1–10.

ROOSEN-RUNGE, Edward, C. & SZOLLOSI, Daniel, 1965. On biology and structure of the testis of *Phialidium* Leuckhart (Leptomedusae). *Z. Zellforsch. mikrosk. Anat.* **68**, 597–610, figs. 1–13.

RUSSELL, F. S., 1956a. On a new medusa, *Amphinema krampi* n.sp. *J. mar. biol. Ass. U.K.* **35**, 371–3, figs. 1–2.

—— 1956b. On two new medusae, *Merga reesi* n.sp. and *Tiaropsidium atlanticum* n.sp. *J. mar. biol. Ass. U.K.* **35**, 493–8, figs. 1–3.

—— 1957. On a new medusa, *Krampella dubia* n.g., n.sp. *J. mar. biol. Ass. U.K.* **36**, 445–7, figs. 1–2.

—— 1958. Notes on the medusa *Amphinema krampi* Russell. *J. mar. biol. Ass. U.K.* **37**, 81–4, figs. 1–3.

RUTHERFORD, Dorothy J., 1962. A rare hydromedusan, *Staurophora mertensi*, at St Andrews. *Ann. Mag. nat. Hist.* ser. 13, **5**, 199–204, figs. 1–6.

SHIMOMURA, Osamu, JOHNSON, Frank H. & SAIGA, Yo, 1962. Extraction, purification and properties of Aequorin, a bioluminescent protein from the luminous hydromedusan, *Aequorea. J. Cell. comp. Physiol.* **59**, 223–39, figs. 1–11.

—— 1963. Microdetermination of calcium by Aequorin luminescence. *Science, N.Y.* **140**, 1339–40.

TAMBS-LYCHE, Hans, 1964. *Gonionemus vertens* L. Agassiz (Limnomedusae). A zoogeographical puzzle. *Sarsia*, no. 15, pp. 1–8, figs. 1–3.

THIEL, Max Egon, 1968. Die Einwanderung der Hydromeduse *Nemopsis bachei* L.Ag. aus dem ostamerikanischen Küstengebiet in die westeuropäischen Gewässer und die Elbmündung. *Abh. Verh. naturw. Ver. Hamburg*, N.F. Bd. XII, 1967, pp. 81–94, figs. 1–6.

TIFFON, Y., 1957. Présence de *Nemopsis bachei* (Agassiz) dans les eaux saumâtres de la Gironde (Anthomedusae). *Vie Milieu*, **7**, 550–3.

TODD, Eric S., KIER, Andrew & EBELING, Alfred, W. 1966. *Gonionemus vertens* L. Agassiz (Hydrozoa: Limnomedusae) in southern California. *Bull. So. Calif. Acad. Sci.* **65**, 205–10, fig. 1.

UCHIDA, Tohru, 1951a. On the frequent occurrence of freshwater medusae in Japan. *Journ. Fac. Sci., Hokkaido Univ.* ser. VI (Zool.), **10**, no. 2, pp. 157–60, figs. 1–2.

—— 1951b. A brackish-water medusa from Japan. *Journ. Fac. Sci., Hokkaido Univ.* ser. VI (Zool.), **10**, no. 2, pp. 161–2, fig. 1.

—— 1955. Dispersal in Japan of the freshwater medusa, *Craspedacusta sowerbyi* Lankester, with remarks on *C. iseana* (Oka & Hara). *Annotnes. Zool. Jap.* **28**, no. 2, pp. 114–20, figs. 1–5.

—— 1964a. Medusae of *Eugymnanthea*, an epizoic hydroid. *Publ. Seto mar. biol. Lab.* **12**, no. 1, pp. 101–7, figs. 1–6.

—— 1964b. A new hydroid species of *Cytaeis*, with some remarks on the inter-relationships in the Filifera. *Publ. Seto mar. Biol. Lab.* **12**, no. 2, pp. 133–44, figs. 1–8.

—— 1968. The two Leptomedusae *Aequorea coerulescens* and *Aequorea vitrina. J. Fac. sc. Hokkaido Univ.* (ser. VI), Zool. **16**, 359–68, figs. 1–6.

UCHIDA, Tohru & NAGAO, Zen, 1960. The life-history of the Hydromedusa, *Bougainvillia superciliaris* (L. Agassiz). *Annotnes zool. Jap.* **33**, 249–53, figs. 1–7.

VALKANOV, A., 1954. Revision der Hydrozoenfamilie Moerisiidae. *Arb. Biol. Meerestat. St. Stalin*, **18**, (1953), 33–47, figs. 1–6.

VANNUCCI, M., 1956. Notes on the Hydromedusae of the Clyde sea area with new distribution records. *Glasg. Nat.* **17**, 243–9.

—— 1957. On Brazilian Hydromedusae and their distribution in relation to different water masses. *Bolm Inst. Oceanogr., S. Paulo*, **8**, 23–109, figs. 1–31.

—— 1958. Considerações em torno das Hydromedusae da região de Fernando de Noronha. *Bolm Inst. Oceanogr., S. Paulo*, **9**, 3–12.

—— 1963. On the ecology of Brazilian medusae at 25° Lat. S. *Bolm Inst. Oceanogr., S. Paulo*, **12**, 143–84, figs. 11–12, graphs 1–9.

VANNUCCI, M. & REES, W. J., 1961. A revision of the genus *Bougainvillia* (Anthomedusae). *Bolm Inst. Oceanogr., S. Paulo*, **11**, 57–100.

WERNER, Bernhard, 1954a. On the development and reproduction of the Anthomedusan *Margelopsis haeckeli* Hartlaub. *Trans. N.Y. Acad. Sci.* ser. II, **16**, 143–6.

—— 1954b. Über die Fortpflanzung der Anthomeduse *Margelopsis haeckeli* Hartlaub durch Subitan- und Dauereier und die Abhängigkeit ihrer Bildung von aüsseren Faktoren. *Verhandl. Deutsch. Zool. Gesellschaft in Tübingen 1954*, pp. 124–33, figs. 1–5.

—— 1956a. Der zytologische Nachweis der parthenogenetischen Entwicklung bei der Anthomeduse *Margelopsis haeckeli* Hartlaub. *Naturwiss.* (XLIII. Jahrg.), Heft 23, pp. 541–2.

—— 1956b. Über die entwicklungsphysiologische Bedeutung des Fortpflanzungswechsels der Anthomeduse *Rathkea octopunctata* M. Sars. *Zool. Anz.* **156**, Heft 5/6, pp. 159–77, figs. 1–5.

—— 1958. Die Verbreitung und das jahreszeitliche Auftreten der Anthomeduse *Rathkea octopunctata* M. Sars, sowie die Temperaturabhängigkeit ihrer Entwicklung und Fortpflanzung. *Helgoländer wiss. Meeresunters.* **6**, Heft 2, pp. 137–70, figs. 1–13.

—— 1959a. The Hydromedusae of Port Erin Bay in May and June 1957. *Rep. mar. biol. Sta. Port Erin*, **71** (1959), 32–8.

—— 1959b. Dauerstadien bei marinen Hydrozoen. *Naturwiss.* Heft 7, p. 238.

—— 1961. Morphologie und Lebensgeschichte, sowie Temperaturabhängigkeit der Verbreitung und des jahreszeitlichen Auftretens von *Bougainvillia superciliaris* (L. Agassiz) (Athecatae–Anthomedusae). *Helgoländer wiss. Meeresunters.* **7**, 206–37, figs. 1–5.

—— 1962. Verbreitung und jahreszeitliches Auftreten von *Rathkea octopunctata* (M. Sars) und *Bougainvillia superciliaris* (L. Agassiz) (Athecatae–Anthomedusae). Ein Beitrag zur kausalen marinen Tiergeographie. *Kieler Meeresforsch.* **18**, Heft 3 (Sonderheft.), pp. 55–66, Taf. 1–2.

—— 1963a. Effect of some environmental factors on differentiation and determination in marine Hydrozoa, with a note on their evolutionary significance. *Ann. N.Y. Acad. Sci.* **105**, 461–88, figs. 1–15.

—— 1963b. Experimentelle Beobachtungen über die Wirksamkeit von Aussenfaktoren in der Entwicklung der Hydrozoen und Erörterung ihrer Bedeutung für die Evolution. *Verh. Inst. Meeresforsch. Bremerhaven*, Sonderband, pp. 153–77, figs. 1–10.

—— 1965. Die Nesselkapseln der Cnidaria, mit besonderer Berücksichtigung der Hydroida. I. Klassifikation und Bedeutung für die Systematik und Evolution. *Helgoländer wiss. Meeresunters.* **12**, 1–39, figs. 1–23.

—— 1968a. Polypengeneration und Entwicklungsgeschichte von *Eucheilota maculata* (Thecata–Leptomedusae). Mit einem Beitrag zur Methodik der Kultur mariner Hydroiden. *Helgoländer wiss. Meeresunters.* **18**, 136–68, figs. 1–20.

—— 1968b. Polypengeneration und Entwicklung von *Eutonina indicans* (Thecata–Leptomedusae). *Helgoländer wiss. Meeresunters.* **18**, 384–403, figs. 1–15.

WERNER, Bernhard & AURICH, Horst, 1955. Über die Entwicklung des Polypen von *Ectopleura dumortieri* van Beneden und die Verbreitung der planktischen Stadien in der südlichen Nordsee (Athecatae–Anthomedusae). *Helgoländer wiss. Meeresunters.* **5**, Heft 2 (List (Sylt) 1955), pp. 234–50, figs. 1–8.

WIDERSTEN, Bernt, 1968. On the morphology and development in some cnidarian larvae. *Zool. Bidr. Upps.* **37**, 139–82, pl. 1–3, text-figs. 1–16.

YAMADA, Mayumi, 1959. Hydroid fauna of Japanese and adjacent waters. *Publs Akkeshi mar. biol. Stn*, no. 9, pp. 1–101.

Plate I s

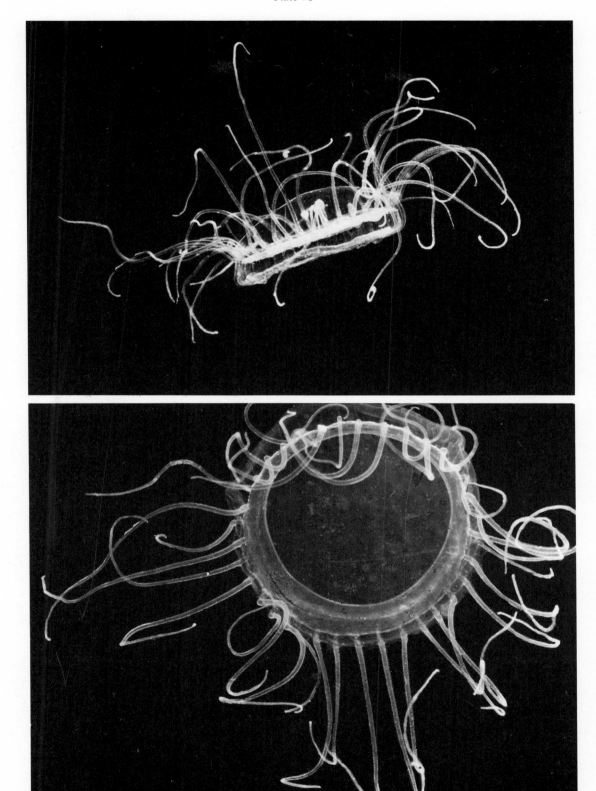

I. SUBJECT INDEX TO VOLUME II

I. SUBJECT INDEX

II. SYSTEMATIC INDEX TO VOLUME II

Synonyms are in italics; the principal page references to descriptions are in black type

II. SYSTEMATIC INDEX

III. SUBJECT INDEX
TO SUPPLEMENT

IV. SYSTEMATIC INDEX TO SUPPLEMENT